KEY EXPERIMENTS IN PRACTICAL DEVELOPMENTAL BIOLOGY

This unique resource presents twenty-seven easy-to-follow laboratory exercises for use in student practical classes, all of which are classic experiments in developmental biology. These experiments have provided key insights into developmental questions, and many of them are described by the leaders in the field who carried out the original pioneering research. This book intends to bridge the gap between state-of-the-art experimental work and the laboratory classes taken at the undergraduate and postgraduate levels. All chapters follow the same logical format, taking the students from materials and methods, through results and discussion, so that they learn the underlying rationale and analysis employed in the research. Chapters also include teaching concepts, discussion of the degree of difficulty of each experiment, potential sources of failure, as well as the time required for each experiment to be carried out in a practical class with students. The book will be an invaluable resource for graduate students and instructors teaching practical developmental biology courses.

Manuel Marí-Beffa is a Lecturer in Developmental Biology at the University of Málaga.

Jennifer Knight is an Instructor in the Department of Molecular, Cellular and Developmental Biology at the University of Colorado, Boulder.

Key Experiments in Practical Developmental Biology

Edited by

Manuel Marí-Beffa

University of Málaga

Jennifer Knight

University of Colorado

CAMBRIDGE
UNIVERSITY PRESS

CAMBRIDGE UNIVERSITY PRESS
Cambridge, New York, Melbourne, Madrid, Cape Town, Singapore,
São Paulo, Delhi, Dubai, Tokyo, Mexico City

Cambridge University Press
The Edinburgh Building, Cambridge CB2 8RU, UK

Published in the United States of America by Cambridge University Press, New York

www.cambridge.org
Information on this title: www.cambridge.org/9780521179768

First published 2005
First paperback edition 2010

A catalogue record for this publication is available from the British Library

Library of Congress Cataloguing in Publication data

Key experiments in practical developmental biology / edited by Manuel Marí-Beffa,
    Jennifer Knight.
        p.   cm.
    Includes bibliographical references and index.
    ISBN 0-521-83315-9
    1. Developmental biology – Experiments.   I. Marí-Beffa, Manuel, 1961–
II. Knight, Jennifer, 1969–
    QH491.K485   2004
571.8–dc22   2004049265

ISBN 978-0-521-83315-8 Hardback
ISBN 978-0-521-17976-8 Paperback

Additional resources for this publication at www.cambridge.org/9780521179768

This book is dedicated to our families

"…causes and effects are discoverable, not by reason but by experience,…"
(David Hume [1748] An Enquiry Concerning Human Understanding.
Section IV. Part I.)

CONTENTS

SECTION VIII. TRANSGENIC ORGANISMS

SECTION IX. VERTEBRATE CLONING

SECTION X. CELL CULTURE

SECTION XI. EVO–DEVO STUDIES

SECTION XII. COMPUTATIONAL MODELLING

CONTRIBUTORS

A. Baonza
MRC Laboratory of Molecular Biology
Hills Road
Cambridge CB2 2QH
UK

G. Begemann
Department of Biology
University of Konstanz
Universitätsstr. 10
D-78464 Konstanz
Germany

M. Blum
University of Hohenheim
Institute of Zoology (220)
Garbenstrasse 30
D-70593 Stuttgart
Germany

H. R. Bode
Developmental Biology Center and
    Department of Developmental and
    Cell Biology
University of California at Irvine
5205 McGaugh Hill
Irvine, California 92697
USA

D. Bueno
Departament de Genètica
Facultat de Biologia
Universitat de Barcelona
Av. Diagonal 645
08028 Barcelona
Spain

S. Campuzano
Centro de Biología Molecular "Severo Ochoa"
Universidad Autónoma de Madrid
Cantoblanco
E-28049 Madrid
Spain

S. Canevascini
Friedrich Miescher Institute for Biomedical
    Research
Maulbeerstrasse 66
4058 Basel
Switzerland

J. Castelli-Gair Hombría
Department of Zoology
University of Cambridge
Downing Street
Cambridge CB2 3EJ
UK

F. J. Díaz-Benjumea
Centro de Biología Molecular "Severo Ochoa"
Universidad Autónoma de Madrid
Cantoblanco
E-28049 Madrid
Spain

J. B. Duffy
Department of Biology
A504/A502 Jordan Hall
Indiana University
101 E. 3rd Street
Bloomington, Indiana 47405-3700
USA

D. Echevarría
Instituto de Neurociencias de Alicante
   (UMH-CSIC)
University of Miguel Hernández
Campus de San Juan
Carretera de Valencia, Km. 87
E-03550 Alicante
Spain

M. Ecke
Max-Planck-Institut für Biochimie
Am Klopferspitz 18a
D-82152 Martinsried
Germany

C. A. Ettensohn
Department of Biological Sciences
Science and Technology Center for Light
   Microscope Imaging and Biotechnology
Carnegie Mellon University
4400 Fifth Avenue
Pittsburgh, Pennsylvania 15213
USA

D. Foronda
Centro de Biología Molecular "Severo Ochoa"
Universidad Autónoma de Madrid
Cantoblanco
E-28049 Madrid
Spain

Y. Gañán
Área Anatomía y Embriología Humanas
Departamento de Ciencias Morfológicas y
   Biología Celular y Animal
Facultad de Medicina
Universidad de Extremadura
E-06071 Badajoz
Spain

R. J. Garriock
Department of Cell Biology and Anatomy,
   LSN 444
University of Arizona College of Medicine
P.O. Box 245044
1501 N. Campbell Avenue
Tucson, Arizona 85743
USA

G. Gerisch
Max-Planck-Institut für Biochimie
Am Klopferspitz 18a
D-82152 Martinsried
Germany

J. B. Gurdon
Wellcome Trust/CRC Cancer UK Institute
Institute of Cancer and Developmental Biology
University of Cambridge
Tennis Court Road
Cambridge CB2 1QR
UK

J. C. Izpisúa-Belmonte
Gene Expression Laboratories
The Salk Institute for Biological Studies
10010 North Torrey Pines Road
La Jolla, California 92037-1099
USA

C. Karcher
University of Hohenheim
Institute of Zoology (220)
Garbenstrasse 30
D-70593 Stuttgart
Germany

C. Klämbt
Institut für Neurobiologie
Universität Münster
Badestrasse 9
D-48149 Münster
Germany

J. Knight
MCD Biology
University of Colorado
Boulder, Colorado 80309-0347
USA

P. A. Krieg
Department of Cell Biology and Anatomy,
   LSN 444
University of Arizona College of Medicine
P.O. Box 245044
1501 N. Campbell Avenue
Tucson, Arizona 85743
USA

D. Macías
Área Anatomía y Embriología Humanas
Departamento de Ciencias Morfológicas y
  Biología Celular y Animal
Facultad de Medicina
Universidad de Extremadura
E-06071 Badajoz
Spain

M. Maden
MRC Centre for Developmental Neurobiology
4th floor New Hunt's House
King's College London
Guy's Campus
London Bridge
London SE1 1UL
UK

M. Marí-Beffa
Department of Cell Biology, Genetics and
  Physiology
Faculty of Science
University of Málaga
E-29071 Málaga
Spain

S. Martínez
Instituto de Neurociencias de Alicante
  (UMH-CSIC)
University of Miguel Hernández
Campus de San Juan
Carretera de Valencia, Km. 87
E-03550 Alicante
Spain

H. Meinhardt
Max-Planck-Institut für Entwicklungsbiologie
Spemannstr. 35
D-72076 Tübingen
Germany

L. de Navas
Centro de Biología Molecular "Severo Ochoa"
Universidad Autónoma de Madrid
Cantoblanco
E-28049 Madrid
Spain

N. Perrimon
Department of Genetics and Howard Hughes
  Medical Institute
Harvard Medical School HHMI
200 Longwood Ave
Boston, Massachusetts 02115-6092
USA

Á. Raya
Gene Expression Laboratories
The Salk Institute for Biological Studies
10010 North Torrey Pines Road
La Jolla, California 92037-1099
USA

J. L. Riechmann
Gene Expression Center
Biology
California Institute of Technology
102B Kerckhoff M/C 156-29
Pasadena, California 91125
USA

C. Rodríguez Esteban
Gene Expression Laboratories
The Salk Institute for Biological Studies
10010 North Torrey Pines Road
La Jolla, California 92037-1099
USA

J. Rodríguez-León
Instituto Gulbenkian de Ciência
Rua da Quinta Grande nº 6, Apt. 14
2780-901 Oeiras
Portugal

A. Rolletschek
In Vitro Differentiation Group
Dept. of Cytogenetics
Institute of Plant Genetics and Crop Plant
  Research (IPK)
Corrensstr. 3
D-06466 Gatersleben
Germany

R. Romero
Departament de Genètica
Facultat de Biologia
Universitat de Barcelona
Av. Diagonal 645
E-08028 Barcelona
Spain

S. Roth
Institut für Entwicklungsbiologie
Universität zu Köln
Gyrhofstr. 17
D-50923 Köln
Germany

L. M. Salgado
CINVESTAV-IPN
Dpto. Biochemistry
Apartado Postal 14-740
07000 México, D.F.
México

E. Saló
Departament de Genètica
Facultat de Biologia
Universitat de Barcelona
Av. Diagonal 645
E-08028 Barcelona
Spain

E. Sánchez-Herrero
Centro de Biología Molecular "Severo Ochoa"
Universidad Autónoma de Madrid
Cantoblanco
E-28049 Madrid
Spain

A. Schweickert
University of Hohenheim
Institute of Zoology (220)
Garbenstrasse 30
D-70593 Stuttgart
Germany

P. Simpson
Department of Zoology
University of Cambridge
Downing Street
Cambridge CB2 3EJ
UK

N. Skaer
SkyLab, Department of Zoology
University of Cambridge
Downing Street
Cambridge CB2 3EJ
UK

S. Sotillos
Centro de Biología Molecular "Severo Ochoa"
Universidad Autónoma de Madrid
Cantoblanco
E-28049 Madrid
Spain

M. Suzanne
Centro de Biología Molecular "Severo Ochoa"
Universidad Autónoma de Madrid
Cantoblanco
E-28049 Madrid
Spain

H. Vaessin
Neurobiotechnology Center
Dept. of Molecular Genetics
Comprehensive Cancer Center
The Ohio State University
176 Rightmire Hall
1060 Carmack Road
Columbus, Ohio 43210
USA

C. Wiese
In Vitro Differentiation Group
Dept. of Cytogenetics
Institute of Plant Genetics and Crop Plant
    Research (IPK)
Corrensstr. 3
D-06466 Gatersleben
Germany

A. M. Wobus
In Vitro Differentiation Group
Dept. of Cytogenetics
Institute of Plant Genetics and Crop Plant
    Research (IPK)
Corrensstr. 3
D-06466 Gatersleben
Germany

Preface

Manuel Marí-Beffa

This handbook of laboratory exercises was first conceived at the Third Congress of the Spanish Society of Developmental Biology held in Málaga, Spain, in 2001. At the time, Professor Antonio García-Bellido suggested including collaborators from the United States and the rest of Europe to give the project a more international scope. The resulting book is a handbook intended to provide a bridge between top scientific researchers and practical laboratories taught at both the undergraduate and postgraduate level. Each chapter introduces a short, inexpensive, and, for the most part, straightforward laboratory project designed to be carried out by students in a standard lab environment. The book uses some of the most popular and best studied model organisms to examine the processes of development. Each chapter is written by specialists in the field describing, in most instances, original pioneering experiments that profoundly influenced the field. The book also demonstrates a historical bridge from classical embryological concepts, using Aristotle and Driesch's entelechia concept (Driesch, 1908) (Chapters 2 and 15) or morphogenetic gradient concept (i.e., Wolpert, 1969) (Chapters 1 and 16) to modern cellular, genetic, and molecular analyses of development such as homeotic genes (Chapters 11 and 20), compartmentalization (Chapter 14), or cell–cell interactions (Chapters 2, 13, and 22). In addition, the high-impact techniques of vertebrate cloning (Section IX) and embryonic stem cells (Section X), as well as the emerging discipline of evolution and development (Evo–Devo, Section XI), are also considered. Finally, although there is much still to learn in this field, Section XII is devoted to computational modelling in the search for a link between genotype and phenotype. During each laboratory exercise, it is our intent that the students imagine themselves working with these highly respected scientists, traveling the same road pioneered by the authors of each chapter.

The format of each chapter is intended to merge the format of standard scientific papers and practical laboratory protocols – a format inspired by texts with similar intent (Stern and Holland, 1993; Halton, Behnke, and Marshall, 2001). Each chapter also includes parts called "Alternative Exercises" and "Questions for Further Analysis" that will permit laboratory instructors or advisors to carry out an "inquiry-based" lab format as supported by the National Research Council of the United States (NRC, 2000). With the guidance provided in each chapter, students can design and carry out their own, related experiments, potentially culminating in the writing of original papers. For most of the laboratory exercises described, the standard laboratory safety protocols maintained in all labs are sufficient; where necessary, more information is given about the controlled use of hazardous substances. IN GENERAL, CAUTIONS MUST BE TAKEN. MANY OF THE CHEMICALS USED IN THESE LABORATORY EXERCISES ARE HAZARDOUS. TO PREVENT EXPOSURE TO THESE CHEMICALS, YOU SHOULD WEAR GLOVES AND SAFETY GLASSES AND WORK WITH THE CHEMICALS IN A FUME HOOD. THIS IS PARTICULARLY IMPORTANT WHEN WORKING WITH SUBSTANCES LIKE PARAFORMALDEHYDE, GLUTARALDEHYDE, RETINOIC ACID, DEAB, DAB XYLENE, OR CHLORAL HYDRATE. MORE DETAILED INFORMATION ON PROPER HANDLING OF THESE CHEMICALS CAN BE OBTAINED FROM MATERIAL SAFETY DATA SHEETS (MSDS), WHICH ARE SUPPLIED BY THE CHEMICAL MANUFACTURERS. The animals used in each laboratory exercise can be obtained from the curators of many international stock centers around the world. In most countries, Home Office approvals are required so that appropriate responsibilities must be taken by receiving departments.

REFERENCES

Aristoteles, De Anima. In *Aristotle De Anima, with Translation, Introduction and Notes.* ed. R. D. Hicks (1965). Amsterdam: Adolf M. Hakkert Publ.

Driesch, H. (1908). *The Science and Philosophy of the Organism. Gifford Lectures in 1908.* London: A. and C. Black.

Halton, D. W., Behnke, J. M., and Marshall, I. (eds.) (2001). *Practical Exercises in Parasitology.* Cambridge: Cambridge University Press.

National Research Council (2002). *Inquiry and the National Science Education Standards: A Guide for Teaching and Learning.* Center for Science, Mathematics and Engineering Education. p. 202 Washington, DC: National Academy Press.

Stern, C. D., and Holland, P. W. H. (eds.) (1993). *Essential Developmental Biology. A Practical Approach.* New York: Oxford University Press.

Wolpert, L. (1969). Positional information and the spatial pattern of cellular differentiation. *J. Theor. Biol.,* 25, 430–1.

Introduction

Jennifer Knight

Experiencing the process of scientific discovery is part of training to be a scientist. This book of laboratory exercises is designed to give students an opportunity to explore and carry out experiments that have each made significant contributions to the fields of Experimental Embryology and Developmental Biology over the past 100 years. It is our hope that students will experience the initial thrill of discovery as they learn how to do each experiment, analyze each outcome, and grasp the significance of each conclusion. However, science is not solely about the end discovery but also about the process. This process cannot be appreciated by reading textbooks or scientific journals alone. Rather, a budding scientist must experience firsthand the myriad pitfalls of each experiment. Despite the way this laboratory manual is designed (with step-by-step instructions to accomplish each experiment), students will encounter unforeseen problems in carrying out the experiments. If they are not already intimately familiar with experimental science, students will undoubtedly discover that this process demands a meticulous approach. Designing, setting up, and executing experiments cannot be accomplished in a haphazard way. For this reason, every student must keep a laboratory notebook, a task that many initially regard as "busy work." In fact, keeping careful record of everything one does in the laboratory is the only way to experience success. At the other end of this process is presenting a finished piece of work to the scientific community. Again, the only way to learn this aspect is to assemble data into a mock scientific "paper," ready for publication in a journal. If possible, verbally presenting the data to an audience is also a valuable learning experience. Below, we give some suggestions for these two important aspects of scientific discovery: keeping a laboratory notebook and writing a laboratory report.

KEEPING A LABORATORY NOTEBOOK

A laboratory notebook is a day-to-day record of plans, procedures, results and interpretations. When a scientist refers back to his/her notebook, the notes on procedures, pitfalls and outcomes should help him/her to easily repeat the experiment. In the scientific community, a notebook is essential both for demonstrating integrity and for helping to keep track of each step of an experiment. In a student laboratory setting, a notebook can be just as useful, provided the student takes the time to make it so. A notebook must be bound, and the first two pages should be reserved for a Table of Contents that can be filled out as experiments are completed. When students begin an experiment, they should write down their thoughts on the experiments, questions they may have, and finally, their objective and hypothesis. Next, as students perform an experiment, the actual steps should be recorded, although often this can be done in an outline format rather than in great detail, since the procedures have already been described in the lab manual. It is most important that students make note of problems encountered during the experiment, or of deviations taken from the laboratory manual. These notes will help with trouble shooting if part of the experiment does not succeed. Next, students should record all of their observations and all of their data, both raw and calculated (graphs, tables, etc.). Finally, when the experiment is complete, the students should summarize their results, conclusions, and interpretations in the notebook, before moving on to the next experiment. It is essential that students realize that a laboratory notebook is a work in progress. It is only useful if it is used during the experiment to record the process.

WRITING A LABORATORY REPORT

A laboratory report should follow the standard format for a scientific paper, described below.

➤ Abstract: a 4–6 sentence summary of the entire paper, including a brief statement of the hypothesis, the methods used, the outcome, and the relevance of the experiment.
➤ Introduction: a well-researched description of the topic addressed by the experiment. The introduction gives the reader the context of the experiment. This section should also restate the hypothesis and describe the predictions and goals of the experiment.
➤ Materials and Methods: a detailed section in which reagents and protocols are clearly described. Often, in a classroom setting, since these details are provided to students in the lab manual, instructors suggest a summary of the materials and methods used. It is still important to write in complete sentences and to accurately state how the experiment was carried out.
➤ Results: a description of the outcome of the experiments. This section of the paper includes only a description of the data and their presentation – figures, tables, and graphs – but does not discuss the interpretation of the findings.
➤ Discussion: an interpretation of the experimental data and how it compares to published information about this topic. In this section, students should discuss what their results mean, the implications or significance of these results, and how they might expand or clarify the results. Ultimately, it is important that students put their experiment into the context of other work on this topic.
➤ References: a detailed citation of each journal article used in writing the paper. There are many different possible formats for references. Students may choose a specific

format by consulting a journal and using that standard format for each article referenced (instructions for authors, often found at the beginning of journals, usually include instructions for referencing other journal articles).

By following the suggestions above, we hope that as instructors and students alike perform the experiments presented in this book, you will find yourselves engaged in and enticed by this exploration of Developmental Biology.

SECTION I. GRAFTINGS

1 Two developmental gradients control head formation in hydra

H. R. Bode

OBJECTIVE OF THE EXPERIMENT Two developmental gradients are involved in the axial patterning of the head and the body column of a hydra. One is a morphogenetic gradient of head activation [= head formation capacity], and the other is a gradient of head inhibition. The objective of the experiment is to demonstrate the presence of these two gradients in the body column of adult hydra using transplantation experiments.

DEGREE OF DIFFICULTY The experiments involve the isolation of a piece of the body column and transplantation to the body column of a second animal. Although this appears difficult at first sight, with a little practice, almost all students learn to carry out these grafts at the rate of 6–10 successful grafts/hour.

INTRODUCTION

In animals, the developmental processes associated with axial patterning occur during early stages of embryogenesis. One example involves the processes governing head formation at the anterior end and tail formation at the posterior end of the anterior–posterior axis. In hydra, a primitive metazoan, this type of axial patterning occurs not only during embryogenesis, but also in the adult. This is due to the tissue dynamics of an adult hydra.

As shown in Figure 1.1a, a hydra has the shape of a cylindrical shell. Along the single axis are the head, body column and foot. The head at the apical end consists of a mouth region, the hypostome, and beneath that the tentacle zone, from which tentacles emerge. The protrusions on the lower part of the body column are early [left] and advanced [right] stage buds, hydra's mode of asexual reproduction. The wall of the shell is composed of two epithelial layers, the ectoderm and endoderm, which extend throughout the animal. Among the epithelial cells (not shown in Figure 1.1a) are smaller cells such as neurons, secretory cells and nematocytes, the stinging cells of cnidaria.

The tissue dynamics is the following. The epithelial cells of both layers are continuously in the mitotic cycle (e.g. Bode, 1996). Yet, despite the ever-increasing number of

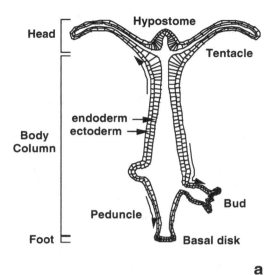

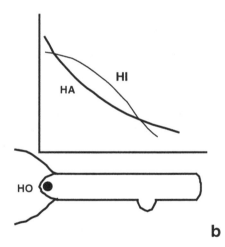

Figure 1.1. (a) Cross section of an adult hydra showing the three regions and the two tissue layers as well as two stages of bud formation. (b) Diagram of the developmental elements that control head formation. HO = head organizer; HA = head activation gradient; HI = head inhibition gradient. (a) is adapted from Amer. Zool., 41, 621–8 (2001).

epithelial cells, the animal remains constant in size. This occurs because the tissue of the upper body column is apically displaced onto the tentacles and eventually sloughed at the tentacle tips (Bode, 1996). Tissue of the lower body column is displaced down the body column and sloughed at the foot (Figure 1.1a). Tissue from the middle of the column is primarily displaced into developing buds, which eventually detach from the adult. Thus, the animal is in a steady state of production and loss of tissue.

As tissue is displaced apically, it is converted into head tissue, whereas tissue displaced basally becomes foot tissue. What are the axial patterning processes that control the changes in the fate of these moving epithelial cells? A body of transplantation and regeneration experiments have provided insight into these processes (Browne, 1909; Wolpert, 1971; MacWilliams, 1983*a*, *b*; Bode and Bode, 1984). Bisection of the body column leads to the regeneration of a head at the apical end of the lower half. This indicates that body column tissue has the capacity to form a head. Transplantation experiments have shown that a head organizer region is located in the hypostome (Figure 1.1b) (Broun and Bode, 2002). This organizer transmits a signal, or morphogen, to the body column which sets up a gradient of head formation capacity, commonly referred to as the Head Activation Gradient (Figure 1.1b) (MacWilliams, 1983*a*). With this capacity, what prevents regions of body column tissue from forming heads? The head organizer also produces and transmits an inhibitor of head formation, which is also graded down the body column (Figure 1.1b), thereby preventing body column tissue from forming heads (Wolpert, 1971: MacWilliams, 1983*b*). These two gradients control the fate of the body column tissue as it is displaced apically. When the tissue reaches a point where [HA] > [HI], the body column tissue is converted into head tissue. This mechanism maintains the axial patterning at the upper end in the context of the tissue dynamics of the animal. These gradients and their behavior have been incorporated into a model that provides a useful overall view of axial patterning in hydra (Meinhardt, 1993; see Chapter 27).

MATERIALS AND METHODS

In this section the equipment and materials required for carrying out transplantation experiments are described using a procedure developed by Rubin and Bode (1982). The culture of hydra and the source of specific pieces of equipment or materials are presented in the Appendix.

EQUIPMENT AND MATERIALS
Per student
> Dissecting microscope with 10 × oculars and, optimally, variable magnification of 1–4 ×.
>
> Pasteur pipette with rubber bulb (Fisher Scientific).
>
> Two pairs of fine-tipped forceps (Fine Science Tools) to handle pieces of fish line and "sleeves."
>
> Scalpel (Fine Science Tools). An ordinary razor blade will work equally well.
>
> Medium-sized [60-mm diameter] plastic or glass petri dishes (Fisher Scientific).

Per practical group. If available, an 18 °C incubator with a light that can be set with a timer so that it is on a cycle for 12 h on and 12 h off. If an incubator is not available, experimental samples can be left in the lab if the temperature is in the 15–25 °C range.

Biological material. One-day–starved adult hydra (see Appendix) without buds. Two adult hydra are needed for each graft: one is the donor, and the other is the host. Determine how many grafts will be made and obtain twice that number of adult animals.

For each transplantation choose two adults that are the same size. Thus, 250–300 hydra are necessary for all four experiments [see Appendix on Maintenance of a Hydra Culture].

REAGENTS

$CaCl_2$	$NaHCO_3$
$MgCl_2$	KNO_3
$MgSO_4$	Glutathione

PREVIOUS TASKS FOR STAFF
Preparation of fish lines and "sleeves"

1–1.5 cm pieces of fish line [8 lb; diameter = 0.3 mm; local store for fishing supplies]. Using a scalpel and forceps, cut 1–1.5 cm long pieces of fish line.

"Sleeves": 2–3 mm pieces of polyethylene tubing (VWR Scientific). [As the ends should be pointed, cut the fish line at a 45 degree angle. Cut as many as are needed for an experiment. For the "sleeves," cut two for each piece of fish line. Make the cuts perpendicular to the axis of the tubing.]

Maintenance of hydra culture. During this and the previous experiment the hydra must be maintained under standard conditions (see Appendix).

Solutions

Hydra medium: 1.0 mM $CaCl_2$, 1.5 mM $NaHCO_3$, 0.1 mM $MgCl_2$, 0.08 mM $MgSO_4$, and 0.03 mM KNO_3; pH 8.0.

1 mM glutathione (Sigma) in hydra medium.

TRANSPLANTATION PROCEDURE

An individual transplantation, or graft, involves the following:

Isolation of a ring of body column tissue (Figure 1.2: Step A). Usually the ring of tissue isolated consists of 1/6 – 1/8 of the body column. Figure 1.2 shows a body column divided into 8 regions. To isolate a region do the following: Place a hydra in a medium-sized petri dish containing hydra medium, and let it stretch out. Determine the location of a region to be isolated. For example, for the 3-region, let the animal stretch out and estimate the location of the middle of the body column. Then estimate the location of the point half way between the middle and the top of the body column [where the tentacles emerge]. This location would be the top of the 3-region. With a pair of forceps in one hand, cradle the hydra. Using the scalpel in the other hand, gently bisect the animal at the apical end of the region you intend to isolate. Let the contracted animal extend, and bisect once more at the point below the apical end which will result in a ring of tissue approximating 1/8 of the length of the body column.

Thread the ring of tissue onto a piece of fish line (Figure 1.2: Step B). When grafting a ring of tissue into a host, it is important that the basal end of the isolated ring be

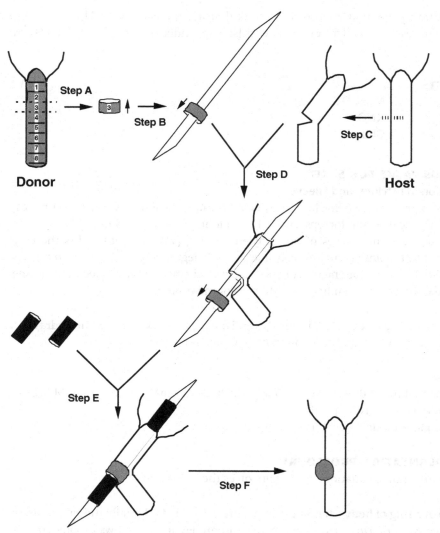

Figure 1.2. Detailed procedure for transplanting a ring of tissue from the body column of a donor hydra to the body column of a host hydra. The six steps for the procedure are described in the text.

brought into contact with the host. To ensure that this happens be certain that the ring of tissue is threaded onto the fish line in the appropriate orientation (as indicated by the arrow in Figure 1.2: Steps A and B). Using two pairs of forceps, gently cradle the ring with one pair, and holding a piece of fish line with the second pair, gently slide the piece of fish line through the ring. Make sure that the apical end of the ring of tissue is facing the end of the fish line. Slide the ring along the fish line until it is about 3–4 mm from the end.

Graft the ring of tissue to the host (Figure 1.2: Steps C and D). Place an adult hydra, which will serve as the host, into the petri dish with the ring of tissue and let it stretch out. Using the scalpel make a cut perpendicular to the body axis that extends about 1/2 way through the body column (Figure 1.2: Step C). For all the experiments described below the location of where the cut will be made will be indicated in terms of the body length [BL]. Thus, when grafting into a location that is 75% of the distance down the body column from the head, the location will be identified as "75% BL."

After cutting, a gap will appear (Figure 1.2: Step C). Using two pairs of forceps, cradle the host with one pair and slide the fish line, holding the ring of tissue into the cut, up through the gastric cavity, and out the mouth (Figure 1.2: Step D). When reaching the mouth, gently push, and the animal will open its mouth. Then slide the ring of tissue along the fish line so that it is in firm contact with the cut edges of the host.

Thread "sleeves" onto the ends of the fish line (Figure 1.2: Step E). It is important to keep the ring of tissue firmly in place as well to keep the animal from moving along the fish line. To do this, pieces of polyethylene tubing, referred to as sleeves, are threaded onto the two ends and brought into contact with the ring of tissue and the head respectively (Figure 1.2: Step E). The 2–3 mm pieces of polyethylene tubing are the "sleeves." With one pair of forceps hold the fish line extending out of the mouth. Use the second pair of forceps to slide a sleeve onto the piece of fish line extending from the ring of tissue, and use it to push the ring of tissue so that it is firmly in contact with the host tissue. Repeat this step with a second sleeve so that it is firmly in contact with the hypostome. Do not push so hard that the tissue folds.

Healing of the graft and removal of the fish line (Figure 1.2: Step F). With a pair of forceps gently transfer the graft to another medium-sized petri dish containing hydra medium. It is not a problem if the graft and fish line float on the surface. When all the grafts for a sample have been completed and transferred to this dish, place the dish [as is, or on a tray] in the 18 °C incubator, or on the lab bench at 15–25 °C.

The cut edges of the ring of tissue and the host will fuse together and heal within 1–2 h. At any time thereafter, remove the sleeves from each graft. Do this by holding one end of the fish line firmly with a pair of forceps, and gently removing the sleeve from the opposite end. Repeat this step for the second sleeve. Then, firmly holding the end of the fish line protruding from the mouth with one pair of forceps, place the other pair of forceps so that it gently cradles the fish line extending from the mouth. Now, slowly pull the fish line through the mouth until it is free of the host animal and the grafted ring of tissue. Or, gently push the animal down the fish line until the animal and the fish line are separated.

Examination of the grafts. Once the sleeves and fish line have been removed from all the grafts in the sample, the grafts should be incubated at 18 °C. Thereafter, the grafts should be examined daily to determine the fate of the grafted ring of tissue.

OUTLINE OF THE EXPERIMENTS

Two pairs of experiments can be carried out to demonstrate the presence of the head activation and head inhibition gradients in hydra.

A. A HEAD ACTIVATION GRADIENT IN THE BODY COLUMN

These simple experiments demonstrate that tissue of the body column has the capacity to form a head and that this property, termed head activation, is graded down the body column.

1. Tissue of the body column has head formation capacity. Head formation capacity can be shown simply by bisecting an animal in the middle of the body column and letting the lower half regenerate a head at its apical end (Figure 1.3a).

PROCEDURE

1. From the stock culture pick out 10 1-day-starved adults of similar size.
2. Place them in a 60-mm petri dish containing about 10 ml hydra medium so that the dish is half full with medium.
3. Using a scalpel, bisect each animal in the middle of the body column resulting in an upper half with a head, and a lower half with a foot.
4. Using a Pasteur pipette with a bulb, remove the lower halves, and transfer to a second 60-mm petri dish with hydra medium.
5. Incubate the dish at 18°C.

DATA RECORDING. The head of a hydra consists of two parts (see Figure 1.1a): The dome-shaped upper half is the hypostome, which contains the mouth. The lower half is the tentacle zone from which a ring of tentacles emerge. Head regeneration will occur as follows (Figure 1.3a): Following bisection, the wound at the apical end of the lower half heals over. At an early stage, a ring of small protrusions, or tentacle bumps, forms below the apical cap. Subsequently, the bumps grow into short tentacles, and later into long tentacles. As the tentacles are forming, the mouth is developing in the hypostome. A fully formed mouth will open widely in response to glutathione treatment, which provides an easy way to assay the formation of the mouth. The analysis of head regeneration should be carried out in the following steps:

➤ Examine each of the 10 regenerating lower halves daily for 4–5 days with respect to tentacle formation and mouth formation using a dissecting microscope. When the daily analysis is complete, return the samples to the incubator.
➤ Determine the extent of tentacle formation. Start this examination on the day after decapitation, and carry out every 1–2 days until the end of the experiment.
   (a) Number of regenerates with a healed apical cap.
   (b) Number of regenerates with a ring of small protrusions, or tentacle bumps, which form a ring at the base of the apical cap.

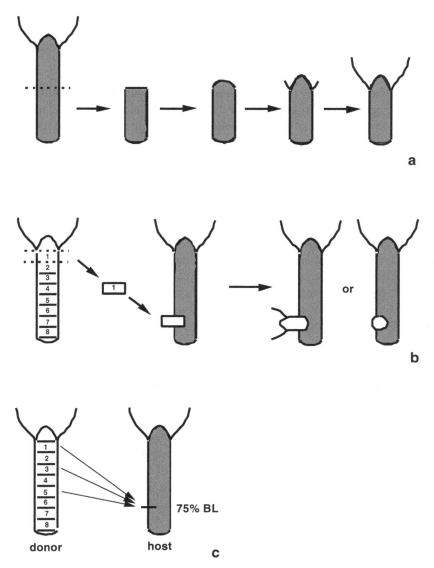

Figure 1.3. Two experiments demonstrating (a) head activation and (c) the head activation gradient. (b) Illustrates the transplantation procedure.

    (c)   Number of regenerates with short tentacles.
    (d)   Number of regenerates with full-length tentacles.
➤ Determine if a mouth has formed. Start this assay when tentacle bumps have appeared, and carry out every 1–2 days until the end of the experiment.
    (a)   Add 100 μl 1 mM of glutathione to the 10 ml of hydra medium in the dish.
    (b)   With the dissecting microscope, observe the animals 5 and 10 minutes later.
    (c)   Determine the number of hydra that open their mouths.

(d)   Determine the extent of opening: small, medium, or wide. The extent of mouth opening indicates the extent of completion of mouth formation.

(e)   When analysis is complete remove the hydra medium containing glutathione, rinse with hydra medium, and then add 10 ml of fresh hydra medium.

2. The head formation capacity, or head activation, is graded down the body column.
Another way to demonstrate that tissue of the body column has head formation capacity is the following. Isolate a piece of the body column from a donor animal and transplant it to a lower location on the body column of a second, or host, animal (Figure 1.3b). For example (Figure 1.3b), when the 1-region of a donor is transplanted to an axial level of 75% BL in many samples, it will form a second axis with a head at the apical end (see Figure 1.3b).

To examine the distribution of head activation along the body column, one can carry out this transplantation experiment using regions from successively more basal parts of the body column (see Figure 1.3c). By determining the fraction of transplants of each kind that form a second axis with a head, the distribution of head activation along the body column can be determined. For this experiment, compare the fraction of transplants of the 1-region, 3-region and 5-region that form second axes. The procedure for carrying out the transplants is described in Materials and Methods.

Carry out 15–20 grafts for each of the following type of transplantation:

(a) Transplant the 1-region of the donor into a host at 75% BL.
(b) Transplant the 3-region of the donor into a host at 75% BL.
(c) Transplant the 5-region of the donor into a host at 75% BL.

DATA RECORDING. Examine each animal in each set of transplants daily and determine the following:

(a) Number of samples with a normally developing head using the criteria for tentacle formation in the head regeneration experiment described in the previous section.
(b) Number of samples developing an attempted head. Such transplants form a single tentacle at the apical end of the transplant.
(c) Number of samples forming a foot. The apical end of the developing transplant will form a blunt end that is sticky. The stickiness can be tested by touching the end with a pair of forceps and seeing if the forceps remain attached to the tissue.
(d) Number of samples forming neither a head nor a foot. The apical end of the transplant remains round and smooth.
(e) The sets of transplants should be examined every 1–2 days until no further changes take place in the type of result formed. For example, some transplants may form a head later than others.

B. A HEAD INHIBITION GRADIENT IN THE BODY COLUMN
Head inhibition is produced by the head organizer in the hypostome and transported to the body column, where it prevents body column tissue from forming heads. The existence of head inhibition and its axial distribution can be demonstrated with the following experiments.

1. Absence of the head reduces head inhibition in the body column. If head inhibition is produced in the head and transmitted to the body column, then removal of the head should reduce the level of head inhibition in the body column. The experiment providing evidence for this statement, as shown in Figure 1.4a, is a variation of the second experiment in the previous section.

Carry out 15–20 grafts for each of the following types of transplantation:

(a) Experimental: decapitate the host and transplant the 1-region of the donor into this host at 50% BL.
(b) Control: transplant the 1-region of the donor into a normal host at 50% BL.

DATA DECORDING. Examine each animal in each set daily using the same criteria as described in the experiment for the distribution of head activation (see previous section, The Head Formation Capacity, or Head Activation, Is Graded Down the Body Column).

2. Head inhibition is graded down the body column. To examine the distribution of head inhibition along the body column, carry out an experiment similar to the one described above for the head activation gradient, as shown in Figure 1.4b. Again, the overall transplantation procedure is as described in Materials and Methods.

The graded distribution of the head inhibition gradient can be demonstrated by grafting 1-regions of donors to different locations (25% BL, 50% BL, and 75% BL) in a host.

Carry out 15–20 grafts for each of the following type of transplantation:

(a) Transplant the 1-region of the donor to 25% BL of the host.
(b) Transplant the 1-region of the donor to 50% BL of the host.
(c) Transplant the 1-region of the donor to 75% BL of the host.

DATA RECORDING. Examine each animal in each set daily using the same criteria as described in the experiment for the distribution of head activation (see previous section, The Head Formation Capacity, or Head Activation, Is Graded Down the Body Column).

EXPECTED RESULTS AND DISCUSSION

The expected results from the two experiments are relatively straightforward. The first experiment of each set demonstrates the existence of the property, whereas the second experiment of each set demonstrates that the property is distributed as a gradient along the body column.

To determine if there are statistically significant differences between the percentage of transplants forming a second axis in, for example, the control and decapitated hosts in the experiment illustrating head inhibition, the following analysis can be carried out (e.g., Zar, 1974): Each student will carry out 15–20 grafts for each of the control and decapitated hosts. A percentage of each type of graft will form a second axis. Calculate the average value for the percentage +/− standard deviation for each of

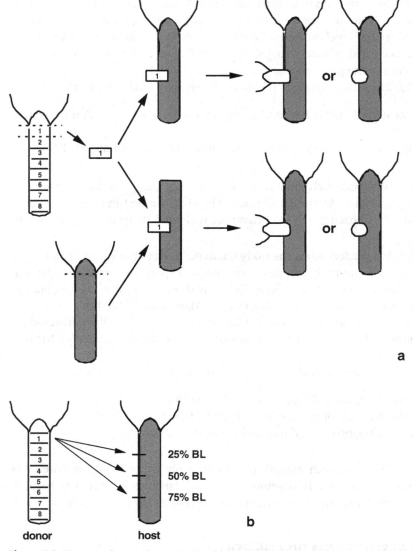

Figure 1.4. Two experiments demonstrating (a) head inhibition and (b) the head inhibition gradient.

the two types of grafts. If these values do not overlap, the difference is statistically significant.

DEMONSTRATION OF THE HEAD ACTIVATION PROPERTY

Bisection of the animal will result in the regeneration of a head at the apical end of the lower half. This indicates that the tissue of the body column is capable of head formation and contains head activation in some molecular form. Because the body column can be bisected anywhere along its length and the lower piece will regenerate a head, the head activation property is distributed all along the body column.

DEMONSTRATION OF THE DISTRIBUTION OF HEAD ACTIVATION

If the distribution of head activation is graded down the body column, one would expect to see the number of grafts forming a second axis with a head decreasing along the lower body column, the source of the isolated ring of tissue. That is, comparing the number of heads formed by each of the three regions, one would expect the 1-region to form more heads than the 3-region. In turn, the 3-region would form more than the 5-region.

DEMONSTRATION OF THE HEAD INHIBITION PROPERTY

Decapitation removes the source of head inhibition. Accordingly, one would expect a lower level of head inhibition in the body column and an increase in the proportion of grafts that form a second axis in the decapitated hosts compared to the grafts in the normal hosts. This is expected if one assumes that head inhibition decays rapidly so that the level is reduced. This is, in fact, the case as the half-life of head inhibition is 2–3 h (MacWilliams, 1983*b*).

DEMONSTRATION OF THE DISTRIBUTION OF HEAD INHIBITION

Here one would expect the reverse of the results in the experiment demonstrating the distribution of head activation. Assume that head inhibition is maximal at the upper end of the body column and graded down the body column. If so, one would expect to see more transplants that form a second axis with a head the farther down the body column the 1-region is transplanted into the body column. That is, the number of heads formed by the 1-region would be higher when transplanted to the 5-region compared to the 3-region. In turn, the number of heads would be higher when transplanted to the 3-region compared to the 1-region.

These results illustrate that the two developmental gradients play a major role in determining the pattern of structures formed along the axis of the body column in hydra. In the instance examined here, the two gradients – the morphogenetic gradient of head activation and the head inhibition gradient – control where a head is formed. Morphogenetic gradients also play a role in other animals, usually during very early stages of embryogenesis when the axes are being set up (see Chapter 19). Hydra is unusual in that the gradients are continuously active in the adult animal.

TIME REQUIRED FOR THE EXPERIMENTS

The execution of these experiments involves learning how to carry out the transplantation procedure. Usually a student will need 1–3 h to learn, become comfortable with, and then, successful with the grafting process. Thereafter, the student will usually be able to carry out 6–10 grafts/h.

To get enough data to obtain a clear result, it is necessary to carry out at least 10 grafts (preferably 15–20) for each type of transplantation. Then, the amount of time required for each experiment would be the following:

➤ Regeneration experiment demonstrating head activation in the body column. This experiment does not require much time. The manipulations, which are the bisecting and handling of the animals, would take 15–20 min. Using a dissecting microscope to examine the extent of regeneration of each animal in a sample requires about 15–20 min/day.

➤ Grafting experiment illustrating head inhibition. As there are two types of grafts involved in this experiment, the total number of grafts would be 20 if 10/type are carried out, or 40 if 20/type are carried out. Assuming that one can do 6–10 grafts/h, then 2–3 h would be required for carrying out 20 grafts and 4–6 h for 40 grafts. Analysis of the grafts would most likely require 30–60 min/day.

➤ Grafting experiments demonstrating the head activation gradient or the head inhibition gradient require a similar amount of time. In both experiments there are three different types of grafts. Thus, the total number of grafts would be either 30 at 10 grafts/type of graft, or 60 at 20 grafts/type. This would require 3–5 h for 30 grafts, or double that for 60 grafts. Analysis of the grafts requires 30–60 min.

One way to reduce the time required for the experiments would be to divide the experiment among several students. For example, for the head inhibition experiment, students could work in groups of 4, each carrying out 5 control grafts and 5 experimental grafts. For the two experiments demonstrating the presence of the head activation and head inhibition gradients, groups of 6 students each doing 10 grafts of one type in an hour would provide the 60 grafts needed for acquiring a reasonable amount of data for each experiment.

POTENTIAL SOURCES OF FAILURE

As the only manipulations involved are the isolation and transplantation of a ring of tissue into a host, the only significant source of failure is a failure of the ring of tissue to graft onto and heal to the host. Practice usually takes care of this problem.

TEACHING CONCEPTS

The major concept illustrated with these experiments is that the pattern along the axis of an animal can be controlled by a morphogenetic gradient. When the morphogen concentration is above a threshold, such as for head formation, then the tissue becomes committed to forming a head. The inhibition gradient illustrates a second process common in embryogenesis and developing systems. Once a piece of tissue, or region of the embryo, has become committed to forming a particular cell type, or a structure, then an inhibitory mechanism, commonly referred to as lateral inhibition, is initiated to prevent that same cell type or structure from forming in the vicinity of the first one (see Chapter 22).

ALTERNATIVE EXERCISES

Two additional experiments can be carried out which extend the information gained from the experiments described above. They would also begin to provide insight into the molecular basis of the head activation gradient.

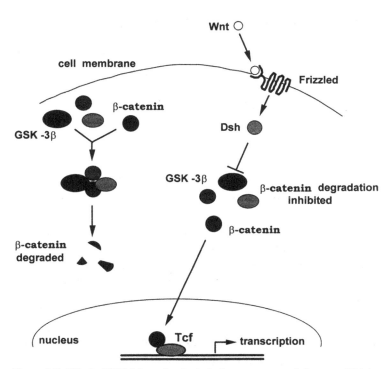

Figure 1.5. Effect of GSK-3β on β-catenin in the presence and absence of Wnt.

PROPOSED EXPERIMENTS

A major pathway that affects a number of developmental events, or processes, during early embryogenesis is the Wnt pathway (Cadigan and Nusse, 1997). On the outer surface of a cell, the pathway consists of Wnt, a signaling molecule, and Frizzled, a receptor for Wnt on the cell surface (Figure 1.5). Inside the cell, the pathway, for the sake of simplicity, consists of Disheveled, GSK-3β, β-catenin and Tcf. As shown in Figure 1.5, when Wnt is absent, GSK-3β causes the degradation of β-catenin. When Wnt is present, the activated form of Disheveled blocks GSK-3β. This in turn prevents the degradation of β-catenin. Then β-catenin coupled with Tcf enters the cell nucleus and acts as a transcription factor, stimulating the transcription of genes required for a specific developmental process. In hydra, *HyWnt*, and *HyTcf*, the hydra homologues of the *Wnt* and *Tcf* genes are expressed in the hypostome (Hobmayer et al., 2000), suggesting the Wnt pathway has a role in the formation and/or activity of the head organizer. LiCl is known to block the activity of GSK-3β (Phiel and Klein, 2001), thereby allowing β-catenin to enter the nucleus, and with Tcf initiate a new developmental process. If so, one might expect treatment of hydra with LiCl to result in the formation of head structures such as tentacles, or complete heads, along the body column; such results have been obtained (Hassel, Albert, and Holfheinz, 1993). Further information on the Wnt pathway can be obtained from http://www.stanford.edu/~rnusse.

The following pair of experiments illustrate this possibility.

Effect of 2 mM LiCl on the body column. Tentacles that form on the body column are called ectopic tentacles. To demonstrate that treatment with 2 mM LiCl will cause the formation of ectopic tentacles, the following experiment can be carried out. This experiment is ideally carried out with a strain of *Hydra vulgaris* or *Hydra littoralis*.

1. Treat 20 animals with 2 mM LiCl in hydra medium for 2 days and then return to hydra medium.
2. Examine animals every 1–2 days for 4–7 days and determine:
   (a) How soon after end of treatment do ectopic tentacles appear on the body column?
   (b) What fraction of the treated animals produce ectopic tentacles?
   (c) Where along the body column do the tentacles appear?

Effect of 2 mM LiCl on the head activation gradient. The formation of ectopic tentacles suggests that the head activation level has risen in the body column, surpassing the level of head inhibition, thereby permitting the formation of head structures. To directly determine if 2 mM LiCl affects the head activation gradient, a transplantation experiment of the type described in a previous section can be carried out as follows:

1. Treat 20 animals with 2 mM LiCl in hydra medium for 2 days.
2. Transplant the 4-region of a LiCl-treated animal to 50% BL of an untreated host using the usual transplantation experiment. Carry this out for all 20 animals.
3. As a control, transplant the 4-region of an untreated animal to 50% BL of an untreated host. Carry out this experiment for 20 animals.
4. Compare the number of animals in the treated and control samples that formed a second axis, or head.

Expected results. One would expect treatment with 2 mM LiCl to result in the formation of ectopic tentacles along the body column. Presumably this reflects a rise in head activation in the body column. If so, one would expect the fraction of transplants using LiCl-treated donors to form a higher fraction of 2nd axes than the controls. The first experiment has been done several times (e.g., Hassel et al., 1993). However, there are no published data concerning the second experiment.

QUESTIONS FOR FURTHER ANALYSIS

These three questions probe the nature and effects of the gradients a little further:

➤ What kind of grafting experiment would you carry out to demonstrate that the head produces the signal that sets up the head activation gradient?
➤ An early step during bud formation involves the initiation of head formation, or the formation of the head organizer. How would you show that the head inhibition gradient has a role in the initiation of bud formation?
➤ The head activation gradient is said to confer a polarity on the tissue in terms of a head forming at the upper end and a foot forming at the lower end of the body column. How would you demonstrate this polarity?

REFERENCES

Bode, H. R. (1996). The interstitial cell lineage of hydra: A stem cell system that arose early in evolution. *J. Cell Science*, 109, 1155–64.

Bode, P. M., and Bode, H. R. (1984). Patterning in hydra. In *Primers in Developmental Biology*, vol. I., *Pattern Formation*, eds. G. M. Malacinski and S. V. Bryant, pp. 213–41. New York: Macmillan Press.

Broun, M., and Bode, H. R. (2002). Characterization of the Head Organizer in Hydra. *Development*, 129, 875–84.

Browne, E. N. (1909). The production of new hydranths in hydra by insertion of small grafts. *J. Exp. Zool.*, 7, 1–37.

Cadigan, K. M., and Nusse, R. (1997). Wnt signalling: A common theme in animal development. *Genes Dev.*, 11, 3286–305.

Hassel, M., Albert, K., and Hofheinz, S. (1993). Pattern formation in *Hydra vulgaris* is controlled by lithium-sensitive processes. *Dev. Biol.*, 156, 362–71.

Hobmayer, B., Rentsch, F., Kuhn, K., Happel, C. M., Cramer von Laue, C., Snyder, P., Rothbacher, U., and Holstein, T. W. (2000). Wnt signalling molecules act in axis formation in the diploblastic metazoan, *Hydra. Nature*, 407, 186–9.

MacWilliams, H. K. (1983*a*). Head transplantation phenomena and the mechanism of *Hydra* head regeneration. II. Properties of head activation. *Dev. Biol.*, 96, 239–57.

MacWilliams, H. K. (1983*b*). Head transplantation phenomena and the mechanism of *Hydra* head regeneration. I. Properties of head inhibition. *Dev. Biol.*, 96, 217–38.

Meinhardt, H. (1993). A model for pattern formation of the hypostome, tentacles and foot in Hydra: How to form structures close to each other, how to form them at a distance. *Dev. Biol.*, 157, 321–33.

Phiel, C. J., and Klein, P. S. (2001). Molecular targets of lithium action. *Ann. Rev. Pharmacol. Toxicol.*, 41, 789–813.

Rubin, D. I., and Bode, H. R. (1982). The Aberrant, a morphological mutant of *Hydra attenuata*, has altered inhibition properties. *Dev. Biol.*, 89, 316–31.

Wolpert, L. (1971). Positional information and pattern formation. *Curr. Topics Dev. Biol.*, 6, 183–224.

Zar, J. H. (1974). *Biostatistical Analysis*. Englewood Cliffs, N.J.: Prentice-Hall, Inc.

APPENDIX: MAINTENANCE OF A HYDRA CULTURE

A total of 250–300 hydra will be needed to carry out all four experiments. The number of hydra needed per student will depend on the number of students as well as which experiments are selected. For practical purposes it is useful to obtain the hydra 3–4 weeks before they are used in experiments. In this way, the number can be increased to reach a level required for the class. When hydra are fed 3 times/week, the population size will double because of asexual reproduction by bud formation in 7–10 days. To obtain a faster doubling time, hydra can be fed 5 times a week. If experiments are to be carried out for a few weeks, it is worthwhile maintaining a culture containing enough hydra so that no more than 40% of them are used each week. With the indicated doubling time, this should permit maintenance of a steady-state culture of animals. In the following, the materials, equipment, and procedures for maintaining a hydra culture will be described. Any species of brown hydra is appropriate for these experiments. If available, a strain of *Hydra vulgaris* or *Hydra magnipapillata* is preferable as most of the work on developmental gradients has been done with these species. *Hydra littoralis* is very closely related to these two species and will work equally well.

MATERIALS AND EQUIPMENT (PER CLASS OF STUDENTS)

50–200 hydra of a single species [*Hydra littoralis*: Carolina Biological Supply Co.].

Dishes for culturing hydra: 150-mm petri dishes (200 hydra/dish; Fisher Scientific), plastic containers, or glass baking dishes (1000 hydra/dish).

Hydra medium: 10 liters. Composition of hydra medium: 1.0 mM $CaCl_2$, 1.5 mM $NaHCO_3$, 0.1 mM $MgCl_2$, 0.08 mM $MgSO_4$, and 0.03 mM KNO_3, pH 7.5–8.0.

Pasteur pipettes and rubber bulbs for the pipettes (Fisher Scientific).

A one-liter glass bottle with rubber stopper and glass tubing for hatching brine shrimp cysts.

NaCl for hatching brine shrimp cysts: 40 g/liter (least expensive option is to obtain it from a supermarket or store for home aquarium supplies).

1000 × antibiotic stock solution: 2.5 g Penicillin (Sigma) + 2.5 g Streptomycin (Sigma) in 50 ml water.

1 can brine shrimp eggs: these are cysts (= desiccated fertilized eggs) of the brine shrimp, *Artemia salina* (Great Salt Lake Artemia Cysts; Sanders Brine Shrimp Co).

A one-liter beaker.

A light source (fluorescent light or incandescent light bulb).

Shrimp net: mesh attached to a circle of plastic (6–8 cm in diameter) attached to a plastic handle – similar to a net for catching butterflies, but smaller. Mesh should be fine enough to retain the hatched shrimp larvae (local store for fishing supplies or pet store).

Round glass bowl (~25 cm in diameter at top of bowl).

Container for hydra medium: 20 liters carboy with spigot (Fisher Scientific).

Fish tank air pump (this kind of pump is commonly used to bubble air into a small home aquarium, or fish tank, and is available in pet stores).

RAISING AND HANDLING HYDRA

In the laboratory, hydra are grown in any convenient transparent container with lids. These include 150-mm plastic petri dishes, plastic boxes, or glass baking dishes covered with a lid of available material. Plastic films such as Saran Wrap should not be used as they may be covered with a reagent or compound that dissolves in hydra medium and damages the animals.

Hydra medium. Hydra medium consists of a dilute salt solution (see Materials and Equipment for composition) made up in fresh water. Use tap water if it is free of high levels of compounds, such as chlorine, meant to reduce the level of micro-organisms. Otherwise, it is wise to use water that has undergone reverse osmosis, or is distilled. For convenience, it is useful to make up 5–20 liters of hydra medium at a time in a large plastic container with a spigot. Store at room temperature [15–25°C].

Handling of hydra. To transfer hydra from one dish to another, use a Pasteur pipette with a rubber bulb attached to the end. With such a pipette one can suck up one or more hydra and some of the medium in one dish, and expel the animals into a second dish. If the hydra are floating in the medium simply use the pipette to suck them up and transfer them. If the animals are attached to the floor of the dish, one can detach them from the dish in three ways: (1) sucking them up directly; (2) expelling fluid at their feet forcing them to be released from the dish; (3) placing the tip of the pipette against the bottom of the dish next to the animal and gently pushing at the foot.

Growth conditions. Hydra are normally grown at 18°C in an incubator with light that is controlled by a timer. The light cycle consists of 12 h on and 12 h off. In case an incubator is not available, hydra can be grown in the laboratory as long as the temperature does not rise above 24–25°C. Above that temperature, the animals suffer, and exposure to 30°C leads to death within an hour. Exposure to lower temperatures [10–18°C] has no effect on the animals although they may grow more slowly at lower temperatures. The density of hydra per dish is optimally 1–2 animals/cm². If grown at double that density, it is difficult to keep the animals clean. Unclean animals become ill and damaged. When using 150-mm petri dishes, enough hydra medium should be used to fill to a depth of 10–12 mm.

FEEDING AND WASHING HYDRA

Hydra catch food with the nematocytes in their tentacles. When a piece of food, such as one or more shrimp larvae, bumps into a tentacle, nematocytes are discharged which capture and kill the larvae. Then the hydra moves the tentacle towards the hypostome, or mouth, and ingests the dead larvae.

Food for hydra. The simplest and most convenient form of food available for hydra is the hatched larvae, or nauplii, of the brine shrimp, *Artemia salina*. Embryos of *Artemia salina* in the form of stable dormant cysts are commercially available (see list of reagents). Once a can of cysts has been opened distribute the cysts to 50 ml or 100 ml plastic tubes with caps and store at 4°C. This will provide enough shrimp for several years without loss of viability.

Hatching of brine shrimp eggs. Dissolve 40 g NaCl in one liter of hydra medium in a one-liter glass bottle. To minimize bacterial growth, add 1 ml of the stock solution of antibiotics. Then add 25 ml of brine shrimp cysts. Firmly insert a rubber stopper containing two holes with a glass tube through one of the holes into the opening of the bottle. The glass tube should extend about 90% of the distance along the length of the bottle in the solution and several centimeters outside the bottle. Attach a rubber or plastic tube to the outer end of the tube and to a fish tank air pump. Let air bubble through the bottle for about 2 days at room temperature (15–25°C).

Collection of hatched shrimp. The hatched shrimp larvae will be bright orange while the unhatched cysts will be brown. To collect the larvae, pour the contents of the bottle into a beaker, and place the beaker on the lab bench. Then place a light source next to the bottom of the beaker. The shrimp larvae migrate towards light and will accumulate near the light source. With a Pasteur pipette or regular pipette withdraw the larvae into the pipette. Transfer the larvae to a shrimp net with a mesh that is fine enough to retain the hatched larvae. Rinse the shrimp by holding the net under a faucet and letting water flow over the shrimp and through the net. Insert the net into a beaker containing hydra medium.

Feeding hydra. To feed animals, use a Pasteur pipette with a bulb to transfer hatched larvae to the dish containing hydra. Scatter the larvae around so that all hydra have a chance of catching larvae. The larvae will swim around until they run into the tentacles of the hydra and are captured. A good meal for a hydra amounts to 4–8 larvae. Larvae that manage to escape the hydra will swim around for less than an hour and die because they are salt water animals and cannot tolerate the freshwater environment of the hydra.

Washing hydra. Two to six hours after feeding, the animals are washed by pouring off the hydra medium and remaining shrimp into a round glass bowl. This also removes any food remains expelled by the hydra. To remove remaining larvae, rinse the hydra bowl with hydra medium and pour into the round bowl.

Most of the hydra will stick to the bottom of the culture dish, and are covered with a film of hydra medium. Add fresh hydra medium to the culture dish. Pouring the medium plus larvae into the glass bowl will most likely result in some hydra being carried along. Gently stir the medium in the bowl in a circular motion which brings all hydra into one place in the middle of the bottom of the bowl. Recover these animals with a Pasteur pipette, and return them to the culture dish.

2 Embryonic regulation and induction in sea urchin development

C. A. Ettensohn

OBJECTIVE OF THE EXPERIMENT Cell–cell interactions play an important role in the early patterning of animal embryos. Polarity inherent in the oocyte or established soon after fertilization entrains subsequent cell signaling events that subdivide the early embryo into distinct territories of gene expression and cell fate.

The objective of the experiments described in this chapter is to illustrate the role of cell–cell signaling in patterning early animal embryos. The sea urchin, a deuterostome that relies extensively on cell interactions to specify blastomere fates, is used as a model system. Two major experiments are described: (1) Analysis of the development of individual blastomeres isolated from early cleavage stage embryos, illustrating the phenomenon of regulative development. (2) Recombination of micromeres with animal blastomeres, illustrating the process of embryonic induction. In addition, a third experiment is described that involves the use of molecular markers (antibodies) to analyze cell fates, an approach that can be applied to either of the first two experiments.

DEGREE OF DIFFICULTY Experiment 1 requires students to collect sea urchin gametes, fertilize eggs, and dissociate embryos. All these skills can be learned relatively easily. Experiment 2 is moderately difficult as it involves micromanipulation, a technique that demands dexterity and patience. Both experiments require stereomicroscopes. The results of Experiments 1 and 2 can be assessed morphologically or by staining embryo whole mounts with antibodies that label specific tissue types (Experiment 3). The latter approach is technically straightforward but requires a compound microscope, preferably one equipped for epifluorescence.

INTRODUCTION

The earliest studies in experimental embryology involved splitting embryos into separate parts. W. Roux (1888) destroyed one blastomere of the 2-cell amphibian embryo and found that each half formed a half-larva. Similarly, embryologists working with ascidian embryos found that individual blastomeres isolated from cleavage stage embryos differentiated according to the particular fates they would have adopted in the intact embryo (Conklin, 1905). Although Roux's experiments were later reinterpreted (McClendon, 1910), these results suggested that in some embryos the prospective potency of blastomeres (i.e., their range of possible fates) was equivalent to their normal fates. Such embryos were described as "mosaic" because they appeared to develop as a mosaic of self-differentiating parts.

Sea urchin embryos have been very useful for experimental embryology because they are easy to obtain, develop externally, and are physically manipulable. Moreover, their optical transparency means that cell behaviors and fates can be observed directly. H. Driesch (1892) found that individual blastomeres isolated from sea urchin embryos at the 2- or 4-cell stages could give rise to complete, well-patterned (although miniature) larvae. This remarkable property, termed "regulation," demonstrated that the prospective potency of each blastomere exceeded its prospective fate. Rather than developing as a mosaic of self-differentiating parts, Driesch's findings showed that the context of blastomeres play an important role in regulating their fates. This, in turn, implied the existence of signals between blastomeres.

S. Hörstadius, one of the great figures of experimental embryology, carried out a series of blastomere isolation and recombination experiments using late cleavage stage sea urchin embryos (reviewed in Hörstadius, 1939). These remarkable experiments provided a wealth of information concerning cellular interactions that pattern the early sea urchin embryo. As part of this work, Hörstadius discovered the potent inductive properties of micromeres – small blastomeres that form at the vegetal pole at the 16-cell stage. These studies provided some of the earliest evidence of embryonic induction.

Today, developmental biologists have a better understanding of the cellular and molecular mechanisms of early fate specification in animal embryos. Eggs are seldom classified as "mosaic" or "regulative." A more modern view is that all eggs are endowed with inherent polarity through the localization of maternal mRNAs or proteins, or acquire such polarity at fertilization. All embryos subsequently rely on cellular interactions, at least in part, to regulate cell fates. "Mosaic" eggs have a relatively high degree of ooplasmic specialization and a spatial pattern of cleavage that segregates maternal regulatory factors into different cell lineages very early in development. In contrast, "regulative" eggs have a less regionalized cytoplasm and a pattern of cleavage that does not segregate maternal regulatory factors into different cell lineages until later in development.

Our understanding of early patterning in sea urchins has advanced greatly since the time of Driesch and Hörstadius. For example, the inductive properties of micromeres have been studied in detail (Ransick and Davidson, 1993; McClay et al., 2000; Ishizuka, Minokawa, and Amemiya, 2001), and at least one molecular signal, Delta, has been identified (Sweet et al., 2002) (see also Chapter 22). The current view of the molecular and cellular basis of early patterning in the sea urchin embryo is described in several recent reviews (Ettensohn and Sweet, 2000; Brandhorst and Klein, 2002; Angerer and Angerer, 2003).

MATERIALS AND METHODS

BIOLOGICAL MATERIAL

Information on obtaining and handling adult sea urchins can be found in Böttger, Unnma, and Walker (2004) and Runft, Adams, and Foltz (2004). Note that many different species are available commercially from suppliers in different parts of the world. Each species has a characteristic reproductive season (usually several months long) and it is only during this season that animals can be obtained in a gravid state. If the season is appropriate, 30 animals should provide sufficient gametes for 20 students to carry out the experiments. Also note that adults of different species must be maintained at different temperatures and their embryos develop only within a certain temperature range. In general, the embryos of most warm water species develop well between 18°C and 25°C, and cold water species between 8°C and 16°C. Within the acceptable range for any particular species, embryos will develop more rapidly at warmer temperatures.

In the absence of marine aquariums suitable for maintaining adult sea urchins, it is possible to make arrangements to have commercial suppliers ship animals to arrive on the day of the experiment. Animals can be held in the original packing materials for at least a day, and usually overnight, so that experiments may be carried out over two days. In addition, as noted below, the spawned eggs of some cold water species (e.g., *Strongylocentrotus purpuratus*) can be stored overnight at 4°C and fertilized the following day. Therefore, even without saltwater aquariums, judicious choice of species, timing of the shipment, and storage of the adults and eggs can allow experiments to be conducted over a 3–4 day period.

Normal stages of sea urchin development are shown in Figure 2.1.

LABORATORY EQUIPMENT AND SUPPLIES

Note: All labware that comes in contact with eggs or embryos (or that is used to make solutions that will come in contact with eggs or embryos) must be free of detergents and toxic contaminants. It is best to keep such labware separate from general laboratory labware and to wash it with water (only) after each use. If it is necessary to use general labware, rinse it thoroughly with hot water followed by a rinse with distilled or deionized water.

Syringe and needle (almost any sizes will work, but a 10-cc syringe with 21-gauge needle works well)

Plastic or glass beakers (50–250 ml)

Pasteur pipettes and rubber bulbs

1.5-ml microfuge tubes

Nylon screen cloth (74 μm or 53 μm mesh opening) (Small Parts, Inc.)

Glass culture bowls (2–10 cm diameter)

Mouthpipettes

Glass microscope slides and coverslips

Plastic Petri dishes (35 mm and 60 mm diameter)

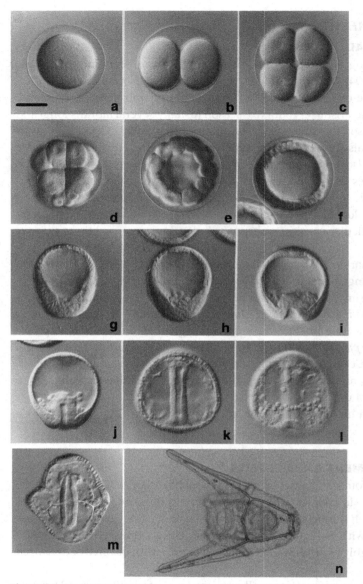

Figure 2.1. Normal stages of sea urchin development (*Lytechinus variegatus*, 23 °C). The time after fertilization is in parentheses. (a) Fertilized egg (10 min). (b) 2-cell stage (1 h). (c) 4-cell stage (1.5 h). (d) 16-cell stage (3 h). (e) 64-cell stage (4.5 h). (f) Pre-hatching blastula (6.5 h). (g) Hatched blastula (8.5 h). (h) Mesenchyme blastula (11 h). (i) Early gastrula (14 h). (j) Mid-gastrula (16 h). (k, l) Late gastrula (18 h), two different focal planes. (m) Prism (21 h). (n) Pluteus larva (36 h). Scale bar in (a) represents 50 μm.

Plastic, round-bottom 96-well plates (e.g., Falcon 353911 flexible U-bottom plates)
Scotch No. 665 double-sided tape (3M Company)
Stereomicroscopes
Compound microscopes, preferably with epifluorescence capability

CHEMICALS AND SOLUTIONS

Sea urchins are marine organisms, and adults and embryos must be kept in natural or artificial seawater (SW). Seawater salts (i.e., Instant Ocean) can be obtained from marine aquarium suppliers and dissolved in distilled or deionized water. There are also many artificial seawater formulae:

MBL formula seawater (from the Marine Biological Laboratory, Woods Hole, MA, USA):

NaCl	24.72 g
KCl	0.67 g
$CaCl_2.2H_2O$	1.36 g
$MgCl_2.6H_2O$	4.66 g
$MgSO_4.7H_2O$	6.29 g
$NaHCO_3$	0.18 g

Dissolve the salts in ~800 ml deionized or distilled water then bring to a final volume of 1 liter. Adjust pH to 8.3.

Calcium-free seawater (CF-SW) + EGTA

NaCl	28.32 g
KCl	0.77 g
$MgCl_2.6H_2O$	5.41 g
$MgSO_4$ (anhyd)	3.48 g
EGTA	0.019 g
$NaHCO_3$	0.20 g

Dissolve the salts in ~800 ml deionized or distilled water then bring to a final volume of 1 liter. Adjust pH to 8.3.

Calcium-free seawater (CF-SW) − EGTA. Same formulation as above, but omit EGTA
para-Aminobenzoic acid (PABA), and sodium salt (SIGMA)
0.5 M KCl
1 M $CaCl_2$
Agarose
Methanol
Phosphate-buffered saline (PBS), pH 7.4
Tween-20
Goat serum
Hybridoma tissue culture supernatants 6a9 and Endo1 (available from the Developmental Studies Hybridoma Bank, University of Iowa; http://www.uiowa.edu/~dshbwww).
Fluorescently conjugated, anti-mouse secondary antibody (broad spectrum, anti-IgG + IgM) (e.g., affinity-purified, fluorescein-conjugated goat anti-mouse IgG + IgM (H+L), Jackson Immunoresearch Laboratories).

OUTLINE OF THE EXPERIMENTS

EXPERIMENT 1. REGULATIVE DEVELOPMENT OF ISOLATED BLASTOMERES

1. Methods for obtaining sea urchin gametes from adults and for fertilizing eggs can be found in Runft et al. (2004) and at http://www.stanford.edu/group/Urchin/first.htm. Briefly, gravid, adult sea urchins are induced to shed gametes by injecting the animals with 0.5 M KCl (1–3 ml total) at several points around the mouth. In most species, males and females cannot be distinguished based on external morphology, but only by the color of the shed gametes. Eggs, which are orange-yellow in color, are collected by inverting the female over a beaker of sea water (SW). Sperm are collected "dry" using a Pasteur pipette to remove the whitish fluid from the surface of the male. The sperm is transferred to a 1.5-ml microfuge tube and kept on ice or at 4°C until use. Eggs of most species should be washed gently with SW 2–3 times before fertilization. This can be accomplished by allowing the eggs to settle gently in a beaker and then aspirating or decanting the liquid. Eggs should generally be fertilized within 2–3 hours of shedding, although eggs of some cold water species (e.g., *Strongylocentrotus purpuratus*) can be stored overnight at 4°C. Sperm can be stored for several days at 4°C, provided they are not diluted with SW. A drop of sperm is activated by mixing with SW and then must be used within 5–10 minutes to fertilize eggs.

2. Fertilize 0.1–0.5 ml eggs in a plastic 100-ml beaker in PABA-SW (0.16 g PABA/100 ml SW, pH 8.3, prepared just before use). PABA irreversibly inhibits an ovoperoxidase released by the fertilized egg that catalyzes the cross-linking and hardening of the fertilization envelopes (Hall, 1978), making it possible to subsequently rupture the envelopes and dissociate the embryos. Allow the fertilized eggs to settle by gravity and wash them at least five times with CF-SW + EGTA. The eggs should be washed by aspirating or decanting as much of the liquid as possible, then pouring fresh CF-SW + EGTA into the beaker. Resuspend the eggs by stirring gently with a Pasteur pipette. Change the solution as soon as the eggs settle. If the eggs remain piled in a dense layer on the bottom of the beaker for a long period of time, their development will be adversely affected. Note that after a few rinses, the eggs will settle quite quickly because the jelly coat will have been washed away. All washing steps should be carried out gently to avoid rupturing the fertilization envelopes (softened by the PABA). If the eggs are inadvertently demembranated they will become sticky, clump together, and will generally be much more difficult to work with.

3. After rinsing with CF-SW + EGTA, culture the embryos in CF-SW + EGTA in clean, detergent-free plastic beakers or glass bowls. Ensure that the embryos are not overcrowded; there should be less than a monolayer of embryos on the surface of the dish.

4. At the desired stage (2-, 4-, or 8-cell), aspirate or decant as much of the CF-SW as possible and resuspend the embryos in 10 ml of fresh CF-SW (*without* EGTA). Pour the embryos through nylon screen cloth. The nylon screen can be draped over a beaker or fastened with a rubber band over the end of a plastic 50-cc syringe from which the tip has been removed with a razor blade. Nylon screen cloth is available in different mesh opening sizes. The eggs and embryos of most commonly used

sea urchin species vary in size from 80 to 120 μm. A mesh opening slightly smaller than the diameter of the embryo is ideal for dissociation. 74-μm mesh works well for many species, although smaller eggs may require 53-μm mesh. After passing the embryos through nylon cloth one time, place a drop on a microscope slide and determine whether the dissociation has been successful. If more than 50% of the embryos remain undissociated, pass them through the cloth a second time. After dissociation, pour an aliquot of the cell suspension into a 60-mm Petri dish that has been coated with 2% agarose in complete SW. (These dishes are prepared by heating a 2% wt/vol solution of agarose in complete SW until the agarose dissolves, then pouring a thin layer of the hot liquid into the dishes, and allowing it to cool.) Do not overload the cells; to avoid clumping they should cover only 5–10% of the dish.

5. Using a stereomicroscope and mouthpipette, select isolated blastomeres and transfer them to complete SW in 35- or 60-mm Petri dishes coated with 2% agarose. One must work quickly to isolate large numbers of blastomeres before they cleave. Knowing the subsequent pattern of cleavage, however, allows one to select desired cell lineages even after additional cleavages have occurred (see Figures 2.2 and 2.3). It is also possible to slow cleavage by lowering the temperature, either by placing the dish containing the cell suspension into a shallow dish of ice water or by adding cold (4 °C) SW to the cell suspension.

6. Periodically observe the development of the isolated blastomeres and determine whether they give rise to pluteus larvae. The larvae can also be fixed and immunostained with tissue-specific antibodies (Experiment 3).

EXPERIMENT 2. INDUCTION OF ENDODERM BY MICROMERES

1. Using the procedure described in Experiment 1, dissociate embryos at the 8-cell stage. Allow the isolated blastomeres to cleave once and then identify and isolate mesomere–mesomere pairs (Figures 2.2 and 2.3) using a mouthpipette. Dissociate a separate cohort of embryos at the 16-cell stage and isolate the smallest blastomeres (micromeres) (Figures 2.1 and 2.2). To allow sufficient time to carry out the mesomere pair isolations, it may be useful to fertilize two batches of eggs 1–2 h apart. The first batch can then be used for the 8-cell separations and the second batch for the 16-cell separations. An alternative approach is to fertilize a single batch of eggs, dissociate the embryos at the 16-cell stage, and then isolate single mesomeres and micromeres to use for recombinations. While this seems simpler, in practice it can be difficult to distinguish single mesomeres from single macromeres, whereas mesomere–mesomere pairs can be unambiguously distiguished from macromere–micromere pairs.

2. Place mesomere–mesomere pairs and micromeres in the same agarose-coated dish, and restore calcium levels by adding a volume of 1 M $CaCl_2$ sufficient to bring the final concentration to 10 mM. Restoring calcium will allow the blastomeres to adhere to one another.

3. Using a fine glass needle, tungsten needle, or finely pulled mouthpipette, attempt to maneuver several mesomere–mesomere pairs into an aggregate along with one or more micromeres. Even a single micromere should be sufficient. It may be helpful to

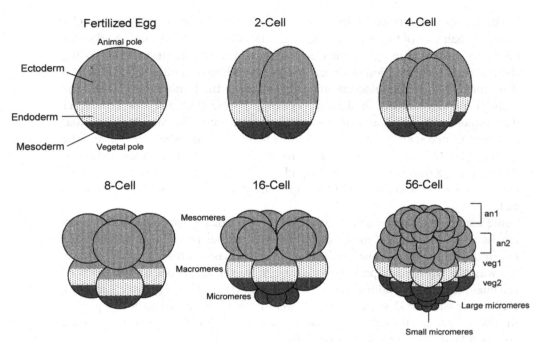

Figure 2.2. Normal cleavage pattern and fate map of the sea urchin embryo. The animal and vegetal poles are indicated at the fertilized egg stage. The prospective ectoderm, endoderm, and mesoderm are shown by different densities of stippling. At the 8-cell stage, the four animal blastomeres are slightly larger than the four vegetal blastomeres. At the 16-cell stage, the eight blastomeres of the animal hemisphere are called mesomeres, the four large blastomeres of the vegetal hemisphere are called macromeres, and the four small blastomeres at the vegetal pole are called micromeres. The micromeres divide more slowly than the remaining cells, resulting in a transient 56-cell stage. The mesomeres give rise to the an1 and an2 tiers of blastomeres, the macromeres to the veg1 and veg2 tiers, and the micromeres divide unequally to produce large and small micromeres.

make shallow grooves or depressions in the agarose and pile the blastomeres inside. As a control, combine several mesomere–mesomere pairs without micromeres.

4. After several minutes, ensure that the blastomeres have adhered to one another by gently agitating the aggregates with a stream of SW dispelled from a mouthpipette. If the aggregates do not fall apart, carefully transfer them with a mouthpipette to a fresh, agarose-coated dish containing complete SW.
5. Periodically observe the development of the aggregates. Do they form a gut (endoderm)? The aggregates can also be fixed and immunostained with tissue-specific antibodies (Experiment 3).

EXPERIMENT 3. IMMUNOSTAINING OF EMBRYO WHOLE MOUNTS

This experiment uses tissue-specific monoclonal antibodies to immunostain embryo whole mounts. It serves as a companion to Experiment 1 or 2, providing students with a simple molecular assay to confirm their morphological findings. Experiment 3 can also be used as a stand-alone experiment (i.e., using unmanipulated embryos) to illustrate the important, general strategy of using molecular probes to assess cell fates.

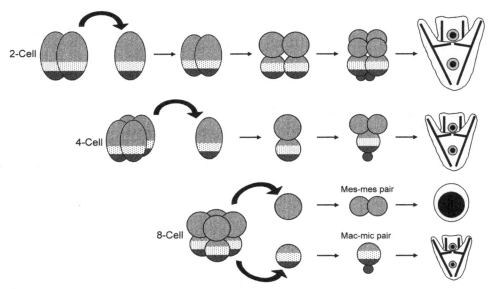

Figure 2.3. Cleavage and development of blastomeres isolated at the 2-, 4-, and 8-cell stages. The isolated cells cleave as they would in the intact embryo. Blastomeres isolated at the 2-cell stage give rise to half-sized pluteus larvae; those isolated at the 4-cell stage give rise to quarter-size pluteus larvae. Single animal blastomeres isolated from 8-cell stage embryos divide equally to produce mesomere–mesomere (mes–mes) pairs. Single vegetal blastomeres divide unequally to produce macromere–micromere (mac–mic) pairs. Typically, animal blastomeres isolated at the 8-cell stage give rise to ectodermal spheres, while vegetal blastomeres give rise to all three germ layers. Note, however, that single mes–mes and mac–mic pairs do not develop as fully as larger aggregates of animal or vegetal blastomeres; i.e., isolated animal or vegetal hemispheres.

1. Using a mouthpipette or Pasteur pipette, transfer embryos into a well (or wells) of a flexible, round-bottomed 96-well plate. Place the plate on ice for 5–10 min to stop the embryos from swimming and to allow them to settle to the bottom of the wells. Observe the embryos in the wells using a stereomicroscope. There should be no more than a monolayer covering the bottom of each well. Remove excess embryos with a mouthpipette.

2. Working under the stereomicroscope and observing the embryos continuously, use a mouthpipette to remove most of the SW. Take care not to aspirate the embryos or allow them to dry out. Add 100 μl of 100% methanol (−20°C) to each well and place the plate on ice for 10 min.

3. Remove as much methanol as possible. *Do not try to remove all the methanol*, as it is very volatile and may evaporate completely, drying the embryos. Wash the embryos 3 × 5 min with phosphate-buffered saline (PBS) at room temperature (RT), 100–200 μl/wash. The embryos should settle to the bottom of the wells within 5 min. Note that in the first rinse or two, the embryos will have a tendency to float on the surface of the liquid rather than settle to the bottom. Once the embryos become rehydrated this will no longer be a problem.

4. Remove as much PBS as possible and add 50–100 μl of primary antibody (full-strength 6a9 or Endo1 monoclonal antibody tissue culture supernatant). Incubate

1–2 h at RT or overnight at 4°C. For these longer incubations, place the 96-well plate in a humid chamber (e.g., a plastic food storage container lined with a damp paper towel) to keep the embryos from drying out.

5. Wash 5 × 5 min with PBS-T (PBS with 0.05% Tween-20), 1 × 5 min with PBS, and 1 × 5 min with PBS + 4% normal goat serum (GS-PBS), all at RT.

6. Remove as much GS-PBS as possible and add 50–100 μl secondary antibody (fluorescently labelled anti-mouse IgG+IgM). The secondary antibody should be diluted 1:50 in GS-PBS. Incubate for 1–2 h at RT or overnight at 4°C.

7. Wash 5 × 5 min with PBS-T and 1 × 5 min with PBS at RT. Add a glycerol-based anti-photobleaching agent (e.g., ProLong Antifade Kit, Molecular Probes) to each well and allow the embryos to settle to the bottom overnight at 4°C. Mount the embryos on a slide, using small strips of Scotch No. 665 double-sided tape as a spacer to prevent the coverslips from crushing the embryos. Alternately, the coverslips can be supported on small daubs of stopcock grease placed in the corners. Seal the edges of the coverslips with fingernail polish. Store the slides horizontally and in the dark at −20°C. Examine them by fluorescence microscopy (conventional epifluorescence or confocal laser scanning microscopy).

8. OPTIONAL: If a fluorescence microscope is not available, it is possible to use an enzyme-conjugated secondary antibody (e.g., an alkaline phosphatase–conjugated antibody). Many companies sell histochemical kits that can be used to generate a dark reaction product visible with transmitted light.

EXPECTED RESULTS

EXPERIMENT 1

Isolated blastomeres will cleave in the same pattern as they would in the intact embryo (Figures 2.2 and 2.3). Blastomeres isolated from 2-cell and 4-cell embryos will give rise predominantly to well-formed pluteus larvae that contain all recognizable tissues (i.e., gut, skeleton, pigment). These larvae will be miniature, as embryonic development in the sea urchin occurs in the absence of cell growth. Blastomeres isolated from 8-cell stage embryos will show different patterns of development depending on whether they are derived from the animal or vegetal region. Animal blastomeres will give rise to small, hollow ectodermal spheres. Vegetal blastomeres will give rise to all germ layers and all major cell types. These "embryoids" will be very small and their morphology may be difficult to interpret. To monitor cell fates more clearly it will be useful to stain with antibodies (see Experiment 3).

EXPERIMENT 2

As described previously, animal blastomeres make a hollow ball of ectoderm that lacks mesodermal (i.e., skeleton) and endodermal (gut) derivatives. When animal blastomeres are combined with micromeres, all major tissues will form and a complete pluteus larva can result. Note that cell-marking experiments have shown that, under these experimental conditions, micromeres give rise only to the skeleton. All other tissues are derived from the animal blastomeres. The endoderm that forms arises as a consequence of inductive signals from the micromeres. As in Experiment 1, it may be useful to stain with antibodies to assay cell fates more clearly (see Experiment 3).

EXPERIMENT 3

Antibody 6a9 recognizes MSP130, a PMC-specific cell surface glycoprotein (Ettensohn and McClay, 1988). It labels the surfaces of all primary mesenchyme cells (skeleton-forming cells) beginning at the late blastula stage and throughout later development. Antibody Endo1 recognizes a determinant expressed specifically by endoderm cells of the mid- and hindgut (Wessel and McClay, 1985). Immunoreactivity is not detectable until the prism stage and becomes much more pronounced at the pluteus larva stage.

DISCUSSION

These results allow the following conclusions to be drawn:

1. Individual blastomeres isolated from 2- or 4-cell stage embryos exhibit regulative development and give rise to complete, miniature larvae. Cell interactions are important in controlling the development of these blastomeres. Blastomeres isolated from 8-cell stage embryos show different patterns of development depending upon whether they are derived from the animal or vegetal region. This reflects the fact that the sea urchin egg is polarized along the animal–vegetal axis and the third cleavage division (a horizontal division) separates animal blastomeres from vegetal blastomeres. Therefore, the sea urchin egg also exhibits "mosaic" properties, at least with respect to the animal–vegetal axis.
2. Micromeres have powerful endoderm-inducing properties. Note that they also induce several kinds of mesodermal cells, but the experiments described here do not examine these cell types.
3. Antibodies can be used to assess the expression of cell-type specific gene products that are a hallmark of cell specification. For example, in these experiments, expression of MSP130 (recognized by monoclonal antibody 6a9) is an indication of skeletogenic primary mesenchyme cell specification, while expression of the Endo1 protein (recognized by monoclonal antibody Endo1) is an indication of endoderm (mid/hindgut) specification.

POTENTIAL SOURCES OF FAILURE

Fertilization of eggs may fail if the adult sea urchins are not in a gravid state or if they are obtained very late or early in their normal spawning season. Poor egg batches are generally indicated by variability in egg size, "roughness" of the egg periphery, rupturing of eggs upon shedding, and/or very small egg volumes. Sperm quality can generally be assessed by diluting a small aliquot and observing the sperm with a compound microscope to ensure they are motile. If both eggs and sperm appear to be of good quality but fertilization does not occur, check to ensure that sufficient numbers of sperm have been added. There should be several to several tens of sperm swimming actively around each egg, but not hundreds. Addition of excess sperm can lead to polyspermy, which results in abnormal cleavage.

Blastomere separations will be more difficult if the embryos are treated roughly and become prematurely demembranated, which will cause clumping. Clumping of blastomeres can also occur following dissociation if the cells are cultured too densely.

All the experiments described in this chapter require harvesting embryos at late (post-hatching) developmental stages. Healthy embryos swim to the top of the seawater after hatching and embryos should be collected from the surface. Embryos on the bottom of the dish are generally less healthy.

For immunostaining studies, ensure that the embryos are fixed at a developmental stage when the antigen recognized by the primary antibody is expressed (late blastula-pluteus stages for 6a9, prism pluteus stages for Endo1). Also, ensure that the secondary antibody used is appropriate for the primary antibody. 6a9 is an IgM while Endo1 is an IgG; therefore, a broad-spectrum secondary antibody is required to recognize both primary antibodies. Bright, nonspecific fluorescence indicates that the secondary antibody was not diluted properly or that the embryos were allowed to dry during the immunostaining procedure.

TIME REQUIRED FOR THE EXPERIMENTS

These experiments typically require 3–5 days to complete, depending on the specific protocols used and the number of times students attempt the experiments. It is critical to have a working knowledge of the developmental timetable of the species being used. Many warm water sea urchins develop from fertilization to the pluteus larva stage within 36 h. Early stages of cleavage will be available from 1–8 h after fertilization, again depending on the temperature of development and the precise stages desired. Some species develop much more slowly.

Here is a reasonable work plan for a fast-developing, warm-water species (e.g., *Lytechinus variegatus*):

➤ Day 1. AM: collect gametes and carry out several fertilizations at 1-h intervals. Staggering the fertilizations ensures that embryos of several different stages will be available in the afternoon. Set aside cultures of normal (undisturbed) embryos for immunostaining experiments on Day 2. PM: perform blastomere separations on various cleavage stages (Experiment 1).

➤ Day 2. AM: collect fresh gametes and fertilize eggs for cell recombination experiments (Experiment 2). Again, several batches can be fertilized at different times to allow for multiple trials. Collect and fix control embryos for antibody staining (Experiment 3). PM: perform blastomere recombination experiments (Experiment 2). As time permits, process fixed embryo whole mounts for immunostaining (Experiments 1 and 3). Depending on other student tasks it is possible to complete the immunostaining protocol on Day 2, but it is more likely that the embryos will be stored overnight in primary or secondary antibody and the protocol completed on

Day 3. Observe embryos derived from separated blastomeres. If desired, fix these embryos for whole mount immunostaining.

➤ Day 3. Complete immunostaining of control embryos and embryos derived from separated blastomeres (Experiments 1 and 3). Observe and fix aggregates from the blastomere recombination experiments (Experiment 2). Repeat Experiments 1 or 2 if needed (note that this may require a Day 5 to complete immunostaining experiments).

➤ Day 4. Complete immunostaining of cell aggregates from blastomere recombination experiments (Experiment 2) and any remaining immunostaining experiments.

ADDITIONAL EXPERIMENTS

Testing the competence of the responding cells at different developmental stages. The concept of competence – the ability of a cell or tissue to respond to an inductive signal – is a central one in developmental biology. In the sea urchin, animal blastomeres are responsive to inductive signals from the micromeres during cleavage and early blastula stages, but not at later stages. Students who become proficient with micromere recombinations may expand their experiments by recombining micromeres with animal fragments that have been isolated from progressively older embryos. Alternatively, animal fragments can be isolated from cleavage stage embryos as described above and then allowed to develop in isolation for various periods of time before being recombined with micromeres.

TEACHING CONCEPTS

Regulative development. The developmental potential of isolated blastomeres can be greater than their normal fates. Fragments of embryos can exhibit regulative development and restore a complete pattern. The regulative properties of early blastomeres are attributable to the fact that the first two cleavage planes do not bisect the maternal (animal–vegetal) axis.

Embryonic induction. Signals from one cell type or tissue (inducing cells) can influence the developmental fate of another cell type or tissue (responding cells). Inducing cells produce secreted or membrane-associate signaling molecules that interact with cell surface receptors expressed by the responding cells. A relatively small number of conserved signaling pathways underlie embryonic induction.

The use of molecular markers to assess cell fates. Cell differentiation is associated with the expression of cell-type–specific gene products; i.e., mRNAs and the proteins they encode. Molecular probes, including antibodies and nucleic acid probes, can be used to monitor the expression of cell-type–specific gene products. This often provides a better assessment of cell fate than simple cell or tissue morphology.

Monoclonal antibodies. These are antibodies of a single specificity produced by a hybridoma cell clone; i.e., a cell line produced from a founder cell derived from the fusion of a B-lymphocyte and a myeloma cell.

REFERENCES

Angerer, L. M., and Angerer, R. C. (2003). Patterning the sea urchin embryo: Gene regulatory networks, signaling pathways, and cellular interactions. *Curr. Top. Dev. Biol.*, 53, 159–98.

Böttger, S. A., Unuma, T., and Walker, C. W. (2004). Care and maintenance of adult echinoderms. In *The Development of Sea Urchins, Ascidians, and Other Invertebrate Deuterostomes: Experimental Approaches.* Methods in Cell Biology, vol. 74, eds. C. Ettensohn, G. Wessel and G. Wray, in press.

Brandhorst, B. P., and Klein, W. H. (2002). Molecular patterning along the sea urchin animal-vegetal axis. *Int. Rev. Cytol.*, 213, 183–232.

Conklin, E. G. (1905). Mosaic development in ascidian eggs. *J. Exp. Zool.*, 2, 145–223.

Driesch, H. (1892). The potency of the first two cleavage cells in echinoderm development. Experimental production of partial and double formations. In *Foundations of Experimental Embryology*, eds. B. H. Willier and J. M. Oppenheimer, 1974, pp. 38–55. New York: Hafner.

Ettensohn, C. A., and McClay, D. R. (1988). Cell lineage conversion in the sea urchin embryo. *Dev. Biol.*, 125, 396–409.

Ettensohn, C. A., and Sweet, H. C. (2000). Patterning the early sea urchin embryo. *Curr. Top. Dev. Biol.*, 50, 1–44.

Hall, H. G. (1978). Hardening of the sea urchin fertilization envelope by peroxidase-catalyzed phenolic coupling of tyrosines. *Cell*, 15, 343–55.

Hörstadius, S. (1939). The mechanics of sea urchin development studied by operative methods. *Biol. Rev.*, 14, 132–79.

Ishizuka, Y., Minokawa, T., and Amemiya, S. (2001). Micromere descendants at the blastula stage are involved in normal archenteron formation in sea urchin embryos. *Dev. Genes Evol.*, 211, 83–8.

McClay, D. R., Peterson, R. E., Range, R. C., Winter-Vann, A. M., and Ferkowicz, M. J. (2000). A micromere induction signal is activated by beta-catenin and acts through Notch to initiate specification of secondary mesenchyme cells in the sea urchin embryo. *Development*, 127, 5113–22.

McClendon, J. F. (1910). The development of isolated blastomeres of the frog's egg. *Am. J. Anat.*, 10, 425–30.

Ransick, A., and Davidson, E. H. (1993). A complete second gut induced by transplanted micromeres in the sea urchin embryo. *Science*, 259, 1134–8.

Roux, W. (1888). Contributions to the developmental mechanics of the embryo. On the artificial production of half-embryos by the destruction of one of the first two blastomeres and the later development (postgeneration) of the missing half of the body. In *Foundations of Experimental Embryology*, eds. B. H. Willier and J. M. Oppenheimer, 1974, pp. 2–37. New York: Hafner.

Runft, L. L., Adams, N. L., and Foltz, K. R. (2004). Echinoderm eggs and embryos: Procurement and culture. In *The Development of Sea Urchins, Ascidians, and Other Invertebrate Deuterostomes: Experimental Approaches.* Methods in Cell Biology, vol. 74, eds. C. Ettensohn, G. Wessel and G. Wray, in press.

Sweet, H. C., Gehring, M., and Ettensohn, C. A. (2002). LvDelta is a mesoderm-inducing signal in the sea urchin embryo and can endow blastomeres with organizer-like properties. *Development*, 129, 1945–55.

Wessel, G. M., and McClay, D. R. (1985). Sequential expression of germ-layer specific molecules in the sea urchin embryo. *Dev. Biol.*, 111, 451–63.

3 The isthmic organizer and brain regionalization in chick embryos

D. Echevarría and S. Martínez

OBJECTIVE OF THE EXPERIMENT The aim of the experimental techniques detailed here is to study the process of regionalization and histogenesis of specific areas in the vertebrate brain at early stages of development, using the chick embryo as an experimental model. We will deal with the morphogenetic properties of particular areas of the neural tube known as "secondary organizers." We will then describe the quail–chick chimaera system developed by Le Douarin (1973) in the discovery and analysis of the isthmic organizer (IsO). This in ovo system has the particular advantage of accessibility to the developing embryo: it is easy and relatively cheap, and the embryo can be monitored for a relatively long period of time as it develops. On the contrary, mouse embryo models, now extensively used in the functional genetic analysis of development, lack these features.

DEGREE OF DIFFICULTY This project is difficult. Experimental embryology requires some experience with vertebrate embryos and knowledge of neuroanatomy. These experiments also demand skillful manipulation and relatively sterile conditions.

INTRODUCTION

The central nervous system is composed of defined neuronal structures interconnected by a complex network of nerve fibers. The potential to construct this complexity is contained in the primary neural anlage: the neural plate. This pseudostratified neuroepithelium, produced during gastrulation, undergoes the process of neurulation by which the neural plate folds and is converted into the neural tube. At the neural tube stage, the brain shows a series of ring-like constrictions that mark the approximate boundaries between the primordia of the major brain regions: the forebrain, the midbrain, and the hindbrain. As development proceeds, additional transverse constrictions subdivide the brain into segments that are called neuromeres (Lumsden and Krumlauf, 1996; Puelles and Rubenstein, 1993; Rubenstein et al., 1998; Martínez and Simeone, 1999). The process of regionalization occurs along the anterior–posterior axis (A/P axis) and the ventro–dorsal axis (V/D axis) and is carried out by specific transcription factors, known as "selector genes," that are expressed in overlapping domains of neuro-epithelial cells through development. Localized areas of the neural plate and tube have been demonstrated by experimental embryology to regulate the further histogenesis of competent neighboring tissue by producing inductive signals. These local patterning centers are called "secondary organizers" and differ from the "primary organizer," often referred to as the Spemann and Mangold organizer (Spemann and Mangold, 1924; 2001; reviewed by Lemaire and Kodjabachian, 1996). The secondary organizers act to *refine* the established A/P and V/D axes (see also Chapters 14 and 16).

Three regions in the neural plate and tube have been identified as secondary organizers: the anterior neural ridge (ANR) at the anterior end of the neural plate, the zona limitans intrathalamica (ZLI) in the middle of the diencephalon, and the isthmic organizer (IsO) at the mid-hindbrain transition (Martínez, 2001). Topologically the IsO is found in the isthmus (a transversal ring-like constriction that delimits the mid-hindbrain boundary). The group of Alvarado Mallart in Paris first demonstrated the morphogenetic properties in the IsO (Martínez, Wassef, and Alvarado-Mallart, 1991). They used the quail–chick chimaera system for transplanting the IsO into an ectopic position in the midbrain. The result was the duplication of mesencephalic and cerebellar structures of the brain. Today we know that the candidate molecule for the inductive signal in the IsO is a member of the fibroblast growth factor family, FGF8. FGF8 alone is capable of reproducing many of the inductive properties of the IsO, including induction of the tectum and cerebellum (Crossley, Martínez, and Martin, 1996; Garda, Echevarría, and Martínez, et al., 2001).

MATERIALS AND METHODS

Cell lineage determination and fate map analysis of the neuroepithelium are the primary techniques used to study how cells determine their destinies in a developmental process. There are a variety of ways to carry out these techniques: classical vital staining (e.g., Nile Blue), fluorescent dyes (carbocyanine dyes or dextrans), infection with retro-viral vectors encoding a specific marker, and the generation of transgenic lines with the Cre-LoxP technology (see Song et al., 1996). All of these methodologies have advantages and disadvantages, primarily with respect to stain properties, durability, specificity, and economy. An alternative approach that does not suffer from the problem of label dilution and durability, and that does not require expensive material or equipment, is to make chimerics between closely related species: i.e., heterospecific grafts of neural tissue using quail–chick chimaeras (Le Douarin, 1973; Alvarado-Mallart and Sotelo, 1993).

EQUIPMENT AND MATERIALS
Per student

Dissecting microscope (Leica MS5)

A grid with concentric circles (35 mm between each two concentric circles, at our working magnification: 40 ×) (Electron Microscopy Sciences)

Surgical scissors (Dumont)

Fine forceps (Dumont #55)

Pasteur pipettes (Brand GmbH)

Silicon rubber tube for holding the glass capillaries (ID 1 mm, OD 3 mm, wall 1 mm) (World Precision Instruments (WPI))

Tungsten needles (thickness: 0.125 mm) (WPI)

India ink (Pelikan)

Tyrode's solution

Syringe of 5 ml (Becton Dickinson S.A.)

Syringe needles (0.8 mm × 25 mm) (Becton Dickinson S.A.)

Ethanol 80% in a plastic bottle dispenser

Per practical group. Egg incubator set at 37°C with 80% humidity atmosphere (Stromberg's Chicks)

Biological material. Eggs (Quail and chick fertilized eggs can be purchased from commercial sources.)

EGG PURCHASE AND STORAGE. Fertilized eggs must be taken as fresh as possible from the commercial sources and can be stored at 12°C for up to 2 weeks.

PREVIOUS TASKS FOR STAFF
Solutions
TYRODE'S SOLUTION
1. Prepare in 900 ml of distilled water Solution (A):

NaCl	12 g
KCl	0.3 g
$CaCl_2 + 2H_2O$	0.39 g
$NaH_2PO_4 + 2H_2O$	0.083 g
$MgCl_2 + 6H_2O$	0.066 g

   Autoclave the solution. Store at 4°C.
2. Prepare separately the following solutions (B and C):

   (B) Glucose   1.5 g in 300 ml of distilled H_2O

   (C) $NaHCO_3$ 1.5 g in 300 ml of distilled H_2O

   Autoclave the solution. Store at 4°C.
3. Prepare on the day of the experiment:

Add: 90 ml of solution (A), 30 ml of solution (B), 30 ml of solution (C), and 1% Penicillin-Streptomycin (Gibco, Invitrogen Corporation).

This solution can be maintained at 4°C for one month.

INDIA INK SOLUTION. Dilute india ink 1:10 in Tyrode's solution.

Preparing glass capillaries. Pasteur pipettes will be used for creating thin glass capillaries. *Wear fire gloves.*

1. Break the thin part of a Pasteur pipette and hold it at the ends.
2. Put the middle of the capillary on a Bunsen burner and maintain it there until it starts melting.
3. Then pull quickly but gently from both sides in a horizontal position 15 to 20 cm away before breaking the glass filament.
4. Use scissors to cut the glass tube to the desired length and tip diameter.
5. Store in Petri dish.

Preparing tungsten (wolfram) needles. Use fire protective gloves.

1. Tungsten needles will be cut in small pieces, approximately 3 cm long. Use tungsten needle handles or the thick part of the broken Pasteur pipettes as holders for the tungsten needles.
2. To make a Pasteur pipette handle, put the broken part in the fire beaker until it starts melting.
3. Hold the stick of tungsten wire with forceps and "glue" it into the melting Pasteur pipette tip.
4. Wait for the glass to cool.
5. Sharpen tungsten needle tips in an electrolytic solution (e.g., NaCl 2.5 M).
6. Use a power supply for an external microscope light (e.g., Bausch and Lomb, 125 V).
7. Attach one side of the output terminal to the tungsten wire (using an alligator clip) and the other side to a reference carborundum rod (the content of a DC battery) that is inserted into the solution.
8. Apply 5 V. You will see bubbles coming out from the tip.
9. Place the tungsten tip into the bath solution, checking periodically under the dissecting microscope, until the correct sharpness is achieved.

Maintenance of the laboratory. Sterile conditions should be maintained in the laboratory to increase the success of the experiments. Use a laminar flow hood when possible. Sterilize the surgical material by either autoclaving or leaving them under the U.V. light chamber for 1 hour.

OUTLINE OF THE EXPERIMENT

The aim of this experiment is to understand the concept of morphogenetic activity and secondary organizers within the context of developmental biology. The properties of the isthmic organizer will be analyzed using surgical manipulations and quail–chick chimaeras (Figure 3.1). We will dissect the IsO using tungsten needles from the quail embryo and transplant this piece of tissue into the chick embryo where a diencephalic hole was made previously. Thereafter, we will use a technique for identifying the grafted

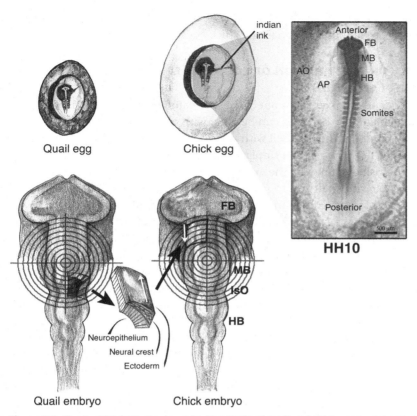

Figure 3.1. Quail–chick heterotopic grafts. The shells of quail and chick eggs were opened and india ink was injected under the blastodermic membrane into the vitelline cavity. The embryos, neural tube and somites were then visible. The developmental stage of the embryos is determined at this point. The graft: The alar plate of the isthmic region, from the right side was excised in the quail embryo and grafted, after 180° inversion along the A–P axis (white arrows) in the left side of the chick embryo, in substitution of the diencephalic alar plate. AO: area opaca, AP: area pellucida, FB: forebrain, HB: hindbrain, IsO: isthmic organizer, MB: midbrain. Bar = 500 μm.

tissue (quail tissue) by means of an immunohistochemistry technique done in toto (the whole embryo).

SETTING EGGS INTO THE INCUBATOR

1. Place about 20 chick eggs and 20 quail eggs per student in the incubator.
2. Place the eggs horizontally in a humidified incubator set at 37°C. Do not tilt the eggs.
3. Mark the upper surface of the eggs: this is where the embryos will be after incubation.
4. For this experiment the embryos must be at stage HH 9–10 (Hamburger and Hamilton, 1951). Synchrony of staging is important when beginning the experiment;

thus, the quail eggs, which have a faster rate of development, must be placed in the incubator 3 to 4 hours later than the chick eggs.

PROTOCOL FOR PREPARATION OF EGGS AND EMBRYOS FOR TRANSPLANTATION

1. Take chick and quail embryos out of the incubator and place them marked side up on a homemade egg bed (this can be done with clay or plasticine).
2. Carefully puncture the eggshell with a syringe (5 ml) at the larger pole of the egg, and remove 1–1.5 ml of the cytoplasm (Albumin or egg white). This step helps to reduce the risk of damaging the embryo when cutting through the eggshell.
3. Gently scrape the shell surface until the placenta membrane is viewed (a white membrane, just underneath the eggshell). Put a drop of Tyrode's solution on top of the eggshell and make a small hole with the tip of the scissors. The embryo and blastodermic membrane will separate from the shell due to atmospheric pressure and the Tyrode's solution will enter the cavity. Then cut gently an ellipse-like hole about 2.5 cm wide and 3 cm long (for a quail egg 1.5 cm diameter). Remove any small pieces of shell that have fallen into the cavity with forceps. Add a few drops of Tyrode's solution. Distinguish under the dissecting microscope a small ring-like patch with milky color at the surface of the yolk sac; this is the germinal disk. Distinguish the anterior and posterior axes, the area opaca (outer ring) and the area pellucida (inner part), and the shape/stage of the embryo (see Figure 3.1).
4. Take a glass capillary and insert it into the silicon tube. Suck a bit of india ink solution and punch the yolk sac very close to the area opaca. Slide and bring the glass capillary underneath the embryo area. Blow a bit of ink out of the capillary, moving it back and forth as if painting the area with ink (Figure 3.1). Take care not to blow out bubbles. Cover the holes with rounded coverslips to keep the embryo from drying out.

PROTOCOL FOR TRANSPLANTING THE ISO REGION TISSUE INTO THE DIENCEPHALON

Ensure that both chick and quail embryos are at the same developmental stage. Use the chick embryo as host and the quail embryo as donor.

We will make heterotopic (different topological position of donor tissue in the host) and isochronic (same developmental stage of donor tissue and host tissue) quail–chick grafts.

1. Insert a grid with concentric circles in one ocular of the dissecting microscope. This will help guide the proportions and topography of each explant/implant in the donor and host embryos, respectively (Figure 3.1). The grid units counted between different points of the grafts provide their dimensions.
2. To get donor tissue from the quail egg, carefully rip off the amnion sac, covering the embryo with the tip of the tungsten wire, and pull it away to leave the embryo free of this shield. Choose the isthmic constriction area, separate the neural epithelium

from the surrounding tissues (migrating neural crest and lateral mesoderm) with the tip of the tungsten needle, and cut very gently a rectangle from the dorsal midline to the medio-lateral side (a quarter of the tube; Figure 3.1). Take your time in this step and memorize position references in your donor tissue. Once finished, leave the rectangular or wedge-shaped area beneath the quail embryo. Cover the shell hole with a rounded coverslip. Remember the topographic orientation of the graft.

3. In the chick embryo, cut and remove the neuroepithelial rectangle in the dien-cephalic vesicle. It is important that size and shape of this rectangle are as similar as possible to those of the quail (use the concentric grid for this purpose).

4. Transfer the quail rectangle piece to the chick egg. Take a glass capillary and put it in the silicon rubber tube. Fill the tip with Tyrode's solution and suck the rectangle piece from the quail and leave it at one side of the chick embryo. Do not interrupt the visual control of the explant tissue – it is very easy to lose it in the host egg membranes. Use the tungsten needles to transfer the piece of quail tissue onto the rectangular hole made in the chick mesencephalon. Take your time in this step and be sure that the programmed orientation is adequately obtained. In our heterotopic grafts of IsO epithelium, the goal is to invert the graft with respect to its original A–P polarity in order to maintain the correct ventro–dorsal orientation (Figure 3.1).

5. Once the quail piece of IsO tissue is placed "as a patch" in the diencephalic rectan-gular hole, cover the chick shell cavity with Scotch tape or a rounded glass coverslip sealed with paraffin, in order to prevent bacterial/fungal contamination and to keep the cavity and embryo humid. Bring it back to the incubator very carefully. Do not tilt the egg. Discard the quail egg.

COLLECTION OF SURGICALLY MANIPULATED CHICK EMBRYOS

1. For embryos between 2 and 3 days of incubation, cut a paper ring with an inner diameter the size of the chick embryo. Crack the egg open, being careful not to rupture the yolk. Place the paper ring on top of the embryo. The amnion will glue to the paper ring. Hold the ring with the forceps and cut from the adjacent outside of the paper ring the amnion and yolk sac around the embryo. Take the ring containing the embryo out of the egg and place it into a Petri dish with PBS. Wash it briefly and fix with PBS-PFA 4%. Leave it in the fixative solution overnight at 4 °C.

2. For embryos older than 4 days of incubation, remove the embryo with round-tip forceps from the umbilical cord insertion and cut the membranes. The embryo can then be removed from the yolk with a spoon.

3. Wash the embryos in PBS-T (see Appendix) twice for 5 minutes. Fix the embryos in PBS-PFA overnight at 4 °C. Then dehydrate the embryos in ascending methanol (25%, 50%, 75%, 100%) for 10 minutes in each step on an orbital shaker platform. Store them at −20 °C.

IDENTIFICATION OF THE QUAIL TISSUE BY IMMUNOHISTOCHEMISTRY

The chick and quail cells differ in two critical ways. First, the quail heterochromatin in the nucleus is concentrated around the nucleoli. This creates a large, deeply stained mass that is easily distinguishable from the diffuse heterochromatin of chick cells. Second, there are some antigens that are quail-specific and are not present on chicken cells. Both of these phenomena allow one to readily distinguish individual quail cells engrafted in the host brain, even when the majority of the cell population is chick.

We describe the technique for labeling the nuclei of quail cells using immunohisto-chemistry in whole embryos. A similar procedure can be used for cryostat or paraffin sections. This technique can be used for embryos that have developed an additional 2 to 3 days in the incubator after the operation.

1. Hydrate the embryos by descending from pure methanol to PBS-T (100%, 75%, 50%, 25%, and PBS-T) for 10 minutes in each step (on an orbital shaker platform).
2. Block the nonspecific epitopes by incubating the embryos in PBS-T-BSA (1%) for 1 hour (orbital shaker) at room temperature (RT).
3. Incubate the embryos in anti-quail antibody QCPN (Hybridoma Bank, diluted 1:10 in PBS-T-BSA) at 4°C overnight (orbital shaker).
4. Wash 7 to 8 times, for 10 minutes each time, at room temperature in PBS-T (orbital shaker).
5. Incubate the embryos in secondary antibody (Goat anti-mouse IgG Biotinylated (Sigma Aldrich) 1:200 in PBS-T) for one hour at RT (orbital shaker).
6. Wash 6 times, for 10 minutes each time, at RT in PBS-T (orbital shaker).
7. Incubate the embryos in Avidin-Biotin Complex (A-B-C-Kit standard; Vector labs; see manufacturer's instructions for preparation) for one hour in the dark at room temperature.
8. Wash 6 times, for 10 minutes each time, at RT in PBS-T (orbital shaker).
9. Preincubate in diaminobenzidine (DAB; Sigma Aldrich) in Triss-T for 10 minutes without H_2O_2 at RT (orbital shaker). WEAR GLOVES AND MASK, AND DO THIS STEP AND STEPS 10 AND 11 UNDER A FUME HOOD. DAB IS A BIOHAZARD!
10. Discard the preincubation DAB (1 mg/ml), conforming to Environmental Health and Safety protocols, and add new DAB with hydrogen peroxydase (H_2O_2; 0.01%).
11. Monitor the reaction under the dissecting scope. Only the quail tissue will turn brown; the rest of the chick embryo should be white/clear. Stop the reaction in Triss-T. Do several washes in PBS-T (orbital shaker).
12. Store the embryos in PBS-azide 0.1% (Azide; Sigma Aldrich) at 4°C for further documentation.

Data recording. The result of each graft experiment will be annotated and drawn using a schematic drawing as in Figure 3.1. Normally, by 6 to 12 hours after implantation, quail grafts into the chick embryo are integrated with the chick tissue. The correct fusing of both tissues depends on the skill and ability of the student to make the graft. Take notes of the survival rate from the experiments done that day. The survival of the embryo can be detected easily by watching the development of vascularization.

Quail cells can be also detected due to the disposition of heterochromatin in the centronuclear mass. Two classical heterochromatin staining techniques are the Feulgen–Rossenbeck method (Le Douarin, 1973) and the fluorescent dye bisbenzimide

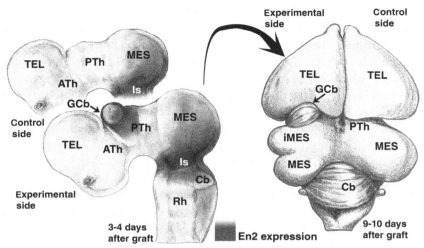

Figure 3.2. Results after IsO heterotopic grafts. At 3 to 4 days of incubation after graft, chimeric embryos show an ectopic vesicle in the diencephalon of the experimental side which was *engrailed* 2 (*En2*) immunopositive. Quail cells formed this vesicle that developed from the grafted isthmic alar plate. The neuroepithelium of the host caudal thalamus (PTh) situated around this vesicle was also induced to express *En2*. The analysis of chimeric embryos at longer survival time, 9–10 days, showed that in the experimental side an ectopic quail cerebellum was generated and an ectopic mesencephalic structure (formed by host cells) developed in substitution of the caudal thalamus, indicating that the diencephalic epithelium induced to express *En2* changed its thalamic fate to develop as mesencephalon. Ath: anterior thalamus, Cb: cerebellum, GCb: grafted cerebellum, iMES: induced mesencephalon, Is: isthmus, Pth: posterior thalamus, TEL: telencephalon, MES: mesencephalon, Rh: rhombencephalon.

method (Hoechst 33258, Sigma). Chick heterochromatin, on the other hand, appears dispersed in the nucleus.

EXPECTED RESULTS

The experimental technique described in this chapter allows the analysis of the function of the IsO, and the dependence of its function on position. The grafted IsO epithelium develops an ectopic cerebellum, made from quail cells, and, moreover, is capable of transforming the presumptive diencephalic epithelium into a supernumerary mesencephalon (Figure 3.2). The A/P polarity of these induced structures is inverted due to the inversion of the graft. Therefore we can also demonstrate an asymmetric induction due to the positional information carried by the grafted tissue. Grafts at different anterior–posterior levels allow the exploration of differential potentiality of each host territory.

DISCUSSION

Distinct neural identities are acquired through progressive restriction of developmental potential under the influence of local environmental signals. Evidence for the localization of such morphogenetic signals at specific locations of the developing neural primordium has suggested the concept of "secondary organizer regions," which regulate the identity and regional polarity of neighboring neuroepithelial areas one step further. In recent years, the most studied secondary neural organizer is the isthmic organizer, which is localized at the mid-hindbrain transition of the neural tube and controls the anterior hindbrain and midbrain regionalization. *Otx2* and *Gbx2* expressions are fundamental for positioning the organizer and the establishment of molecular interactions that induce *Fgf8*. In the isthmic region, an overlapping expression of *Otx2* and *Gbx2* has been demonstrated. This area is the transversal domain where expression of *Fgf8* is induced. The FGF8 protein produced in the isthmus stabilizes and up-regulates *Gbx2* expression, which, in turn, down-regulates *Otx2* expression. The inductive effect of the *Gbx2/Otx2* limit keeps *Fgf8* expression stable and thus maintains its positive role in the expression of *Pax2, En1, En2,* and *Wnt1*. Temporospatial patterns of such gene expressions are necessary for the correct development of the organizer, which, by a planar mechanism of induction, controls the normal development of the rostral hindbrain from r2 to the midbrain–diencephalic boundary. FGF8 appears as the active diffusible molecule for isthmic morphogenetic activity and has been suggested to be the morphogenetic effector in other inductive activities revealed in other neuroepithelial regions (for review see Martínez, 2001).

TIME REQUIRED FOR THE EXPERIMENTS

➤ Getting the quail and chick embryos to operating stage (HH9–10) takes one and a half days in the incubator.
➤ The graft experiment procedure takes about 15 minutes for each operation.
➤ The collection of the operated embryos takes three days.
➤ The immunohistochemistry technique takes two days.

POTENTIAL SOURCES OF FAILURE

It is clear that the experiment requires good and precise handling by the student. When monitoring embryo survival after the operation, it is important not to leave it too long out of the incubator.

To our knowledge, the most important source of failure is the quality of the chick and quail embryos. Ensure fresh eggs from the commercial company. If the eggs are not fresh, incubation time requirements and quality of the embryos will change and successfully completing the experiment will be more difficult. Skills in manipulation of the eggs are a "keep trying" work for the student. Grafting has an initial rate of success of about 5%.

ALTERNATIVE EXERCISES
HISTOLOGICAL ANALYSIS OF THE ECTOPIC STRUCTURES FROM THE GRAFT EXPERIMENTS

The histological study of the chimeric brains and tissues will reveal the exact character of the structures derived from the grafted quail epithelium and their relations with the host brain. A common method used after fixation in PBS-PFA or Clarke fixative (see Appendix) is embedding the tissue in paraffin and then sectioning the embryo in sagittal sections 12 μm thick. Make parallel series of the sections for different detection techniques. Use three glass slides and put the first section on the first slide, the second section on the next slide, the third section on the third slide, the fourth section back on the first slide, and so on. By this mounting procedure one can achieve different detection techniques. For example, use a serie (one slide) for immunohistology against the quail cells (Q¢PN staining), another serie for cresyl violet staining, and another for calcium binding protein (i.e., calbindin). (For more details see Dumesnil-Bousez and Sotelo, 1993; Nieto, Patel, and Wilkinson, 1996.)

IMPLANTATION OF FGF8-SOAKED HEPARIN ACRYLIC BEADS

Local application of specific proteins in the developing embryo has revealed an enormous tool for understanding the molecular cascade events that are taking place during early development. We have seen that FGF8 is the candidate molecule for morphogenetic activity of the IsO. Crossley et al. (1996) have demonstrated this property in the brain of vertebrate embryos. More details are described elsewhere (see also Garda et al., 2001).

IN SITU HYBRIDIZATION IN TOTO

In situ hybridization is a procedure that allows detection of the site(s) of transcription of a given gene at a cellular level within the entire organism. *Fgf8* has a specific expression patterning along the A/P axis in secondary organizers (Martínez, 2001). Therefore, *Fgf8* is a good indicator of sites where morphogenetic activity might occur. The execution of this technique on the quail–chick chimaeras will show the induction of *Fgf8* and other genes ectopically in the tectum. More precise information and description can be obtained elsewhere (see also Rosen and Beddington, 1993).

QUESTIONS FOR FURTHER ANALYSIS

➤ How could you analyze the character, vertical or planar, of the IsO inductive activity?
➤ How could you establish the possible polarity and the limits of competence to IsO induction in the neural tube?

ACKNOWLEDGEMENTS

We would like to thank Dr. Raquel García López and Dr. Claudia Vieira for helpful comments on this chapter. The work presented by the authors has been supported by the following: European Union Grants U.E. QLG2-CT-1999-00793; UE QLRT-1999-31556; UE QLRT-1999-31625; QLRT-2000-02310; Spanish Grants DIGESIC-MEC PM98-0056; FEDER-1FD97-2090; Seneca Foundation; Francisco Cobos Foundation; the Spanish Multiple Sclerosis Foundation and the Fundação para a Ciência e a Tecnologia do Ministério para a Ciência e o Ensino Superior (FCT/MCES) and Generalitat Valenciana CTDIA/2002/91. The authors would like to express their thanks to the technicians Francisca Almagro and Mónica Ródenas for the helpful description of the protocols used in our laboratory.

REFERENCES

Alvarado-Mallart, R. M., and Sotelo, C. (1993). Cerebellar grafting in murine heredodegenerative ataxia. Current limitations for a therapeutic approach. *Adv. Neurol.*, 61, 181–92.

Crossley, P. H., Martínez, S., and Martin, G. R. (1996). Midbrain development induced by FGF8 in the chick embryo. *Nature*, 380, 66–8.

Dumesnil-Bousez, N., and Sotelo, C. (1993). Partial reconstruction of the adult Lurcher cerebellar circuitry by neural grafting. *Neurosci.*, 55, 1–21.

Garda, A. L., Echevarría, D., and Martínez, S. (2001). Neuroepithelial co-expression of Gbx2 and Otx2 precedes Fgf8 expression in the isthmic organizer. *Mech. Dev.*, 101, 111–18.

Hamburger, V., and Hamilton, H. L. (1992). A series of normal stages in the development of the chick embryo. 1951. *Dev. Dyn.*, 195, 231–72.

Le Douarin, N. (1973). A biological cell labeling technique and its use in experimental embryology. *Dev. Biol.*, 30, 217–22.

Lemaire, P., and Kodjabachian, L. (1996). The vertebrate organizer: Structure and molecules. *Trends Genet.*, 12, 525–31.

Lumsden, A., and Krumlauf, R. (1996). Patterning the vertebrate neuraxis. *Science*, 274, 1109–15.

Martínez, S., Wassef, M., and Alvarado-Mallart, R. M. (1991). Induction of a mesencephalic phenotype in the 2-day-old chick prosencephalon is preceded by the early expression of the homeobox gene *en. Neuron*, 6, 971–81.

Martínez, S., and Simeone, A. (1999). Specification and patterning of the rostral neural tube. In *The Development of Dopaminergic Neurons*, eds. U. do Porzio, R. Pernas-Alonso and C. Perrone-Capano, pp. 1–8. Austin: R. G. Landes Company.

Martínez, S. (2001). The isthmic organizer and brain regionalization. *Int. J. Dev. Biol.*, 45, 367–71.

Nieto, M. A., Patel, K., and Wilkinson, D. G. (1996). In situ hybridization analysis of chick embryos in whole mount and tissue sections. *Methods Cell Biol.*, 51, 219–35.

Puelles, L., and Rubenstein, J. L. (1993). Expression patterns of homeobox and other putative regulatory genes in the embryonic mouse forebrain suggest a neuromeric organization. *Trends Neurosci.*, 16, 472–9.

Rosen, B., and Beddington, R. S. (1993). Whole-mount in situ hybridization in the mouse embryo: Gene expression in three dimensions. *Trends Genet.*, 9, 162–7.

Rubenstein, J. L., Shimamura, K., Martínez, S., and Puelles, L. (1998). Regionalization of the prosencephalic neural plate. *Ann. Rev. Neurosci.*, 21, 445–77.

Song, D. L., Chalepakis, G., Gruss, P., and Joyner, A. L. (1996). Two Pax-binding sites are required for early embryonic brain expression of an Engrailed-2 transgene. *Development*, 122, 627–35.

Spemann, H., and Mangold, H. (1924). Über die Induktion von Embryoanlagen durch Implantation artfremder Organisatoren. *Roux' Arch. Entwicklungsmech.*, 100, 599–638.

Spemann, H., and Mangold, H. (2001). Induction of embryonic primordia by implantation of organizers from a different species. 1923. *Int. J. Dev. Biol.*, 45, 13–38.

APPENDIX

➤ *PBS*

NaCl	8 g
KCl	0.2 g
Na_2HPO_4	1.44 g
KH_2PO_4	0.24 g

H_2O to 1 liter and adjust pH to 7.4, store at 4 °C.

➤ *PBS-T*: add 1% Triton X-100 to PBS.

➤ *PBS-T-BSA*: add 1% of Bovine Serum Albumin (Sigma) to PBS-T.
➤ *PBS-PFA*: 4% paraformaldehyde in PBS:
  (a) Mix 1 g paraformaldehyde with 25 ml PBS.
  (b) Heat to 60°C.
  (c) Add 1 drop of 2 N NaOH.
  (d) Mix to dissolve.
  (e) Filter.
  (f) Refrigerate at 4°C and use within 24 hours. WEAR MASK AND WORK UNDER FUME HOOD.
➤ Clarke fixative: 75% ethanol absolute and 25% acetic acid glacial.
➤ DAB solution: 1 mg/ml diaminobenzidine (Sigma) in PBS-T.
  1. Dissolve 10 mg tablet in 10 ml.
  2. Filter.
  3. Aliquot; 1 ml aliquots can be stored at −20°C for months. DAB IS A CARCINOGEN AND SHOULD BE HANDLED WITH GLOVES AND MASK UNDER A FUME HOOD. After inactivation with bleach, decontaminate instruments and work surface with bleach and dispose the used bleach as a biohazard.

4 Chemotaxis of aggregating *Dictyostelium* cells

G. Gerisch and M. Ecke

OBJECTIVE OF THE EXPERIMENT In the course of *Dictyostelium* development, a multicellular organism is established by the aggregation of single cells. In the following experiment, the chemoattractant that guides cell movement in an aggregation field is replaced by treating cells with cyclic AMP diffusing out of a micropipette. By the use of a micropipette that is easily moved by a micromanipulator, the direction of diffusion gradients can be changed quickly enough to study the response of the cells within the first few seconds of reorientation.

DEGREE OF DIFFICULTY The cultivation of *Dictyostelium* and preparation of chemotactically responsive cells is easy. Manipulation of the micropipette requires some skill and endurance.

INTRODUCTION

Prokaryotic as well as eukaryotic cells are capable of responding to concentration gradients of chemical compounds, but the mechanisms of their orientation within these gradients are different. The best-studied examples are bacteria that swim by flagella rotation and eukaryotic cells that migrate by actin-based shape changes on a substrate surface. Bacteria such as *Escherichia coli* or *Salmonella* strains control the rotation of their flagella motors: counterclockwise rotation results in smooth swimming by bundled flagellae, whereas clockwise rotation leads to tumbling caused by separation of the flagellae. These small bacterial cells use temporal changes of chemoattractant to find the source of a gradient: when they swim toward the source, concentration will increase with time and tumbling is suppressed. The tumbling frequency is increased when bacteria swim in the opposite direction. As a result, net movement of the bacteria will be biased toward higher concentrations of the attractant (for a textbook on bacterial and eukaryote cell motility see Bray, 2001). Eukaryotic cells with an actin-based motility system, like *Dictyostelium* amoebae, combine temporal and spatial signal inputs to reorganize the actin network in the cell cortex in a way that pseudopods are preferentially directed toward the source of the gradient.

Common to these two examples is the recognition of chemoattractants by transmembrane proteins that act as receptors on the cell surface. Binding of a ligand to the receptors will activate signal transduction pathways that control the machineries responsible for movement of the cells. The signal transduction pathways and the machineries driving the responses are different. In eukaryotic cells such as neutrophils and *Dictyostelium* amoebae, heterotrimeric G-proteins are essential constituents of this pathway. Central (although not indispensable) effectors in chemotactic signal transduction are PI3 kinases, which generate phosphatidylinositol-3,4,5-triphosphate (PI(3,4,5)P$_3$) (Funamoto et al., 2002), and PTEN, a phosphatase that converts the PI(3,4,5)P$_3$ into PI(4,5)P$_2$ (Iijima and Devreotes, 2002). In cells exposed to a gradient of attractant, PI3 kinases are recruited to the leading edge membrane, whereas PTEN is directed to lateral and posterior positions (Funamoto et al., 2002). Through the action of the antagonistic effectors PI3 kinases and PTEN, high levels of PI(3,4,5)P$_3$ are sustained at the leading edge (Comer and Parent, 2002). Proteins comprising a pleckstrin homology (PH) domain recognize the change in membrane phosphatidyl inositides. One of these proteins is the cytosolic regulator of adenylyl cyclase (CRAC), which is consequently directed to the front of the responding cell (Parent et al., 1998). Other effectors, parallel or downstream of PI(3,4,5)P$_3$ in the signal system, contribute to the robustness, speed, and precision of the responses (Kimmel and Parent, 2003).

The principal target system of chemotactic signal transduction in *Dictyostelium* is the network of actin filaments in the cortex of the cells. This network of polymerized actin is formed from complexes of unpolymerized actin that are located in the cytoplasm. The cortical actin network is continuously reorganized in the presence and also in the absence of chemoattractant. What the attractant gradient actually does is to prime the site where actin is induced to polymerize and to accumulate. At the leading edge, actin filaments form a membrane-anchored network together with the Arp2/3 complex, coronin, and other regulatory proteins that control the polymerization and cross-linking of actin filaments (Insall et al., 2001).

By cell aggregation, a multicellular organism of about 10^5 cells is built in *D. discoideum* from single cells (Figure 4.1). In the course of growth in the single-cell state and of multicellular development, the cells use their capability of orienting in gradients of chemoattractant for different purposes. They recognize folate in searching for food bacteria, they respond to cAMP during cell aggregation, and they sort out in a cell-type specific manner within the multicellular body called a "slug," thereby responding to signals generated at the tip of this organism (Weijer, 2003).

For chemotactic orientation in a stationary gradient, *Dictyostelium* cells need a concentration difference of roughly 2 percent between front and tail along their length of 10–20 μm (Fisher, Merkl, and Gerish, 1989). This requirement is similar to the 1-percent difference detected by neutrophils (Zigmond, 1977). This limit of sensitivity restricts the distance from which cells can be attracted by a single diffusion gradient. *D. minutum*, a *Dictyostelium* species that generates a single gradient around each aggregation center, can organize only small aggregation territories with a radius of several hundred

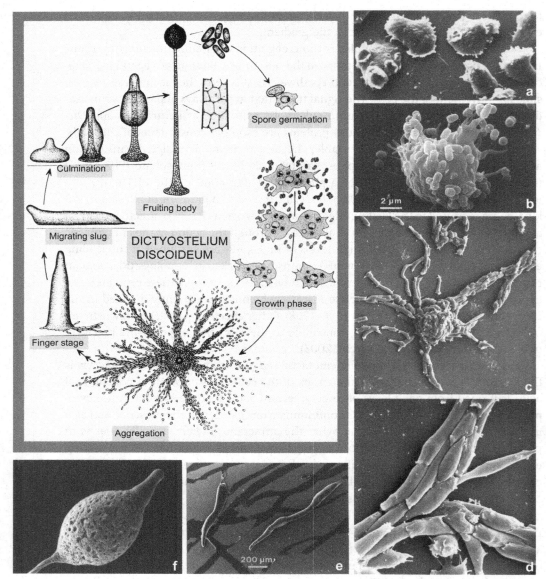

Figure 4.1. Scheme of the developmental cycle of *Dictyostelium discoideum* and scanning electron micrographs exemplifying unicellular and multicellular stages. (a) Growth-phase cells with multiple fronts and crown-shaped protrusions known to be enriched in filamentous actin. (b) A cell growing on bacteria that are attached to the cell's surface and engulfed through circular, actin-rich cell-surface protrusions called phagocytic cups. (c) Aggregating cells streaming toward an aggregation center. (d) Streaming cells that are cohering and strongly polarized. (e) Slugs consisting of some hundred thousand cells, leaving trails of extracellular material when they migrate on a surface. (f) Upper portion of a fruiting body shortly before the spore mass assumes a terminal position at the end of the stalk.

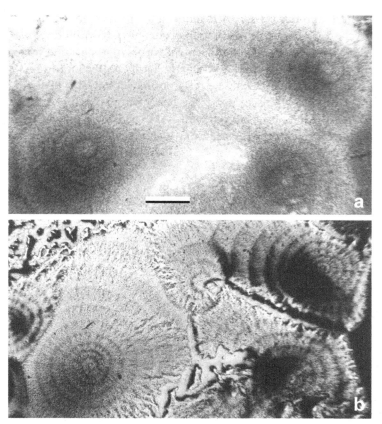

Figure 4.2. Wave patterns of chemotactic responses in aggregation fields of *D. discoideum*. Cells were allowed to aggregate in a dense layer on a phosphate agar plate. The pattern of chemotactic responses to spiral-shaped or concentric cAMP waves was visualized by conventional bright-field microscopy at low-power magnification. (a) Early spiral patterns organizing aggregation territories. (b) At 20 min after (a) cells began to separate from each other at the borders of the territories and to accumulate in the centers. Note that waves are usually prevented from crossing each other by refractoriness of cells just after they have relayed a signal. Bar = 1 mm. From Gerisch, 1971. Copyright of Springer-Verlag.

micrometers (Gerisch, 1964). In aggregates of *D. discoideum*, cells are attracted from a much larger distance by virtue of a relay system. This system is based on the fact that a cell acting as a responder can also act as a producer of the chemoattractant. This is possible because in an aggregation field the synthesis and release of cAMP is phase-shifted relative to the chemotactic response. The cells produce pulses of cAMP at periods of about 7 minutes. In order for the relay system to work, the field is cleaned up between each pulse of cAMP production by cAMP-phosphodiesterases, hydrolyzing enzymes that convert cAMP into 5'-AMP.

Cells that are in the same phase of the signaling and response cycle are organized in a pattern of spiral-shaped or concentric waves that travel from the center to the periphery of an aggregation territory (Figure 4.2). Each cell in the field is capable of responding to cAMP signals in two ways: by chemotactic movement into the direction of the gradient and by the synthesis and release of cAMP, induced by the receptor-mediated activation of adenylyl cyclase. Chemotactic orientation occurs already within the first 30 seconds of arrival of a cAMP wave, and cell movement within a propagating wave is restricted to a window of about 150 seconds at which the local cAMP concentration is increasing (Wessels, Murray, and Sell, 1992). The relay occurs with a delay of about 90 seconds, needed for the cells to synthesize and release cAMP in the form of a pulse (Gerisch and Wick, 1975). As a result of the relay system, cAMP waves are propagating over a field of cells much larger than the distance from which the cells can respond to a single diffusion gradient of the chemoattractant.

TIME REQUIRED FOR THE EXPERIMENT

The cells of *D. discoideum* need 2–4 days to grow on agar plates. The micropipette experiment can be performed within one day.

MATERIALS AND METHODS

EQUIPMENT AND MATERIALS

Per student

Eppendorf Femtotips (glass-microcapillaries). If electrophysiological equipment is available, micropipettes can be self-made on a pipette puller. The pipettes should be pulled like electrodes for intracellular recording with a tip opening of about 0.3 μm diameter.

Eppendorf Microloader Tips, to fill the Femtotips from behind. Alternatively, a filling tool can be produced from a polyethylene tubing. The tubing is gently heated on a burner and, once removed from the burner, drawn into a long, thin extension that can be cut in the middle with a razor blade.

Glass bottom culture dishes (with a coverslip as a bottom). No. 0 uncoated, MatTek Cooperation. Alternatively, a Plexiglas ring, 4 mm high and 40 mm in diameter, is tightened with silicone grease onto a 50 × 50 mm coverslip.

1 scalpel

1 pipette 1 ml and set of tips

1 forceps with a flat tip

Counting chamber for cells (optional; only needed for cultures in HL-5 medium)

Per practical group

Inverted microscope (e.g., Axiovert 200 from Zeiss) with a 100 × oil immersion objective, e.g. Plan-Neofluar for fluorescence recording and phase contrast for imaging of cell shape.

Filter equipment to view GFP fluorescence (e.g., set 41015 for GFP/YFP from Chroma Technology Corp). On a laser scanning microscope the 488-nm line of an argon-ion laser is used for excitation, and a 515–565 bandpass filter is recommended for emission.

12Bit CCD Imaging camera and corresponding software (e.g., SensiCam (PCO)).

Micromanipulator (e.g., Eppendorf InjectMan NI 2) with micropipette holder.

Plastic Petri dishes of 9 cm diameter for agar plate cultures.

Sterilized Drigalski spatulas, self-made from glass rods or purchased as one-way plastic material.

Biological material

Spores of *D. discoideum* strain AX2

Spores of a strain expressing GFP- (or YFP)-ABD120

Spores of a strain expressing coronin-GFP

E. coli B/2

For strain requests from the *Dictyostelium* stock center, contact Jakob Franke (jf31@columbia.edu) or Richard H. Kessin (rhk2@columbia.edu) at Columbia University, New York, NY.

REAGENTS

Adenosine 3′,5′-cyclic monophosphate sodium salt monohydrate (cAMP-Na, Sigma)

KH_2PO_4

$Na_2HPO_4 \times 2H_2O$

Agar-agar (Difco)

Bacto-Peptone (Oxoid)

Glucose-monohydrate

LB or other medium for *E. coli* growth

HL-5 liquid medium (optional; needed only for axenic cultures)

PREVIOUS TASKS FOR STAFF

Preparation of phosphate buffer. For 1 L 17 mM phosphate buffer, weigh out:

KH_2PO_4	2.0 g
$Na_2HPO_4 \times 2H_2O$	0.356 g

Dissolve to 1 L in nanopur or double-distilled water if available, adjust to pH 6.0, and autoclave.

Preparation of nutrient agar plates. For 1 L weigh out:

Agar-agar (Difco)	15 g
Bacto-Peptone (Oxoid)	1 g
Glucose-monohydrate	1 g

Dissolve to 1 L in phosphate buffer and autoclave at 120°C, pour out about 30 ml per Petri dish. Store plates at 4°C in a plastic bag until used.

Liquid medium HL-5. This medium for axenic (bacteria-free) culture can be purchased from Qbiogene as a powder, which is dissolved and autoclaved at not more than 120°C.

Preparation of cAMP solution. Prepare 10 ml of 10^{-3} M cAMP-Na in phosphate buffer, and store frozen at $-20°C$ in 1-ml aliquots. Before filling the micropipettes, dilute to 10^{-4} M with phosphate buffer.

Maintenance of cell strains. *D. discoideum* cells grow by phagocytosis on *E. coli* bacteria. A stock of *E. coli* B/2 is prepared by cultivating the bacteria overnight at room temperature in LB or other nutrient medium. The bacteria are centrifuged and washed in sterile phosphate buffer. After resuspension in fresh buffer the bacteria are shock-frozen and stored at $-20°C$. After thawing, the suspended bacteria can be directly used to inoculate agar plates.

Dictyostelium spores are harvested from cultures with *E. coli* B/2 on nutrient agar plates, shock-frozen like the bacteria in phosphate buffer, and stored at $-20°C$. For the chemotaxis experiment, plate out *E. coli* B/2 on nutrient-agar plates with a Drigalski spatula and allow the surface to dry. Dip a sterile inoculation loop into a suspension of *Dictyostelium* spores and distribute the spores in a straight line on the surface of the plate. At room temperature, the spores will germinate, and the cells will grow into the bacterial lawn and develop within 3 days on both sides of the inoculation line. Spores from these cultures can be picked to inoculate plates for the students' experiments.

For transfer of *Dictyostelium* spores from bacteria-grown cultures to axenic ones, add 100 units of penicillin and 100 μg of streptomycin per ml to HL-5 medium. Cells suspended in the medium are agitated in Erlenmeyer flasks on a rotary shaker at 150 rpm. The cells should be harvested during the exponential growth phase – before they reach a density of 5×10^6 cells per ml.

PROCEDURES

D. discoideum cells cultivated on agar plates with bacteria are rinsed off the agar with phosphate buffer and transferred to a glass bottom culture dish. Cells in a cAMP-sensitive early aggregation stage are stimulated with a micropipette filled with 10^{-4} mM cAMP and fastened on a micromanipulator. The response of the cells is directly observed or video recorded. The chemoattractant-induced redistribution of GFP-tagged cytoskeletal proteins visualized by fluorescence microscopy can also be video recorded. All procedures are performed at room temperature (about $22°C$ and not higher than $25°C$).

OUTLINE OF THE EXPERIMENTS

CHEMOTACTIC STIMULATION

1. Colonies of *D. discoideum* AX2, or transformants derived from this strain that express GFP-fusion proteins, are allowed to grow up to 3 cm from the site of inoculation into the bacterial lawn. At this time all stages of development can be observed on the same plate (Figure 4.3).

    Alternatively, cells of *D. discoideum* AX2 or transformants derived from this strain that express GFP-fusion proteins are harvested from HL-5 medium during

Layer of bacteria [

Starving cells in pre-aggregation phase [

Multicellular development up to fruiting body formation [

Growth zone of *Dictyostelium* cells

Aggregating cells forming streams

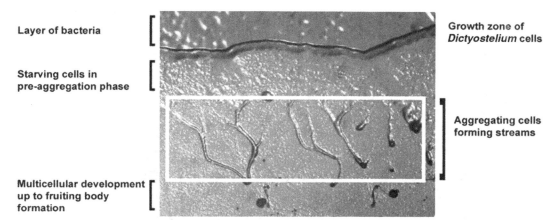

Figure 4.3. Cells of *D. discoideum* growing and developing on an agar plate in a lawn of bacteria. The framed area delineates the field of the agar plate to be cut out in order to harvest cells for the chemotaxis assay.

exponential growth, washed twice in phosphate buffer, and resuspended in the buffer at a density of 1×10^7 cells per ml. After shaking for 5–6 hours under these starving conditions, the cells are optimally suited for chemotactic stimulation.

2. For chemotaxis experiments, students harvest the cells at early aggregation stages. On agar plates, the best stage is recognizable by the beginning of stream formation as in the framed area of Figure 4.3. A piece of agar is cut out with a scalpel and kept with flat forceps. At room temperature, cells are rinsed with phosphate buffer off the agar surface directly into a dish with a glass bottom. There the cells are allowed to settle and spread for about 10 minutes. The cells should be covered by 3–4 mm of buffer.

3. In the meantime, a micropipette is filled free of air bubbles with 10^{-4} M cAMP solution. The pipette should be completely filled to avoid rearward streaming of the solution by capillary forces. The micropipette is fixed on the micromanipulator at a vertical angle of about 30° to the microscope stage.

4. The most critical phase of the experiment is the attempt to place the pipette tip into the microscope field slightly above the cells. First immerse the tip region into the phosphate buffer. By watching from the side, bring the pipette into the light beam of the microscope. DON'T TOUCH THE GLASS BOTTOM OF THE DISH WITH THE PIPETTE, OTHERWISE THE TIP WILL BREAK. Under microscopic inspection with the focus slightly above the cells, the pipette is moved laterally and slowly downward until its out-of-focus image is detectable. Then continue to move the pipette down and backward. When the tip is positioned a few micrometers above the glass surface without touching the cells, the experiment can begin.

5. The chemotactic response is immediately obvious under the microscope because the cells move at a speed of 10–20 μm per minute. Before a cell reaches the tip, the pipette is slightly raised and moved to another position. This can be repeated as many times as is necessary.

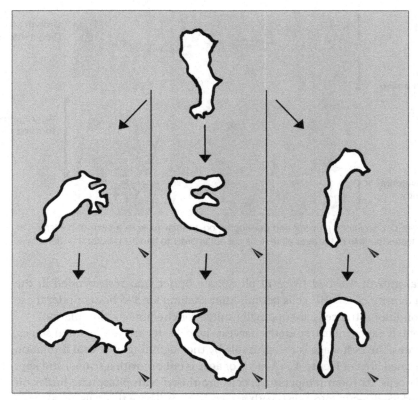

Figure 4.4. Variations in chemotactic response behaviour. Starting with a front pointing to the top of this diagram, a cell can either keep or change its polarity when it is attracted to the source of a gradient (tip of a micropipette indicated by arrowhead). Left: Turn of the initial front. Middle: Induction of several new protrusions, of which one eventually persists as a leading edge (in the example shown the initial rear end). Right: Initiation of a directional response at the most distant portion of a cell, actually at its previously established leading edge. Compiled from Segall and Gerisch (1989) and Weber et al. (2002).

Data recording. The response of *D. discoideum* cells in a diffusion gradient of cAMP (Figure 4.4) can be recorded by DIC or phase-contrast optics using a CCD camera at a frame-to-frame interval of 1 second. For imaging the redistribution of GFP-tagged proteins, a fluorescence microscope is required. Dual-channel imaging allows the user to record fluorescence and cell shape simultaneously. If available, use of a confocal laser scanning microscope is recommended. However, attention should be paid to the fact that the front of a cell is often lifted up with respect to the rest of the cell body. Therefore, the front may no longer be in the focal plane when the cell body is in focus.

POTENTIAL SOURCES OF FAILURE

The cells must be gently harvested at the proper developmental stage and should be seen to move on a glass surface. Cells that are rounded rather than spread are suffering from inappropriate treatment. The optimal temperature is between 21° and 23 °C, and should not be higher than 25 °C; lower temperatures slow down all activities of the cells. *D. discoideum* is an aerobic microorganism. Its cells suffer from loss of ATP within a few minutes of oxygen depletion at room temperature. Cells can be stored for a longer time in an ice bath, but thereafter need about a quarter of an hour to move and respond appropriately.

Light intensity should be reduced to a minimum because the cells tend to contract in response to illumination. In particular, they respond temporarily when light is turned on. This is critical in recording GFP fluorescences, since the cells are most sensitive to the blue light needed for GFP excitation. For phase-contrast recording, a red filter is recommended for illumination.

It is crucial to limit the diffusion of cAMP by a narrow opening of the micropipette. If no attraction to the micropipette is observed, the pipette might be blocked or, more likely, the tip may be broken. The tip should be kept a few micrometers above the glass surface. This protects the tip from breaking and prevents its obstruction by cells committing suicide by hara-kiri.

EXPECTED RESULTS

When cells are stimulated by cAMP through a micropipette from a distance of about 20 μm, the first signs of a chemotactic response may be seen within 10 seconds. If the cell is stimulated at the flank (at an angle of about 90° relative to its axis of polarity) or from behind, it will either turn with its previous front into the direction of the gradient, or protrude one or several new fronts into the correct direction while the previous front is retracting. Turning may occur even when the cell is stimulated from behind, indicating that the response does not necessarily occur at a site where the cell is most strongly stimulated or where the attractant reaches its surface first (Figure 4.4). If polarity is changed in response to the attractant, often multiple protrusions are induced, which compete with each other until a single front guides the directional movement of the cell. The variability in the behavior reflects the contribution of the cytoskeleton, its organization and its flexible rearrangements, to the chemotactic response.

Reversion of polarity is most obvious when cells are already assembled by end-to-end contacts into streams. In Figure 4.5, a micropipette is placed on the side of a stream. Initiation of chemotaxis is accompanied by the local loss of adhesiveness, so that a free tail is produced in one cell and a free front in the other. Both ends are attracted in parallel, which means that the cell that was ahead of the other has reversed its polarity.

By fluorescence imaging of cells expressing GFP-fusions of actin-associated proteins, spontaneous and chemoattractant-induced reorganization of the actin system can be monitored. Local polymerization and depolymerization of actin in the cell cortex is visualized by GFP-tagged ABD120, the actin-binding domain of a 120 kDa filament-like protein from *Dictyostelium* (Pang, Lee, and Knecht, 1998). In response to the chemoat-tractant, a GFP-fusion of the actin-associated protein coronin is recruited within seconds from the cytoplasm to a new leading edge (Gerisch et al., 1995). Figure 4.6 illustrates the redistribution of coronin-GFP during turning of a front, in response to the

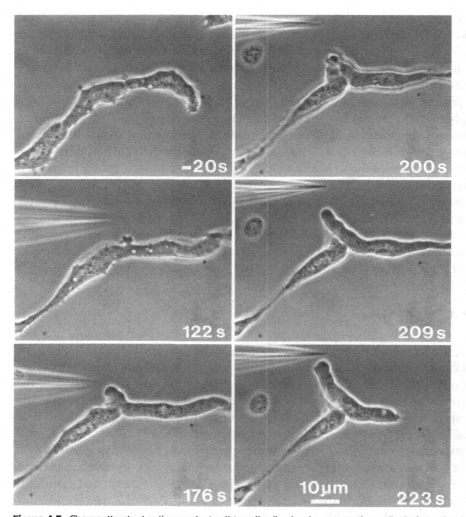

Figure 4.5. Chemoattractant acting against cell-to-cell adhesion in aggregating cells. Before stimulation, the one-row stream of three highly polarized cells moved to the right. Stimulation through a micropipette close to the connection between the second and third cell in the stream firstly induced the second cell to form a lateral protrusion. Subsequently, adhesion was paralyzed and both cells responded in parallel, one with its tail, the other with its tip (200-second frame). (Afterwards the front of the third cell moved out of focus.) Note the hyaline, particle-free front region of responding cells, which coincides with the dense actin network at their tip. Time is indicated in seconds before and after insertion of the micropipette.

induction of new fronts, and upon loss of polarity when a cell approaches the pipette tip, where the cell is surrounded by high cAMP concentrations.

DISCUSSION

There are two networks of molecular interaction that need to be coupled to each other in order for a *Dictyostelium* cell to respond in a gradient of attractant by directional movement: the signal transduction system and the cytoskeleton, primarily

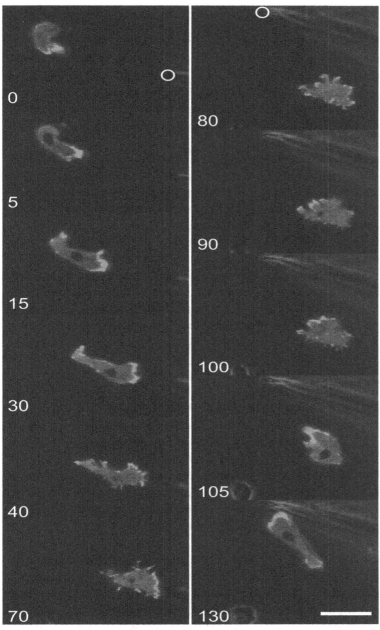

Figure 4.6. Chemotaxis and reorganization of the actin cytoskeleton. This time series shows the response of a *D. discoideum* cell to a micropipette filled with cAMP. Between the left and right row of images the pipette has been moved toward the left and top of the frame. Positions of the pipette tip before and after the move are marked by circles. The cell shown expresses coronin-GFP as a marker. Together with filamentous actin, coronin is recruited from the cytoplasm to sites of chemotactic responses. The redistribution of coronin was recorded by confocal fluorescence microscopy, and the fluorescence images in green are superimposed on phase-contrast images in dark red. The dark area within the cell is the nucleus. The numbers are in seconds. Bar = 10 μm. (See also color plate 1).

the actin network in the cell cortex. In principle, both systems can act independently of each other. When the actin system is paralyzed by latrunculin A, a drug that in vivo blocks the polymerization of actin, cells become immobilized and rounded. Nevertheless, stimulation through a micropipette causes the accumulation of CRAC at the side of cAMP application, a response for which the PH-domain of CRAC is responsible (Parent and Devreotes, 1999). When the micropipette is moved around the cell, the crescent of CRAC recruitment beneath the cell membrane follows that movement of the pipette, indicating that the accumulation is a dynamic response to continuous signal inputs.

The actin system is capable of driving cell movement in the absence of any chemotactic stimulation. In fact, polymerized actin accumulates locally in the cell cortex in mutants lacking the β-subunit of heterotrimeric G-proteins that are required for chemotactic signal transduction. In the mutant cells, chemotactic stimulation does not determine the site where actin is accumulating (Peracino et al., 1998).

In their actual chemotactic behavior cells integrate the inputs from the signal transduction pathways and from the cytoskeletal organization. The variable balance of these different influences determines the repertoire of chemotactic behavior. In some cases the polarity imposed on the cell by the cytoskeletal organization overrides the gradient of receptor occupancy from the external attractant gradient. A striking case is the initiation of a response at a site remote from the source of the gradient (right panel of Figure 4.4).

ALTERNATIVE EXERCISES

STIMULATION OF OTHER CELLS THROUGH A MICROPIPETTE

Stimulation through a micropipette, as described here for *Dictyostelium*, is an efficient method to induce chemotaxis in mammalian cells. For human neutrophils (or differentiated cells of a permanent cell line such as HL-60) the pipette is filled with 10 μM fMetLeuPhe (Gerisch and Keller, 1981). Mammary carcinoma cells can be stimulated similarly by epidermal growth factor (Bailly et al., 1998). By applying the same technique to the stimulation of budding yeast with mating factor α, directed cell growth leading to polarized projections can be induced (Vallier et al., 2002).

Waves of signal relay in aggregation territories. Aggregation territories of *D. discoideum* reach sizes of up to 1 cm in diameter. Two mechanisms enable the cells to reach the center of an aggregate from a long distance. One is cohesion that allows the cells to move in the form of streams (Figures 4.1c and 4.1d). The other mechanism is the relay of cyclic-AMP signals over a field of aggregating cells. These signals are generated in the aggregation centers in a pulsatile manner by periodic adenylate cyclase activation. The signal relay implies three steps: (1) intracellular signal transduction from activated cAMP receptors through a transducer protein (CRAC) to adenylate cyclase, which is activated to produce cAMP; (2) release of the de-novo synthesized cAMP into the medium where it diffuses as a transmitter, stimulating neighboring cells by binding to their cell-surface receptors; (3) hydrolysis of cAMP by phosphodiesterases that are located on the cell surface or released into the medium. These enzymes convert the 3′,5′-cyclic AMP into inactive 5′AMP.

Wave patterns can be visualized at low magnification by bright- or dark-field microscopy. Any microscope equipment available should be adaptable to the time-lapse recording of wave propagation using a video camera at about 4 frames per min. For bright-field illumination, a narrow diaphragm and slightly acentric illumination is helpful, as used in Figure 4.2. Dark-field illumination can be improvised; for high-quality images of wave patterns in wild-type and mutants, special equipment is recommended (Ross and Newell, 1979).

Waves can be seen to propagate along streams of cells aggregating under conditions as shown in Figure 4.3. The wave patterns are more impressive if cells are dispensed as a dense layer on an agar surface. To do that, cells are cultivated in HL-5 medium in a shaken suspension, harvested during exponential growth at a density of not more than 5×10^6 cells per ml, washed twice by centrifugation in phosphate buffer, and resuspended at a density of 1×10^7/ml in the buffer. Droplets of the cell suspension are placed on the surface of phosphate agar plates (1.5% agar in phosphate buffer). After settling of the cells, excess fluid is removed. The generation of wave patterns is recorded for several hours at frame-to-frame intervals of 15–30 seconds.

TEACHING CONCEPTS

The chemotactic response of *Dictyostelium* cells is representative of the orientation of neutrophils, tumor cells, and embryonic cells in fields of signaling molecules that are generated in spatial patterns within an organism. To find a direction along concentration profiles, it is not sufficient for the cells to recognize the signaling molecules: the pattern of occupancy of the chemoattractant receptors over the surface of the cells is important and needs to be translated into a directional response. From a theoretical point of view, chemotaxis exemplifies general principles of biological pattern formation (Meinhardt, 1999). In molecular terms, receptors uniformly distributed on the cell surface feed their pattern of activation into a signal-transducing network that recruits a set of target proteins to different positions beneath the cell surface (Kimmel and Parent, 2003). This pattern of recruitment provides a platform on which the actin cytoskeleton is remodeled.

In principle, actin reorganization is based on rapid and local shifts in the equilibrium between unpolymerized and filamentous actin, which is controlled by a large number of proteins and other regulatory factors, for instance, the lipid phosphatidylinositol-4,5-bisphosphate (PIP_2) (Yin and Janmey, 2003). Actin filaments are the material on which membrane-anchored network structures of various shapes are built (Medalia et al., 2002). Many proteins involved in the organization and protrusion of a leading edge are common to motile eukaryotic cells (Pollard and Borisy, 2003). What makes chemotaxis special is the coupling of these processes to the environmental pattern of specific chemical signals.

Periodic generation of cAMP signals and the relay of signals in the form of waves and spirals is a special case of spatio–temporal pattern formation in excitable media

(Levine et al., 1996; Hess, 1997). A model based on the interaction of known proteins simulates this oscillatory behavior of the biological system (Maeda et al., 2004).

ADDITIONAL INFORMATION AND TEACHING AIDS

Because the chemotaxis field is in rapid progress, updated reviews should be consulted before each course. Information on all aspects of *Dictyostelium* research and on the *Dictyostelium* stock center is available from the Model Organism Database for Dictyostelium (http://www.dictybase.org). The user has to enter a gene name, researcher name, or gene product name to be presented with the relevant data. The search box will search the dictyNews if the tick box is checked. There is also access to chemotaxis movies that can be downloaded.

A 15-minute movie with English sound track covering the entire developmental cycle is available under http://www.iwf.de from the Institut für den Wissenschaftlichen Film, Wissen und Medien gGmbH, Nonnenstieg 72, D-37075 Göttingen, Germany. Movies on special topics of *Dictyostelium* growth and development are distributed by the same institution.

APPENDIX

HISTORICAL OUTLOOK

For those who are interested in early steps of research that have led to our current knowledge, a few references documenting development of the field are given. Cyclic AMP has been identified as the chemoattractant of aggregating cells in John Bonner's group (Konijn et al., 1967). The micropipette technique was invented by Dieter Hülser through an unsuccessful attempt to record intracellularly the effect of cAMP on the membrane potential of *Dictyostelium* cells (Gerisch et al., 1975). Identification of the constituents of the chemotactic signaling cascade began with the cloning of the first cAMP receptor in Peter Devreotes' laboratory (Klein et al., 1988).

The story of signal relay goes back to a pioneering time-lapse study (the first one on the entire development of a multicellular organism) by Arthur Arndt (1937), who discovered the pulsatile activity of aggregation centers and the propagation of stimuli in the form of waves. Spiral waves in aggregation territories were first described by Gerisch (1965), and scroll waves in slugs by Bretschneider, Siegert, and Wiejer (1995). The importance of pulsatile signaling in *D. discoideum* has been stressed by the finding that mutants deficient in phosphodiesterase, which cannot degrade the cAMP after each pulse, are unable to develop (Darmon, Barra, and Brachet, 1978). Spontaneous oscillations of adenylate cyclase activity and release of cAMP from the cells were analyzed in cell suspensions (Gerisch and Wick, 1975; Roos, Scheidegger, and Gerisch, 1977). Devreotes and colleagues have used a flow system to quantitatively analyze signal relay in response to stimuli of defined temporal patterns (see Dinauer, Steck, and Devreotes (1980), the last in a series of five papers). To visualize the spatial pattern of cAMP release in aggregation fields, an assay based on binding of the chemoattractant to an affinity support was invented by Tomchik and Devreotes (1981).

REFERENCES

Arndt, A. (1937). Rhizopodenstudien. III. Untersuchungen über *Dictyostelium mucoroides* Brefeld. *Wilhelm Roux' Archiv für Entwicklungsmechanik der Organismen*, 136, 681–747.

Bailly, M., Yan, L., Whitesides, G. M., Condeelis, J. S., and Segall, J. E. (1998). Regulation of protrusion shape and adhesion to the substratum during chemotaxis responses of mammalian carcinoma cells. *Exp. Cell Res.*, 241, 285–99.

Bray, D. (2001). *Cell Movements. From Molecules to Motility*, 2nd ed. New York: Garland.

Bretschneider, T., Siegert, F., and Weijer, C. J. (1995). Three-dimensional scroll waves of cAMP could direct cell movement and gene expression in *Dictyostelium* slugs. *Proc. Natl. Acad. Sci. USA*, 92, 4387–91.

Comer, F. I., and Parent, C. A. (2002). PI 3-kinases and PTEN: How opposites chemoattract. *Cell*, 109, 541–4.

Darmon, M., Barra, J., and Brachet, P. (1978). The role of phosphodiesterase in aggregation of *Dictyostelium discoideum. J. Cell Sci.*, 31, 233–43.

Dinauer, M. C., Steck, T. L., and Devreotes, P. N. (1980). Cyclic 3′,5′-AMP relay in *Dictyostelium discoideum*. V. Adaptation of the cAMP signaling response during cAMP stimulation. *J. Cell Biol.*, 86, 554–61.

Fisher, P. R., Merkl, R., and Gerisch, G. (1989). Quantitative analysis of cell motility and chemotaxis in *Dictyostelium discoideum* by using an image processing system and a novel chemotaxis chamber providing stationary chemical gradients. *J. Cell Biol.*, 108, 973–84.

Funamoto, S., Meili, R., Lee, S., Parry, L., and Firtel, R. A. (2002). Spatial and temporal regulation of 3-phosphoinositides by PI 3-kinase and PTEN mediates chemotaxis. *Cell*, 109, 611–23.

Gerisch, G. (1964). *E 673/1964 Dictyostelium minutum* (Acrasina). Aggregation. IWF Wissen und Medien GmbH, Nonnenstieg 72, D-37075 Göttingen, Germany. Internet: http://www.iwf.de, last accessed September 2004.

Gerisch, G. (1965). Stadienspezifische Aggregationsmuster bei *Dictyostelium discoideum. Roux' Archiv für Entwicklungsmechanik*, 156, 127–44.

Gerisch, G. (1971). Periodische Signale steuern die Musterbildung in Zellverbänden. *Naturwissenschaften*, 58, 430–8.

Gerisch, G., Albrecht, R., Heizer, C., Hodgkinson, S., and Maniak, M. (1995). Chemoattractant-controlled accumulation of coronin at the leading edge of *Dictyostelium* cells monitored using a green fluorescent protein-coronin fusion protein. *Curr. Biol.*, 5, 1280–5.

Gerisch, G., Hülser, D., Malchow, D., and Wick, U. (1975). Cell communication by periodic cyclic-AMP pulses. *Phil. Trans. R. Soc. Lond. B*, 272, 181–92.

Gerisch, G., and Keller, H. U. (1981). Chemotactic reorientation of granulocytes stimulated with micropipettes containing fMet-Leu-Phe. *J. Cell Sci.*, 52, 1–10.

Gerisch, G., and Wick, U. (1975). Intracellular oscillations and release of cyclic AMP from *Dictyostelium* cells. *Biochem. Biophys. Res. Comm.*, 65, 364–70.

Hess, B. (1997). Periodic patterns in biochemical reactions. *Quart. Rev. Biophys.*, 30, 121–76.

Iijima, M., and Devreotes, P. (2002). Tumor suppressor PTEN mediates sensing of chemoattractant gradients. *Cell*, 109, 599–610.

Insall, R., Müller-Taubenberger, A., Machesky, L., Köhler, J., Simmeth, E., Atkinson, S. J., Weber, I., and Gerisch, G. (2001). Dynamics of the *Dictyostelium* Arp 2/3 complex in endocytosis, cytokinesis, and chemotaxis. *Cell Motility Cytoskeleton*, 50, 115–28.

Kimmel, A. R., and Parent, C. A. (2003). The signal to move: *D. discoideum* go orienteering. *Science*, 300, 1525–26.

Klein, P. S., Sun, T. J., Saxe III, C. L., Kimmel, A. R., Johnson, R. L., and Devreotes, P. N. (1988). A chemoattractant receptor controls development in *Dictyostelium discoideum. Science*, 241, 1467–72.

Konijn, T. M., van de Meene, J. G. C., Bonner, J. T., and Barkley, D. S. (1967). The acrasin activity of adenosine-3′,5′-cyclic phosphate. *Proc. Natl. Acad. Sci. USA*, 58, 1152–4.

Levine, H., Aranson, I., Tsimring, L., and Truong, T. V. (1996). Positive genetic feedback governs cAMP spiral wave formation in *Dictyostelium. Proc. Natl. Acad. Sci. USA*, 93, 6382–6.

Maeda, M., Lu, S., Shaulsky, G., Miyazaki, Y., Kuwayama, H., Tanaka, Y., Kuspa, A., and Loomis, W. F. (2004). Periodic signaling controlled by an oscillatory circuit that includes protein kinases ERK2 and PKA. *Science*, 304, 875–8.

Medalia, O., Weber, I., Frangakis, A. S., Nicastro, D., Gerisch, G., and Baumeister, W. (2002). Macromolecular architecture in eukaryotic cells visualized by cryoelectron tomography. *Science*, 298, 1209–13.

Meinhardt, H. (1999). Orientation of chemotactic cells and growth cones: Models and mechanisms. *J. Cell Sci.*, 112, 2867–74.

Pang, K. M., Lee, E., and Knecht, D. A. (1998). Use of a fusion protein between GFP and an antibinding domain to visualize transient filamentous-actin structures. *Curr. Biol.*, 8, 405–8.

Parent, C. A., Blacklock, B. J., Froehlich, W. M., Murphy, D. B., and Devreotes, P. N. (1998). G protein signalling events are activated at the leading edge of chemotactic cells. *Cell*, 95, 81–91.

Parent, C. A., and Devreotes, P. N. (1999). A cell's sense of direction. *Science*, 284, 765–9.

Peracino, B., Borleis, J., Jin, T., Westphal, M., Schwartz, J.-M., Wu, L., Bracco, E., Gerisch, G., Devreotes, P., and Bozzaro, S. (1998). G protein β subunit-null mutants are impaired in phagocytosis and chemotaxis due to inappropriate regulation of the actin cytoskeleton. *J. Cell Biol.*, 141, 1529–37.

Pollard, T. D., and Borisy, G. G. (2003). Cellular motility driven by assembly and disassembly of actin filaments. *Cell*, 112, 453–65.

Roos, W., Scheidegger, C., and Gerisch, G. (1977). Adenylate cyclase activity oscillations as signals for cell aggregation in *Dictyostelium discoideum*. *Nature*, 266, 259–61.

Ross, F. M., and Newell, P. C. (1979). Genetics of aggregation pattern mutations in the cellular slime mold *Dictyostelium discoideum*. *J. Gen. Microbiol.*, 115, 289–300.

Segall, J. E., and Gerisch, G. (1989). Genetic approaches to cytoskeleton function and the control of cell motility. *Curr. Opin. Cell Biol.*, 1, 44–50.

Tomchik, K. J., and Devreotes, P. N. (1981). Adenosine 3′,5′-monophosphate waves in *Dictyostelium discoideum*: A demonstration by isotope dilution-fluorography. *Science*, 212, 443–6.

Vallier, L. G., Segall, J. E., and Snyder, M. (2002). The alpha-factor receptor C-terminus is important for mating projection formation and orientation in *Saccharomyces cerevisiae*. *Cell Motil. Cytoskeleton*, 53, 251–66.

Weber, I., Niewöhner, J., Du, A., Röhrig, U., and Gerisch, G. (2002). A talin fragment as an actin trap visualizing actin flow in chemotaxis, endocytosis, and cytokinesis. *Cell Motil. Cytoskeleton*, 53, 136–49.

Weijer, C. J. (2003). Visualizing signals moving in cells. *Science*, 300, 96–100.

Wessels, D., Murray, J., and Soll, D. R. (1992). Behavior of *Dictyostelium* amoebae is regulated primarily by the temporal dynamic of the natural cAMP wave. *Cell Motil. Cytoskeleton*, 23, 145–56.

Yin, H. L., and Janmey, P. A. (2003). Phosphoinositide regulation of the actin cytoskeleton. *Annu. Rev. Physiol.*, 65, 761–89.

Zigmond, S. H. (1977). Ability of polymorphonuclear leukocytes to orient in gradients of chemotactic factors. *J. Cell Biol.*, 75, 606–16.

5 Inhibition of signal transduction pathways prevents head regeneration in hydra

L. M. Salgado

OBJECTIVE OF THE EXPERIMENT Development is the result of the coordinated expression, at the right time and in the right place, of a set of genes. The expression pattern of these mostly regulatory genes depends on the ability of cells to interpret different signals from their neighbors and from their position in the embryo. This chapter presents an easy way to analyze some of the signaling systems typically used by cells to acquire their correct fate. By using well-defined chemicals to selectively block key enzymes in signaling pathways, the normal role of these signaling pathways in hydra development can be determined.

DEGREE OF DIFFICULTY Medium.

INTRODUCTION

Hydra is an evolutionary old metazoan. Its tissue organization is simple, but its cells are determined via a complex pattern of cell–cell interactions. A hydra is essentially a tube consisting of two layers of tissue, the ectoderm and the endoderm, separated by an intermediate layer, the mesoglea. There are two cell lineages, the muscular epithelium and the interstitial stem cells, from which all the cells derive. The adult animal has around 100,000 cells that can be grouped into twelve different tissues. Hydra can reproduce either sexually or asexually. Most significantly, hydra has a high capacity for regeneration and is one of the more complex organisms that is able to regenerate from cell aggregates. These features make hydra an ideal model for studying morphogenesis and differentiation in the adult animal (Gierer, 1974).

The development of hydra consists primarily of head and foot formation from tissue of the gastric region. This is a continuous process in the adult animal, allowing for study of an embryonic-like process that persists in the adult. A well-fed animal is continuously involved in growth, pattern formation, morphogenesis, and differentiation (Bode and Bode, 1980). Hydra has a third specialized structure as an outcome of asexual reproduction, the bud. Pattern formation for the new bud starts at approximately

2/3 the distance along the body to the head. The cells form an evagination and new cells are incorporated to the new axis as the bud growths. After three days, the new organism, about 1/4 the size of the mother, is ready to detach (Otto, 1976). An important characteristic of hydra is that the two extremities have organizer capacity, similar to the dorsal lip on the amphibian gastrula, and can induce a secondary axis when transplanted. This capacity implies that the two organizers have the ability to regulate the development of the whole animal (Gierer, 1970).

When a hydra is bisected at any point along the body column, there is an activation of the regeneration machine. Bode (1996) divides the pattern formation in such conditions into four stages. The first stage occurs when the decision to regenerate must be started. The relative position of the cells or the strength of the gradients helps to make a head or a foot. The absence of extremities stimulates the regeneration process.

The characterization of several genes, most of them transcription factors that are expressed at early stages during regeneration, has helped us to gain some information about the molecular basis of head and axis formation. In addition, some of the signal transduction pathways involved in head formation have been elucidated. A transduction system mediated by protein kinase C (PKC) has been implicated for head regeneration. Treating animals with PKC activators, such as diacylglycerol (DAG), 12-O-tetradecanoylphorbol-13-acetate (TPA), and arachidonic acid (AA), results in the formation of ectopic heads (reviewed in Müller, 1996). Treatment with sphingosine, the natural inhibitor of PKC, and two other inhibitors, staurosporine and H7, stops head regeneration (Cardenas et al., 2000). The activation of PKC by TPA produces an enhanced transcription of *Ks1*, a head-specific gene (Weinziger et al., 1994; Endl et al., 1999) and inhibition by sphingosine results in inhibition of the transcription of the same gene (Cardenas et al., 2000). The function of other signal transduction pathways in the regulation of head development has been analyzed using inhibitors against SRC, PI3K, and ERK. The inhibitors used reversibly block head development in *Hydra*, and this indicates that signal transduction pathways involving each of these enzymes are required for head development (Cardenas et al., 2000; Fabila et al., 2002). In treated animals, the inhibition of several genes expressed early during head and bud formation provides additional evidence for their regulatory function. As an example, the selective SRC kinase inhibitors, PP1/AGL1872 and PP2/AG1879, inhibit both head regeneration and the expression of the head specific genes *Ks1, Hybra1* and *Hym301* (Cardenas et al., 2000). Using morphological and gene expression assays, we found that the signaling pathways mediated by SRC, and PI3K play a central regulatory role in bud formation.

MATERIALS AND METHODS

EQUIPMENT AND MATERIALS
Per student

12-well cell culture plates	Eppendorf tubes
Surgical bistoury	15- and 50-ml falcon tubes
Stereoscopic microscope	Pipette tips
Pasteur pipettes	

Per practical group

Micropipettes	Rotator
Incubator at 18–20°C	Thermoblocks
Hybridization oven	Centrifuge
Shaker	Bakers

Biological material

HYDRA STRAINS. *Hydra magnipapillata* (wt strain *105*) or *Hydra vulgaris* (wt strain Basel) cultured using standard procedures (Takano and Sugiyama, 1983) are used for all experiments.

HYDRA REQUEST. There is no hydra stock center. The animals should be requested from the nearest hydra group. I will cite here three of them for reference – one in Europe, one in the United States, and one in Japan – that have the largest hydra collections.

Prof. Charles N. David. Zoological Institute, University of Munich; Luissenstr. 14, D-80333 Munich, FRG.

Prof. Hans R. Bode. Department of Developmental and Cell Biology, University of California, Irvine, CA 92697.

Prof. Toshitaka Fujisawa. Department of Developmental Genetics, National Institute of Genetics. Mishima 411-0801, Japan.

REAGENTS

Milli Q grade water	Forskolin (Sigma)
RNA labeling kit (Roche)	BpV(phen) (Calbiochem)
$CaCl_2$	Me-Me-IBMX (Sigma)
$MgCl_2$	Urethane (Sigma)
KCl	Paraformaldehyde
$NaHCO_3$	Proteinase K (Roche)
NaCl	Glycine
HCl	Acetic anhydride
NaOH	Trietanolamine
Ethanol	Formamide
DMSO	Sodium citrate
Apigenin (Calbiochem)	Ficoll
PP2/AG1879 (Calbiochem)	Bovine serum albumin
LY2940022 (Calbiochem)	Polyvinylpyrrolidone
PD98059 (Calbiochem)	tRNA from yeast
Heparin	Levamisole (Sigma)
Tween 20	Nitro blue tetrazolium
CHAPS	
Maleic acid	BCIP (5-bromo-4-chloro-3-indolyl phosphate) (Roche)
Sheep serum (Invitrogen)	Euparol
Alkaline phosphatase-conjugated anti-digoxygenin Fab fragment (Roche)	

PREVIOUS TASKS FOR STAFF

Maintenance of hydra culture. Media for maintaining hydras contain 1 mM $CaCl_2$, 0.1 mM $MgCl_2$, 0.1 mM KCl, 1 mM $NaHCO_3$, pH 7.8 adjusted with HCl. Hydra are fed daily with recently hatched *Artemia salina*. For the experiments involving inhibition of regeneration, the animals are starved for 48 h before their utilization. Further information can be obtained in Chapter 1.

Solutions

Hydra Medium (HM). Described in Chapter 1.

Saline solution for Artemias. Described in Chapter 1.

Stock solutions of the inhibitors. THESE COMPOUNDS ARE CONSIDERED AS IR-
RITANTS AND IT IS RECOMMENDED TO USE PROTECTIVE CLOTHES. PLEASE
READ CAREFULLY THE INDICATIONS FROM THE PROVIDER.

In situ hybridization solutions

2% urethane in HM

4% paraformaldehyde in HM

PBS (phosphate-buffered saline) (0.15 M NaCl, 0.08 M Na_2HPO_4, 0.021 M NaH_2PO_4,
pH 7.34)

PBT (phosphate-buffered saline with 0.1% Tween 20)

75%–25% ethanol-PBT

50%–50% ethanol-PBT

25%–75% ethanol-PBT

10 mg/ml Proteinase K in PBT

4 mg/ml glycine in PBT

0.1 M triethanolamine pH 7.8

0.25% (v/v) acetic anhydride in 0.1 M triethanolamine pH 7.8

5 × SSC (750 mM NaCl, 75 mM sodium citrate)

2 × SSC (300 mM NaCl, 30 mM sodium citrate)

2 × SSC, 0.1% chaps

Hybridization solution (HS) (50% formamide, 5 × SSC, 0.02% (w/v) each Ficoll,
bovine serum albumin (BSA, Fraction V), and polyvinylpyrrolidone, 200 mg/ml
yeast tRNA, 100 mg/ml heparin, 0.1% Tween 20, and 0.1% chaps)

75%–25% HS – 2 × SSC

50%–50% HS – 2 × SSC

25%–75% HS – 2 × SSC

MAB (100 mM maleic acid, 150 mM NaCl, pH 7.5)

MAB, 1% BSA

80%–20% MAB – heat-inactivated sheep serum

NTMT (100 mM NaCl, 100 mM Tris, pH 9.5, 50 mM $MgCl_2$, 0.1% Tween-20)

3.7% paraformaldehyde

70% ethanol

95% ethanol

PROCEDURES

In situ hybridization. The whole-mount in situ hybridization procedure described is based in the method reported previously by Grens et al. (1996):

1. Animals are "relaxed" in 2% urethane for 30–60 s in hydra medium (HM) and fixed overnight at 18°C in 4% paraformaldehyde in HM.
2. Following fixation, the samples are washed for 5 min each in 100% ethanol, 75%, 50%, and 25% ethanol-PBT, and then rinsed three times for 5 min each in PBT. These washes and subsequent treatments are performed at room temperature on a rotator unless otherwise indicated.
3. The samples are then treated for 10 min with 10 mg/ml Proteinase K in PBT.
4. Proteinase K digestion is stopped by a 10-min wash with 4 mg/ml glycine in PBT, and the glycine is removed by two 5-min washes in PBT.
5. The samples are treated twice for 5 min each with 0.25% (v/v) acetic anhydride in 0.1 M triethanolamine pH 7.8, and re-fixed for 20–30 min in 4% paraformaldehyde in PBT at room temperature.
6. Fixative is removed by five 5-min washes with PBT, after which the samples are heated to 85°C for 30 min to inactivate endogenous alkaline phosphatases.
7. Before hybridization, the samples are washed for 10 min in PBT – hybridization solution (50%–50%) and for 10 min in hybridization solution, and then incubated for at least 2 h in hybridization solution at 55°C.
8. A digoxygenin-labeled probe is added to a final concentration of approximately 0.3 mg/ml and hybridized overnight at 55°C.
9. To remove unhybridized probe, the samples are washed for 10 min each in HS – 2 × SSC, 75%–25%, 50%–50%, and 25%–75%.
10. Two 30-min stringent washes are performed in 2 × SSC with 0.1% chaps. All washes are carried out at 55°C.
11. To prepare the samples for binding of the anti-digoxygenin antibody, they are first washed twice for 10 min at room temperature in MAB and then for 1 h in MAB containing 1% BSA.
12. Samples are then blocked for at least 2 h in a solution of 80% MAB – 20% heat-inactivated sheep serum (Sigma).
13. Alkaline phosphatase-conjugated anti-digoxygenin Fab fragments (Roche) are diluted 1:400 in blocking solution and pre-absorbed, for at least 2 h, against fixed hydra before being used.
14. The pre-absorbed Fab fragments are diluted to a final dilution of 1:1,600 in blocking solution and applied to the samples overnight at 4°C.
15. Unbound anti-digoxygenin Fab fragments are removed by eight 20-min washes in MAB at room temperature.
16. Samples are then equilibrated with the alkaline phosphatase staining buffer NTMT in three 10-min washes, with the final wash also containing 1 mM Levamisole.
17. Substrate solution consisting of NTMT-Levamisole with 4.5 ml/ml NBT (nitro blue tetrazolium) and 3.5 ml/ml BCIP (5-bromo-4-chloro-3-indolyl phosphate) is added, and the samples are incubated in the dark for 3 h or until the development of color.

18. Staining is stopped by two 20-min washes in ethanol.
19. The samples are re-fixed for 30 min in 3.7% paraformaldehyde, rinsed three times in ethanol, and allowed to destain overnight in ethanol.
20. For mounting, the samples are dehydrated by successive 2-min washes in 70% ethanol, 95% ethanol, and twice in 100% ethanol and then mounted in Euparol (ASCO Laboratories).

OUTLINE OF THE EXPERIMENTS

EXPERIMENT: EFFECT OF THE INHIBITOR OF HEAD REGENERATION

Standardization of the parameters of regeneration. To ensure the reproducibility of the inhibition experiments, the place of the cut for the head and the foot should be carefully established and followed. The hydras should be decapitated at 2/3 the body length, measured from the foot. To remove the foot, a cut should be made above the peduncle, around 1/3 the body length measured from the foot. The manipulations are done with a surgical bistoury and under a stereoscopic microscope.

An animal is considered to have regenerated its head when the hypostome and the tips of the tentacles are clearly visible. Foot regeneration is complete when the animals are able to adhere to the surface of the culture box.

Effect of the inhibitors on head regeneration. Decapitate 10–20 starved adult animals without buds and incubate in 5 ml of hydra medium plus 10 μl of the inhibitor. Include two controls as follows: (a) 10 animals incubated in hydra medium and (b) 10 animals incubated in hydra medium plus 10 μl of the solvent in which inhibitors are dissolved. Replace the inhibitor every 24 h and follow the progress of regeneration with a stereomicroscope over the next 3 days. After this time, the control animals should regenerate a head and the inhibitory effect should be clearly observed in the treated animals. The percentage of inhibition will be calculated as the number of animals that do not regenerate the head compared to the controls. After 3 days, the inhibitor can be removed and the animals returned to hydra medium for 3 more days to see if the effect is reversible or not.

DATA RECORDING. The regeneration must be followed during several days under the microscope, taking note of the number and the appearance of animals, at least every day. During the first 48 h differences between the control and the treated animals are not expected. After this time, the controls begin to regenerate and the treated animals do not. At the end of the experiment you should be able to plot your results putting the percentage of regeneration in one axis and the time in the other.

Critical time for the inhibition of head regeneration. Ten adult starved animals without buds are used for each assay. The animals are decapitated and incubated in hydra medium, and the boxes marked as 0, 2, 6, 8, 12, and 24. Ten μl of the inhibitor will be added at times 0, 2, 6, 8, 12, and 24 h after decapitation to each corresponding box. The medium and the inhibitor should be changed every 24 h and the regeneration followed with a stereomicroscope for 3 days.

EXPERIMENT: EFFECT OF THE INHIBITOR ON GENE EXPRESSION

In situ hybridization. In addition to morphological observations, the efficacy of the treatments can be tested by measuring changes in the expression of early genes that are regulated by the signaling pathways in question. Digoxygenin-labeled RNA probes corresponding to the sense and anti-sense strands of the genes to be tested should be prepared using the RNA labeling kit for in vitro transcriptions from Roche Molecular Biochemicals. In situ hybridization should be carried out on samples of control and inhibitor-treated hydras with both sense and antisense probes, following the protocol provided in the Procedures section.

SUGGESTED GENES TO BE USED FOR IN SITU HYBRIDIZATION

➤ *Ks1* is an early marker gene for head formation in hydra (Weinziger et al., 1994; Endl, Lohmann, and Busch, 1999). Its expression is repressed in animals treated with Apigenin, LY294002, and PP2 and delayed in those treated with PD98059 (Cardenas et al., 2000; Fabila et al., 2002; Manuel et al., personal communication).

➤ *Hybra-1* is an early gene involved in the development of the hypostome, the proposed head organizer in hydra (Technau and Bode, 1999; Broun and Bode, 2002). Its expression is inhibited by treatment with Apigenin, LY94002, and PP2 (Cardenas et al., 2000; Manuel et al., personal communication).

➤ *BudHead* is expressed later than *Hybra-1* during development, and is implicated in the formation of the head organizer (Martínez et al., 1997). Its expression is blocked in animals treated with Apigenin, PP2, and LY94002 (Fabila et al., 2002).

➤ *HyAlx* is expressed relatively late during the formation of the head, around 20 h after decapitation, and is a good marker where the tentacles will develop (Smith, Gee, and Bode, 2000). Its expression is inhibited in animals treated with Apigenin and LY294002 but not by PP2 (Manuel et al., personal communication).

SUGGESTED INHIBITORS. The main effects of the inhibitors used in these experiments are the following:

➤ Apigenin: its effect has been attributed to the inhibition of MAP kinase activity.

➤ BpV(phen): it is a potent protein phosphotyrosine phosphatase inhibitor and insulin receptor kinase activator.

➤ LY294002: it is a potent and specific phosphatidylinositol 3-kinase inhibitor that acts on the ATP-binding site of the enzyme.

➤ PD98059: it is a selective inhibitor of MAP kinase kinase (MEK) that acts by inhibiting the activation of MAPK and subsequent phosphorylation of MAPK substrates.

➤ PP2/AG1879: it is a potent and selective inhibitor of the SRC family of protein tyrosine kinases.

Table 5.1. Suggested inhibitors to study the influence of different signal transduction pathways on the head regeneration in hydra

Inhibitor	Target enzyme	IC$_{50}$	Solubility
Apigenin	ERK1/ERK2 (Kuo and Yang, 1995)	4.2 μM	DMSO
Bpv(phen)	Phosphatases P-Tyr residues (Posner et al., 1994)	100 nM	Water
LY294002	PI3K (Vlahos et al., 1994)	6.6 μM	Ethanol
PD 98059	MEK1/MEK2 (Alessi et al., 1995)	40 μM	DMSO
PP2	Src (Hanke et al., 1996)	2 nM	DMSO
Forskolin	Adenylate cyclase (Galli et al., 1995)	5 μM	DMSO
Me-Me-IBMX	Phosphodiesterase (Ahn et al., 1989)	100 nM	Ethanol

➤ Forskolin: its main pharmacological activities are due to its activation of adenylate cyclase resulting in increased cAMP levels.

➤ 8-Methoxymethyl-3isobutyl-1-methylxanthine (Me-Me-IBMX): it is a selective inhibitor of Ca^{+2}-calmodulin-dependent phosphodiesterase.

EXPECTED RESULTS

In the previous table of the suggested inhibitors, we can classify them in two main groups based on their effects. There are compounds that selectively and reversibly inhibit the development of the head: Apigenin, LY294002, and PP2, and one compound that delays its development for about 24 h, PD98059. The decapitated animals treated with these compounds will be inhibited for at least 48 h, and changes in the expression of the indicated genes should be detectable.

The other group (BpV(phen), forskolin, and Me-Me-IBMX) can accelerate the regeneration for at least twelve hours. In some experiments, we observed up to 24 h of acceleration in head regeneration.

DISCUSSION

Head regeneration in hydra is a suitable model to study the signaling systems that regulate development. The use of specific compounds to modify these signals is helping us to understand which signal transduction pathways are involved in the regulation of the head formation. From the results obtained, it must be clear that the inhibition of the signaling systems mediated by SRC and PI3K reversibly block the development of the head in hydra. This inhibition correlates with an inhibition of the expression of some genes involved in the regulation of the early steps of head formation. The fact that the inhibitors did not block the expression of the same set of genes strongly suggests that the two signaling pathways are independent. We believe that there must be a crosstalk point between them, and we have evidence that it must be at the ERK1-2 level because in animals treated with any of the two inhibitors there is a delay in the phosphorylation of these key kinases (Manuel et al., personal communication). Also, we are able to block the expression of all the genes tested in animals treated with Apigenin (the inhibitor of the ERK1-2). Then, the effects of the inhibitors on the expression of the different genes tell us about their regulation and the possible interaction between the signaling systems on the regulation of these genes.

TIME REQUIRED FOR THE EXPERIMENTS

A typical experiment of regeneration, and the effect of the inhibitors on the regeneration, takes a week. The in situ hybridization experiments also take a week.

TEACHING CONCEPTS

Pattern formation is the process by which cells become organized in the embryo to form the rudiments of the body plan.

Signal tranduction pathways are the processes used by the cells to change the extracellular signals (hormones, growth factors, etc.) into different intracellular signals (changes in the level of phosphorylation of proteins, cAMP, Ca^{+2}, etc.) to give an answer to the stimuli.

Regeneration is the ability to replace missing body parts after embryogenesis.

ALTERNATIVE EXERCISES

QUESTIONS FOR FURTHER ANALYSIS
- ➤ Do you think that the expression of the genes tested is regulated by different signaling systems? Or do the signal transduction pathways converge at certain points?
- ➤ Do you believe that the effect of the compounds could be antagonized in certain conditions?

REFERENCES

Ahn, H., Crim, W., Romano, M., Sybertz, E., and Pitts, B. (1989). Effects of selective inhibitors on cyclic nucleotide phosphodiesterases of rabbit aorta. *Biochem. Pharmacol.*, 38, 3331–9.

Alessi, D., Cuenda, A., Cohen, P., Dudley, D., and Saltiel, A. (1995). PD98059 is a specific inhibitor of the activation of mitogen-activated protein kinase kinase *in vitro* and *in vivo*. *J. Biol. Chem.*, 270, 27489–94.

Bode, H. R. (1996). The interstitial cell lineage of hydra: A stem cell system that arose early in evolution. *J. Cell. Sci.*, 109, 1155–64.

Bode, P. M., and Bode, H. R. (1980). Formation of pattern in regenerating tissue pieces of *Hydra attenuata*. I. Head-body portion regulation. *Dev. Biol.*, 78, 484–96.

Broun, M., and Bode, H. R. (2002). Characterization of the head organizer in hydra. *Development*, 129, 875–84.

Cardenas, M., Fabila, Y., Yum, S., Cerbon, J., Böhmer, F. D., Wetzker, R., Fujisawa, T., Bosch, T. C. G., and Salgado, L. M. (2000). Selective protein kinase inhibitors block head specific differentiation in hydra. *Cell Signal.*, 12, 649–58.

Endl, I., Lohmann, J. U., and Bosch, T. C. G. (1999). Head-specific gene expression in hydra: complexity of DNA-protein interactions at the promoter of *ks1* is inversely correlated to the head activation potential. *Proc. Natl. Acad. Sci. USA,* 96, 1445–50.

Fabila, Y., Navarro, L., Fujisawa, T., Bode, H. R., and Salgado, L. M. (2002). Selective inhibition of protein kinases blocks the formation of a new axis, the beginning of budding, in *hydra. Mech. Dev.*, 119, 157–64.

Galli, C., Meucci, O., Scorziello, A., Werge, T., Calissano, P., and Schettini, G. (1995). Apoptosis in cerebellar granule cells is blocked by high KCl, forskolin, and IGF-1 through distinct mechanisms of action: The involvement of intracellular calcium and RNA synthesis. *J. Neurosci.*, 15, 1172–9.

Gierer, A. (1970). A theory of biological pattern formation. *Kybernetik*, 12, 30–9.

Gierer, A. (1974). Hydra as a model for the development of biological form. *Sci. Am.*, 231, 544–59.

Grens, A., Gee, L., Fisher, D. A., and Bode, H. R. (1996). cnNK-2, an NK-2 homeobox gene, has a role in patterning the basal end of the axis in hydra. *Dev. Biol.*, 180, 473–88.

Hanke, J. H., Gardner, J. P., Dow, R. L., Changelian, P. S., Brissette, W., Weringer, E. J., Pollok, B. A., and Connelly, P. A. (1996). Discovery of a novel and Src family-selective tyrosine kinase inhibitor. Study of Lck- and Fyn T-dependent cell activation. *J. Biol. Chem.*, 271, 695–701.

Kuo, M. L., and Yang, N. C. (1995). Reversion of v-H-ras transformed NIH 3T3 cells by apigenin through inhibiting mitogen activated protein kinase and its downstream oncogenes. *Biochem. Biophys. Res. Comm.*, 212, 767–75.

Martínez, D. E., Dirksen, M. L., Bode, P. M., Jamrich, M., Steele, R. E., and Bode, H. R. (1997). Budhead, a fork head/HNF-3 homologue, is expressed during axis formation and head specification in hydra. *Dev. Biol.*, 192, 523–36.

Müller, W. (1996). Pattern formation in the immortal *Hydra. Trends Genetics*, 12, 91–6.

Otto, J. (1976). Budding in *Hydra attenuata*: Bud stages and fate map. *J. Exp. Zool*, 200, 417–28.

Posner, B. I., Faure, R., Burgess, J. W., Bevan, A. P., Lachance, D., Zhang-Sun, G., Fantus, I. G., Hall, D. A., and Lum, B. S. (1994). Peroxovanadium compounds. A new class of potent phosphotyrosine phosphatase inhibitors which are insulin mimetics. *J. Biol. Chem.*, 269, 4596–604.

Smith, K. M., Gee, L., and Bode, H. R. (2000). HyAlx, an aristaless-related gene, is involved in tentacle formation in hydra. *Development,* 127, 4743–52.

Takano, J., and Sugiyama, T. (1983). Genetic analysis of developmental mechanisms in hydra. VIII. Head-activation and head-inhibition potentials of a slow-budding strain (L4). *J. Embryol. Exp. Morphol.*, 78, 141–68.

Technau, U., and Bode, H. R. (1999). *HyBra1*, a *Brachyury* homologue, acts during head formation in *Hydra. Development,* 126, 999–1010.

Vlahos, C. J., Matter, W., Hui, K., and Brown, R. (1994). A specific inhibitor of phosphatidylinositol 3-kinase, 2-(4-morpholinyl)-8-phenyl-4H-1-benzopyran-4-one (LY294002). *J. Biol. Chem.*, 269, 5241–8.

Weinziger, R., Salgado, L. M., David, C. N., and Bosch, T. C. G. (1994). Ks1, an epithelial cell-specific gene, responds to early signals of head formation in *Hydra. Development,* 120, 2511–17.

6 Retinoic acid during limb regeneration

M. Maden

OBJECTIVE OF THE EXPERIMENT The objective of the experiment is to demonstrate the existence of the phenomenon known as positional information. Within an organ such as the regenerating limb, the administration of retinoic acid (RA) at different doses can change this property and induce the regeneration of extra limb segments. This experiment is an excellent introduction to this basic principle in developmental biology.

DEGREE OF DIFFICULTY Easy. The experiment requires the limbs of anaesthetised axolotl larvae to be amputated with a razor blade or scissors at different levels, followed by the application of a powder to the water in which they live. After 4 weeks the regenerated limbs are stained as whole mounts for cartilage to reveal the regenerated structures.

INTRODUCTION

Urodeles (tailed amphibians) have the remarkable ability to regenerate their limbs throughout life, even as adults. Wherever a limb is amputated, the precise parts that were removed are replaced by a series of well-described processes. Firstly, the wound rapidly heals over by epithelial migration and a thickened epithelial cap appears over the stump (Figure 6.1a). Secondly, the cells of the tissues at the amputation plane (muscle, cartilage, nerve, and connective tissue) dedifferentiate and return to a single-celled, embryonic state (Figure 6.1b). Thirdly, these dedifferentiated cells begin to proliferate and together form a conical structure at the tip of the stump, called a blastema (Figure 6.1c). Fourthly, as the blastema elongates, the structures of the limb that were amputated begin to redifferentiate in a proximal to distal direction.

Since there is a perfect match between the regenerated structures and those removed by amputation, that is, no more and no fewer tissues are regenerated (Figure 6.1d), it must be the case that the dedifferentiated cells at the amputation plane have a knowledge of their position relative to the limb as a whole, that is, they have

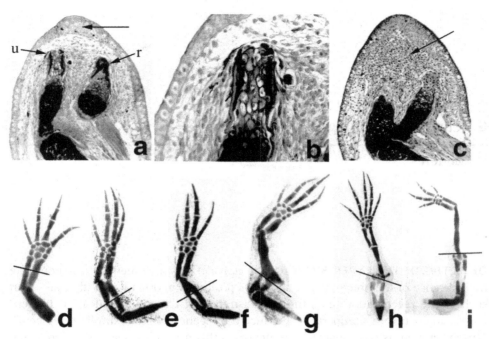

Figure 6.1. (a–c) Histological sections showing several of the features of limb regeneration in axolotls. (a) Section through a limb which had been amputated through the mid-radius and ulna 3 days previously. The wound has healed completely due to migration of the epidermis over the stump and there is a pileup of epithelial cells at the tip of the limb (arrow). The cut radius (r) and ulna (u) have begun to dedifferentiate, as have the cut ends of the muscle fibres. (b) Close-up of the cut end of the ulna showing a loss of staining and an expansion of the chondrocytes as they dedifferentiate. (c) Section through a regenerate after 6 days showing the blastema (arrow) which consists of a mass of rapidly dividing, pluripotent cells which have accumulated at the tip of the limb and have arisen from dedifferentiation of the cut tissues. The ends of the radius and ulna can still be seen. (d–i) The effect of retinoids on limb regeneration in axolotls visualised by Victoria Blue staining of the cartilage elements in regenerated limbs. The amputation plane is marked in each case by the solid line. (d) Control forelimb regenerate amputated through the mid-radius and ulna which regenerated the ends of the radius and ulna, the carpals and 4 normal digits. (e) A forelimb regenerate treated with a low level of retinyl palmitate which regenerated the ends of the radius and ulna and then another radius and ulna, carpals and 4 digits from the amputation plane. (f) A regenerate treated with a medium level of retinyl palmitate which regenerated the distal end of the humerus, then an elbow joint, another radius and ulna and hand. (g) A regenerate treated with a high level of retinyl palmitate which regenerated a complete forelimb beginning with the humerus from the amputation plane. (h) A hindlimb regenerate which had been amputated through the tarsals and treated with a high level of retinyl palmitate. A complete hindlimb has regenerated from the amputation plane (note the 5 digits of the hindlinb) instead of just the tarsals and 5 digits as expected. (i) A forelimb regenerate which had been amputated through the carpals and treated with a high level of retinyl palmitate. A complete forelimb has regenerated from the amputation plane instead of just the carpals and 4 digits as expected.

positional information. The application of retinoids to the regenerating limb alters this positional information of the blastemal cells in a concentration-dependent manner, such that, instead of regenerating only those structures that were removed, extra limb segments appear as well (Niazi and Saxena, 1978; Maden, 1982). These data provide us with a system whereby the nature of positional information can be investigated by comparing normal regenerates with RA-treated regenerates and searching for molecular changes within the regenerating tissues (e.g. da Silva, Gates, and Brockes, 2002).

When a limb is amputated through the mid-radius and ulna, and subsequently treated with 30 IU/ml retinyl palmitate for 10 days, an extra radius and ulna form in tandem with the regenerated one (Figure 6.1e), rather than merely replacing the distal ends of the radius and ulna and then the hand. When the concentration of retinyl palmitate is raised to 75 IU/ml, the regenerate had an extra elbow joint in tandem (Figure 6.1f). When the same concentration (75 IU/ml) is used, but the length of treatment is increased from 10 days to 15 days, then regeneration begins from the shoulder level with a complete humerus in addition to the more distal elements (Figure 6.1g). These results show that as the concentration of retinyl palmitate is increased, regeneration commences from an increasingly more proximal level; that is, the positional information of the blastema cells is gradually respecified in a more proximal direction, towards the shoulder. The same range of results can be obtained from hindlimb amputations instead of forelimb amputations (Figure 6.1h) and after amputating through the hand instead of the mid-forearm (Figure 6.1i), but in the latter case a higher concentration of retinoids is required.

Retinoids are a family of compounds all derived from vitamin A, known as retinol. Retinol is metabolised to retinal, the molecule used in the visual cycle, and then to retinoic acid (RA). RA acts at the level of the nucleus by binding to two classes of ligand-activated transcription factors known as the retinoic acid receptors (RARs) and the retinoid X receptors (RXRs) (Chambon, 1996). When RA binds to these RARs and RXRs, gene transcription is activated. RA is widely used as an inducer of differentiation in embryonal carcinoma cells or as an inhibitor of proliferation in cancer cells. It is also known to have profound effects on embryos where, in excess or deficiency, it is teratogenic to several organ systems. Clearly, the level of RA needs to be carefully regulated in the embryo to ensure correct development, because of the control that RA exerts in patterning embryonic organ systems (see also Chapter 17).

Different retinoids have different efficacies in the regeneration assay described previously. Of the naturally occurring retinoids, *all-trans*-retinoic acid is the most potent, followed by retinol, retinyl palmitate, and retinyl acetate (Maden, 1982; 1983*a*). There are now many synthetic retinoids that have been designed to activate specific RARs; most of these are far more potent in the regeneration assay than *all-trans*-RA. For example, a compound known as TTNPB which specifically activates RARα is at least 100 times more potent than RA (Keeble and Maden, 1989). Increasing the time of treatment with retinoids has the same effect as increasing the dose: The regenerate begins at a more proximal level as the time or dose is increased. However, after a certain point, regeneration is permanently inhibited by too lengthy or too strong a treatment, due to the effect of retinoids on cell proliferation (Maden, 1983*a*).

The preceding effects of retinoids have also been described for a wide range of amphibian species. In the Urodeles, proximalisation has been observed in *Ambystoma mexicanum* (axolotls) (Maden, 1982; 1983*a*; Thoms and Stocum, 1984), *Pleurodeles waltl* (Niazi, Pescitelli, and Stocum, 1985; Lheureux, Thomas, and Carey, 1986), *Triturus vulgaris* (Koussoulakos et al., 1986), *Triturus helveticus* (Koussoulakos, Kiortsis, and Anton,

1986), *Triturus alpestris* (Koussoulakos, Sharma, and Anton, 1988), *Triturus cristatus* (Sharma and Anton, 1986), and *Notophthalmus viridescens* (Thoms and Stocum, 1984). Retinoids also work on the limbs of Anurans if the experiments are done before metamorphosis (for example, in *Bufo andersonii* (Niazi and Saxena, 1978), *Bufo melanosticus* (Jangir and Niazi, 1978), *Rana breviceps* (Sharma and Niazi, 1979), *Xenopus laevis* (Scadding and Maden, 1986), and *Rana temporaria* (Maden, 1983*b*)). In these Anuran species it is often the case that not only are the regenerates proximalised, but more than one limb appears at the site of regeneration, suggesting a respecification of the positional information in the other two axes (anteroposterior and dorsoventral) as well.

MATERIALS AND METHODS

EQUIPMENT AND MATERIALS
Dissecting microscope
Razor blade
Forceps
Small glass sample bottles for fixing tissues

Biological materials. Larval axolotls (*Ambystoma mexicanum*) 3–4 cm long.

REAGENTS
3-aminobenzoic acid ethyl ester (MS-222) (Sigma)
Retinyl palmitate, type VII, dispersed in acacia-starch matrix, 250,000 USP units per gram (Sigma)
10% neutral formalin
Ethanol
Acid alcohol
Methyl salicylate

PREVIOUS TASKS FOR STAFF
Adult axolotls breed in spring and early summer. They lay thousands of eggs that can be grown up for use in these experiments. After hatching, the larval axolotls feed on brine shrimp which must be hatched from eggs by placing them into salt water. When the larvae get larger they feed on *Tubifex* worms. Adult axolotls can be fed on strips of liver. These experiments can also be performed on another European species, the larval *Pleurodeles waltl*, or they can be performed on adult newts. They can also be performed on frog tadpoles (*Rana temporaria*) which are in plentiful supply in ponds in spring throughout Europe.

Solutions
1 in 10,000 dilution of MS-222 in water. Check that the solution is still at neutral pH by adding a few drops of indicator dye.
10% neutral formalin = 100 ml formalin, 900 ml distilled water, 10 g calcium chloride.

1% Victoria Blue B in acid alcohol
Acid alcohol = 100 ml of 70% ethanol with 1 ml of concentrated HCl added.
50% ethanol
70% ethanol
90% ethanol
95% ethanol

PROCEDURES
Victoria Blue staining of fixed limbs to reveal the cartilage elements
1. Fix for at least 24 hours.
2. Place limbs in 50% ethanol for 1 h; acid alcohol for 2 h
3. Stain in 1% Victoria Blue for 45 minutes. The tissues all go blue and the colour is removed from the soft tissues during subsequent dehydration leaving the cartilage elements blue.
4. Place limbs in 70% ethanol for 20 minutes, 90% ethanol for 20 minutes, 95% ethanol for 20 minutes, 100% ethanol for 20 minutes, 100% ethanol for 20 minutes.
5. Clear in methyl salicylate 24 hours.

OUTLINE OF THE EXPERIMENT
1. Give 18 larval axolotls about 3–4 cm in length to each student and anaesthetise them by placing them in MS-222 solution until they have stopped moving (about 3–5 minutes).
2. Place the larvae on a wet tissue and observe under the dissecting microscope. Lay the forelimbs out.
3. To nine of the axolotls, amputate both forelimbs through the mid forearm level using the razor blade.
4. To the other nine larvae, amputate both forelimbs through the hand.
5. Place three of the forearm amputated larva and three of the hand amputated larvae in clean water to recover, and keep them in clean water. These are the controls to confirm that only what was removed will regenerate.
6. Place three of the forearm amputated larvae and three of the hand amputated larvae in 1 litre of water to which 9 mg of retinyl palmitate has been added. The solution will turn milky as the compound dissolves in the water. Keep the larvae in this solution for a total of 15 days.
7. Place the other three forearm amputated larvae and the other three hand amputated larvae in 1 litre of water to which 22.5 mg of retinyl palmitate has been added. This is a more concentrated solution than that in step 6 because it requires a greater concentration of retinoids to produce extra elements from the hand than from the forearm. Keep the larvae in this solution for a total of 15 days.
8. Leave the larvae for about 4–6 weeks until they have completed regeneration. Change the water every 2–3 days and replace the required amount of retinyl palmitate for the required total time (15 days).

9. The controls should look perfectly normal. The experimentals should have fore-limbs that now look too long.
10. Anaesthetise the larvae in MS-222 again. Amputate all the forelimbs at the shoulder and fix them in individual bottles in 10% neutral formalin.
11. On the next day, stain the limbs for cartilage using the protocol for Victoria Blue staining. At the end of the staining procedure, clear the limbs in methyl salicylate.
12. On the next day, look at the limbs through the dissecting microscope and draw or photograph the cartilage elements.

EXPECTED RESULTS

The controls should have regenerated the structures that were removed by amputation, either a hand or a lower forearm (e.g. Figure 6.1d). The forearm amputations treated with retinyl palmitate should have regenerated extra elements between those that were expected, probably an extra radius and ulna in tandem or an extra elbow as well (Figures 6.1e and 6.1f). The hand amputations will have regenerated an extra complete limb from the amputation plane (Figure 6.1i). The effect of differences in concentration between the same amputation plane can also be evaluated.

DISCUSSION

This experiment demonstrates that retinoids can change the process by which blastemal cells measure their positional information and can induce extra elements appearing in the proximodistal axis instead of only those elements that were removed by ampu-tation. Increasing the concentration of retinoids or the time during which they are administered will result in a progressive proximalisation of the positional information. Retinoids are acting like classical morphogens in this system. Knowledge of the exact downstream targets of retinoic acid would give us valuable information concerning the pathways through which RA acts and thus how fields of cells becomes organised into complex patterns of differentiation. We know that RA acts though the RARδ receptor to change the positional information of blastemas cells (Pecorino, Entwistle, and Brocker, 1996), but other target genes have not been identified. We might guess that Hox genes would be targets as they are expressed during limb development and re-expressed dur-ing limb regeneration (Gardiner et al., 1995; Torok et al., 1998). Furthermore, several of the Hox genes have retinoic acid response elements in their enhancer sequences and would therefore be expected to be targets.

TIME REQUIRED FOR THE EXPERIMENTS

To get 3–4 cm axolotl larvae from fertilisation will take about 3 months, but an alter-native is to collect *Rana temporaria* tadpoles from ponds at the correct stage. From amputation to completion of regeneration will take 4–6 weeks and then 2 further days are needed to analyse the results.

POTENTIAL SOURCES OF FAILURE

The axolotls may not respond to the retinyl palmitate or there may be variation in the response so that they do not all behave the same amongst the class. This problem is best overcome by increasing the number of animals that each student uses, for example a group of 10 is better than a single animal. The Victoria Blue staining may be too dark to see the cartilage elements in which case a further round of destaining can be conducted.

TEACHING CONCEPTS

Limb regeneration: some animals such as the Amphibian can regenerate whole organs such as the limbs after amputation. They do this by returning differentiated cells at the amputation plane to an embryonic, pluripotent state. These processes by which regeneration takes place are together known as epimorphic regeneration.

Dedifferentiation: the process by which fully differentiated cells at the amputation plane return to an embryonic, pluripotent state and commence cell division to generate a blastema.

Blastema: a group of dedifferentiated cells which are rapidly proliferating and which will replace the elements that were removed by amputation.

Positional information: a state whereby the cell must have some knowledge of where it is in relation to the whole organ of which it is a part. In developing embryos this state is assumed to be established by gradients of extracellular morphogens. However, in regeneration this process is unfeasible and the concept that it is a cell surface property has been proposed instead.

Morphogen: a substance which is distributed across a field of embryonic cells in a gradient and which acts to induce patterns of spatial differentiation across that field according to differences in the concentration of the morphogen that the cells experience.

ALTERNATIVE EXERCISES

PROPOSED EXPERIMENT

The experiments can be performed on tadpoles of *Rana temporaria* following the scheme described in Maden (1983*b*). In this case, not only will extra elements appear in the proximodistal axis, but several complete limbs will appear – often including the pelvic girdle.

ACKNOWLEDGEMENTS

Figure 6.1(e–i) Reprinted with permission from *Nature* (Maden, 1982). Copyright (1982) McMillan Magazines Limited.

REFERENCES

Chambon, P. (1996). A decade of molecular biology of retinoic acid receptors. *FASEB J.*, 10, 940–54.

da Silva, S. M., Gates, P. B., and Brockes, J. P. (2002). The newt orthologue CD59 is implicated in proximodistal identity during amphibian limb regeneration. *Dev. Cell*, 3, 547–55.

Gardiner, D. M., Blumberg, B., Komine, Y., and Bryant, S. V. (1995). Regulation of *HoxA* expression in developing and regenerating axolotl limbs. *Development*, 121, 1731–41.

Jangir, O. P., and Niazi, I. A. (1978). Stage dependent effects of vitamin A excess on limbs during ontogenesis and regeneration in tadpoles of the toad *Bufo melanosticus* (Schneider). *Indian J. Exp. Biol.*, 16, 438–45.

Keeble, S., and Maden, M. (1989). The relationship among retinoid structure, affinity for retinoic acid-binding protein, and ability to respecify pattern in the regenerating limb. *Dev. Biol.*, 132, 26–34.

Koussoulakos, S., Kiortsis, V., and Anton, H. J. (1986). Vitamin A induces dorsoventral duplications in regenerating Urodele limbs. *IRCS Med. Sci.*, 14, 1093–4.

Koussoulakos, S., Sharma, K. K., and Anton, H. J. (1988). Vitamin A induced bilateral asymmetries in *Triturus* forelimb regenerates. *Biol. Struct. Morphog.*, 1, 43–8.

Lheureux, E., Thoms, S. D., and Carey, F. (1986). The effects of two retinoids in limb regeneration in *Pleurodeles waltl* and *Triturus vulgaris*. *J. Embryol. Exp. Morph.*, 92, 165–82.

Maden, M. (1982). Vitamin A and pattern formation in the regenerating limb. *Nature*, 295, 672–5.

Maden, M. (1983*a*). The effect of vitamin A on the regenerating axolotl limb. *J. Embryol. Exp. Morph.*, 77, 273–95.

Maden, M. (1983*b*). The effect of vitamin A on limb regeneration in *Rana temporaria*. *Dev. Biol.*, 98, 409–16.

Niazi, I. A., and Saxena, S. (1978). Abnormal hind limb regeneration in tadpoles of the toad, *Bufo andersoni*, exposed to excess Vitamin A. *Folia Biol. (Krakow)*, 26, 3–8.

Niazi, I. A., Pescitelli, M. J., and Stocum, D. L. (1985). Stage-dependent effects of retinoic acid on regenerating Urodele limbs. *Roux's Arch. Dev. Biol.*, 194, 355–63.

Pecorino, L. T., Entwistle, A., and Brocker, J. P. (1996). Activation of a single retinoic acid receptor isoforms mediates proximodistal respecification. *Current Biol.*, 6, 563–9.

Scadding, S. R., and Maden, M. (1986). Comparison of the effects of vitamin A on limb development and regeneration in tadpoles of *Xenopus laevis*. *J. Embryol. Exp. Morph.*, 91, 55–63.

Sharma, K. K., and Niazi, I. A. (1979). Regeneration induced in the forelimbs by treatment with vitamin A in the froglets of *Rana breviceps*. *Experientia*, 35, 1571–2.

Sharma, K. K., and Anton, H. J. (1986). Biochemical and ultrastructural studies on vitamin A induced proximalization of limb regeneration in axolotl. In *Progress in Developmental Biology*. Part A, ed. H. C. Slavkin, pp. 105–8. New York: Alan R. Liss Inc.

Thoms, S. D., and Stocum, D. L. (1984). Retinoic acid-induced pattern duplication in regenerating Urodele limbs. *Dev. Biol.*, 103, 319–28.

Torok, M. A., Gardiner, D. M., Shubin, N. H., and Bryant, S. V. (1998). Expression of *HoxD* gnes in developing and regenerating axolotl limbs. *Dev. Biol.*, 200, 225–33.

SECTION III. BEAD IMPLANTATION

7 Experimental manipulations during limb development in avian embryos

Y. Gañán, J. Rodríguez-León, and D. Macías

OBJECTIVE OF THE EXPERIMENT The students will observe the three main events of limb development. They will observe first, how proximo-distal limb outgrowth is achieved; second, how chondrogenic differentiation contributes to the establishment of the future skeleton; and third, how limb shape is sculpted by apoptosis. The avian embryo is a good model for the study of these processes, as eggs are easy to obtain and manipulate, and the resulting phenotypes can be observed at different time points. By using the application of microspheres soaked in the different proteins, students will visualise how the proximo-distal axis is controlled during limb bud development. Students will also observe how the chondrogenic template, which will be substituted by osteogenic tissue, develops to form the different elements and joints of the limb. Lastly, students will use a technique for visualizing cell death that enables the assessment of both how cell death takes place during limb development and how this is correlated with the morphology of the limb bud in different species. The exercises are completed by studying the action of different secreted molecules involved in apoptosis and chondrogenesis.

DEGREE OF DIFFICULTY Manipulations are difficult. They require patience and practice with microsurgery materials. Staining embryos is easy, requiring only some care during the washing steps.

INTRODUCTION

Morphogenesis during limb outgrowth is an excellent model system for studying the mechanisms of development. In our studies, we compare chick and duck embryos because, although they are very similar at the initial steps of development, they present different morphologies at advanced stages (free or webbed digits).

At the beginning of development, limb buds consist of an ectodermal jacket that covers a core of mesodermal cells. In the distal part of the limb bud, the Apical Ectodermal

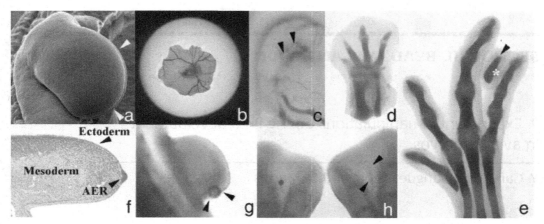

Figure 7.1. (a) Dorsal view of the control leg bud at stage 26 under scanning electron microscope. AER is evident at the distal tip of the limb bud (arrowheads). (b) The eggshell is open in the airspace at the desired stage. (c) The embryo, stage 21, is positioned showing its right flank, head, and limbs. This picture shows the removal of the AER (arrowheads). (d) Control autopod at stage 29 stained with Alcian Green, showing the morphology of the skeletal elements. (e) Leg bud stained with Alcian Green 3 days after the implantation of an Activin A bead in the third interdigit at stage 28, showing extra digit induced. (f) Longitudinal section of leg bud at stage 24, stained with hematoxylin eosin. (g) Wing bud stage 21 treated with a bead of heparin soaked in FGF2 and stained with Neutral Red 24 hours after treatment. (h) Leg bud at stage 30 after application with bead soaked in BMP-2 and control bud at the same stage. Both limbs are stained with Neutral Red. (See also color plate 2).

Ridge (AER), a specialised thickening of the ectoderm, is responsible for its proximo-distal outgrowth (Figures 7.1a and f). Signals emanating from this structure maintain the cells of the underlying mesoderm, the Progress Zone (PZ), in a proliferative and undifferentiated state. The PZ cells contribute to the growth of limbs, eventually forming the future skeleton or shaping the limb by apoptosis (Summerbell and Wolpert, 1972). Surgical removal of the AER truncates the proximo-distal outgrowth and the resulting limb lacks distal skeletal elements (Summerbell, 1974). The truncation obtained is more proximal if the removal is made at early stages whereas distal truncations are obtained when the AER is removed at later stages. These truncations can be rescued with the application of beads soaked in fibroblast growth factor (FGF) after AER removal (Fallon et al., 1994; Niswander et al.; 1993). Detection of expression of different FGFs in the AER and the experiments showing their ability to substitute for the AER make them the best candidates for mediating AER action. The classical view of how cells take on different fates from within the PZ is based on the time that cells spend in the PZ (the "Progress Zone" model). When cells in the PZ are under the influence of AER signals for a shorter period of time, they take on proximal fates, but the longer they spend in the PZ, the more distal their fate. The validity of this model is currently being debated; however, it still serves as a useful baseline in thinking about how the limb forms.

Two of the major differentiation processes that can be observed during limb development are chondrogenesis and apoptosis (programmed cell death). Chondrogenesis leads to the formation of a chondrogenic template, necessary for the establishment of the apendicular skeleton. Apoptosis is responsible for the shape of the limb bud, as well as free digit, and joint formation.

(A) CHONDROGENESIS

In this chapter, we will focus on the signalling proteins (extracellularly secreted molecules) which are responsible for triggering chondrogenesis. One family of proteins involved in this process is the Transforming Growth Factor beta family (TGFβ). It has been shown that, at least in the autopod (carpals/tarsals, metacarpals/metatarsals, and phalanxes), the primary signals for inducing chondrogenesis are Activins. At stage 25–26 (Hamburger and Hamilton, 1951) expression of *activin* β*A* subunit is detected in the position where digits will develop and it is maintained until the last phalanx is formed. Beads soaked in Activin A protein and placed in the interdigital space leads to the formation of an ectopic digit (Merino et al., 1999*a*). Another member of the family, *Tgfβ2*, is also expressed in prechondrogenic condensations later in development (Merino et al., 1998; Millán et al., 1991), and its application in the interdigital mesenchyme is followed by induction of ectopic chondrogenesis (Merino et al., 1998). These two molecules induce chondrogenesis by upregulation of the receptor for other molecules belonging to the TGFβ family, the Bone Morphogenetic Proteins (BMPs). Several BMPs, namely BMP-2, BMP-4, and BMP-7, are expressed in the anterior and posterior parts of the early limb bud, in the AER, in the future joints and in the interdigital space and surrounding chondrogenic condensations when digits appear (Francis et al., 1994; Francis-West et al., 1995; Duprez et al., 1996*a*; Duprez et al., 1996*b*; Macías et al., 1997). Application of Activin A or TGFβ induces expression of BMP receptor 1b (*bmpR-1b*, which is expressed in the digital rays) preceding the appearance of chondrogenic tissue. If a bead of BMP is implanted at the tip of a digit, an enlargement in the phalanx is detected as well as the formation of an extra interdigital space (Gañán et al., 1996).

However, the activity of these molecules belonging to the TGFβ family is controlled by specific antagonists that modulate their actions during limb development. Expression of several of the TGFβ antagonists is detected during limb development, including *noggin, follistatin, chordin,* and *gremlin* (Baleman and Van Hul, 2002). Treatment with these molecules at the tips of digits leads to digit truncation due to inhibition of chondrogenesis (Merino et al., 1998; Merino et al., 1999*b*).

(B) APOPTOSIS

The double phenotype obtained after treatment with BMPs (enlargement of chondrogenic area and formation of an extra interdigital space) can be explained by their dual role during limb development. BMPs are expressed in the areas where cell death will take place, the Anterior Necrotic Zone (ANZ), the Posterior Necrotic Zone (PNZ), interdigital spaces, and joints. Application of beads soaked in BMPs in the anterior or posterior part of the limb bud at early stages, or in the interdigital mesenchyme, induces apoptosis (Gañán et al., 1996; Macías et al., 1997). The action of BMPs in cell death is also controlled by their inhibitors; they are expressed in and at surrounding areas where BMPs are expressed to control the level of cell death induction and to protect different areas from apoptosis. For example, implantation of beads carrying these antagonists in the chick interdigital space inhibits cell death, leading to a phenotype similar to that seen in duck (Merino et al., 1999b).

FGFs also have a dual role during limb bud development. We have shown previously how these proteins are responsible for maintaining cell proliferation during proximo-distal limb outgrowth. In addition, another role for FGFs has been discovered recently. When FGFs are applied to the interdigital mesenchyme, an upregulation of cell death is observed (Montero et al., 2001). Furthermore, application of an FGF inhibitor in the same areas inhibits the onset of apoptosis. On the other hand, the apoptotic effect of BMPs in duck interdigital spaces is enhanced by co-application of beads soaked in FGF protein (Gañán et al., 1998). These results suggest that FGFs are involved in the process of cell death and that their participation is required for the proper establishment of apoptotic regions.

MATERIALS AND METHODS
EQUIPMENT AND MATERIALS

Stereomicroscope
Incubator
Petri dishes
Forceps (thick and thin ones)
Scissors
Microscissor
Tungsten needle
Fertilised eggs

PREVIOUS TASKS FOR STAFF
Tungsten needle. Attach a thin tungsten wire to the tip of a glass tube with glue or by heating the glass. Once the tungsten wire is secure, put sodium nitrate in a spoon and heat it with a Bunsen burner until the powder becomes liquid (use caution and safety glasses). Dip the tip of the wire into the hot sodium nitrate several times to obtain a very thin tip.

SOLUTIONS
PBS
Dissolve 8 g of NaCl, 0.2 g of KCl, 1.44 g of Na_2HPO_4, and 0.24 g of KH_2PO_4 in 800 ml of distilled water.
Adjust to pH 7.4 with HCl.
Add water to 1 liter.

Neutral red stock solution (2:1000)
Dissolve 0.2 g of Neutral Red in 100 ml of distilled water.
Filter the solution.
Store at 4°C.

Formol/Calcium solution

Prepare 100 ml of 4% formaldehyde.

Before bringing final volume to 100 ml, add 1 g of $CaCl_2$ to the formaldehyde solution.

Adjust pH to 7.0.

Add water to 100 ml.

Trichloroacetic acid solution

Dissolve 5 g of trichloroacetic acid in 100 ml of distilled water.

Acid alcohol

Add 1 ml of 1N HCl to 99 ml of 70% ethanol.

Alcian green solution

Dissolve 0.1 g of Alcian Green in 100 ml of acid alcohol.

Stir with a stir bar and filter.

Nile blue solution

Dissolve 1 g of Nile Blue in 100 ml of PBS by stirring.

Filter the solution when completely dissolved.

BIOLOGICAL MATERIAL

Fertilised eggs can be obtained from specialised farms. Once the eggs are received, it is better to maintain them at 16°C before incubation to keep embryos in a non-proliferative state. Eggs should be ordered just prior to incubation (1 to 5 days) because remaining out of the incubator for more than 10 days can result in dead embryos. The eggs should be washed with water to avoid contamination in the laboratory.

OUTLINE OF THE EXPERIMENTS

INCUBATION OF EGGS

Fertilised chick embryos are incubated under rotation, with the airspace up, at 38.5°C and staged according to Hamburger and Hamilton (1951).

MANIPULATION OF EMBRYOS

The day before manipulation, rotation in the incubator is stopped. One hour before surgery, the eggs are sprayed with 70% ethanol. The eggs are windowed using forceps at the air chamber (Figure 7.1b) and then placed on a cylinder (a cap of 3–4 cm in diameter is very useful). For observation of the embryo and blood vessels under the microscope (Figure 7.1c), the membranes immediately surrounding the embryo must be removed. Usually, manipulations are performed on the right limb, leaving the left one as a control. After manipulation, the eggs are sealed with transparent adhesive tape and reincubated without rocking after one hour at room temperature.

Many different kinds of manipulations are possible; we describe two of the most commonly used.

Removal of AER. In general, removing a tissue allows one to deduce the normal function of that tissue during limb development. One of the tissues that is interesting to us is the AER.

1. Once the embryo is exposed, as explained earlier, the AER can be surgically removed.
2. To eliminate the AER, place the tip of a forceps adjacent to the ventral side of the limb to provide contrast to the tissue.
3. Sometimes forceps are not enough to observe the AER; if necessary, the AER can be stained with Nile Blue solution.
4. For that purpose, make a rounded tip at the end of a Pasteur pipette by heating it with a Bunsen burner and soaking it in a liquid solution of 1% agar. This will form a drop of smooth agar at the tip of the pipette that can be soaked in the Nile Blue solution. The AER can be stained by passing the pipette gently along the distal tip of the limb.
5. To remove the AER, use a tungsten needle to make a slit in the anterior region. Then, remove the ridge by scraping with the needle gently towards its posterior end (Figure 7.1c). AER removals must be done between stages 20 and 28.
6. The resulting phenotype should be observed by Alcian Green staining when embryos reach 6–7 days of incubation.

Implantation of beads soaked in different molecules
1. Prepare a 35-mm diameter petri dish with five or six drops (40 μl) of PBS (leave the central area without liquid).
2. Place the microspheres (beads) in one of the drops and let them wash for 30 minutes. Beads made of different materials can be used (for instance heparin or Affi-Gel Blue) depending on the protein or chemical compound to be applied (Table 7.1). The size of the beads must range between 80 and 150 μm in diameter.
3. After this period, place 2 μl of the desired protein or compound solution in the centre of the petri dish. Using thin forceps, wash the beads in each PBS drop. The beads are then soaked in the solution with the molecule of interest for at least 1 hour at 4°C to avoid evaporation of the solution.
4. At this point, the eggs can be opened as described previously.
5. While holding the limb on its ventral side with forceps, a slit can be made with a tungsten needle in the area to receive the bead implant (anterior or posterior border, distal mesenchyme, tip of digit, or interdigital web).
6. Then, using thin forceps, carry a loaded bead from the petri dish to the little hole made in the limb. To do this, take the bead in one of the tips of the forceps in one hand and with the other hand, secure the limb such that its ventral side is outside the liquid albumin. Place the bead at the top of the hole and press it with the forceps to position the bead in the mesenchyme (Figures 7.1g and h).

Table 71. Summary of different bead implantations with different molecules, beads to use, concentration of the protein, time points for implantation and fixation procedure

Protein	Beads	Concentration	Location–Stage	Time period for incubation
Chicken embryos–Chondrogenesis				
Activin A or TGFβ	Heparin acrylic beads or Affi-Gel Blue beads	0.5 to 1 μg/μl	Interdigital mesenchyme St 28 to 29	3 days. Fix for Alcian Green staining
BMP-2, -4 or -7	Heparin acrylic beads	0.1 to 0.5 μg/μl	Tip of digit St 28 to 29	3 days. Fix for Alcian Green staining
Gremlin, Follistatin, Noggin or Chordin	Heparin acrylic beads	0.5 to 1 μg/μl	Tip of digit St 28 to 29	3 days. Fix for Alcian Green staining
Chicken embryos–Cell death				
FGF-2, -4 or -8	Heparin acrylic beads or Affi-Gel Blue beads	0.5 to 1 μg/μl	Interdigital mesenchyme St 28 to 29	24 hours. Fix for Neutral Red staining
BMP-2, -4 or -7	Heparin acrylic beads	0.1 to 0.5 μg/μl	Interdigital mesenchyme St 28 to 29	10 hours. Fix for Neutral Red staining
BMP-2, -4 or -7	Heparin acrylic beads	0.1 to 0.5 μg/μl	Anterior or posterior mesenchyme St 20 to 26	10 hours. Fix for Neutral Red staining
Gremlin, Follistatin, Noggin or Chordin	Heparin acrylic beads	0.5 to 1 μg/μl	Interdigital mesenchyme St 28 to 29	20–24 hours. Fix for Neutral Red staining
Duck embryos–Cell death				
FGF-2, -4 or -8	Heparin acrylic beads or Affi-Gel Blue beads	0.5 to 1 μg/μl	Interdigital mesenchyme 8.5 days	24 hours. Fix for Neutral Red staining
BMP-2, -4 or -7	Heparin acrylic beads	0.1 to 0.5 μg/μl	Interdigital mesenchyme 8.5 days	10–20 hours. Fix for Neutral Red staining
FGF and BMP Double treatment			Interdigital mesenchyme 8.5 days	10–20 hours. Fix for Neutral Red staining

Table 7.1 summarises the different timepoints, stages, and purposes to perform these applications (study of chondrogenesis or apoptosis). In general, hindlimbs are better than forelimbs for manipulations because the interdigital space is bigger and the digits have more phalanxes. For our experiments, we used the third interdigital space and digit number three of the leg bud as targets.

PREPARATION OF EMBRYOS

1. When the embryos reach the desired stage, cut the tape with scissors and secure the vessels on the ventral side of the embryo with forceps. Cut away the tissue surrounding the embryo and remove its head.
2. Put the trunk with the limbs in a petri dish with PBS and dissect the limbs if the embryos are older than stage 27.
3. Wash embryos or limbs two or three times in PBS and proceed with the protocol for cell death detection or cartilage staining.

NEUTRAL RED PROTOCOL FOR DETECTION OF CELL DEATH

1. For Neutral Red staining, the embryo or the limbs are incubated in 2 ml of culture medium (F12 Ham) or PBS with 100 μl of Neutral Red stock solution in a 35-mm petri dish at 37°C.
2. After 20 minutes, look at the embryos or limbs under the microscope and, when they are stained as desired (between 20 and 30 minutes), wash them in PBS two times.
3. Fix the embryos in formol/calcium at 4°C overnight.
4. After this, dehydrate with two washes in isopropyl alcohol (2-propanol) at 4°C 1 hour each.
5. Clear the embryos with xylene (twice, 1 hour each), and observe under the microscope.

This technique can be used after manipulations or in wild-type limbs. By using this staining, students can observe the different areas of cell death (ANZ, PNZ, and interdigital areas). Alternatively, Neutral Red staining can be done in different species (duck and chicken) to compare differences in their patterns.

Data recording. For visualizing ANZ and PNZ, chicken embryos between stage 20 and 26 are required. Interdigital necrotic zones can be observed in 6.5-day-old chicken embryos (Figures 7.1g and 7.1h) and 9.5-day-old duck embryos.

CARTILAGE STAINING

1. Fix the embryos or limbs in 5% trichloroacetic acid overnight.
2. Then remove fixative solution and stain the tissue with 0.1% Alcian Green overnight.
3. In order to remove excess dye, wash embryos with acid alcohol three or four times or until the medium becomes transparent.

4. Dehydrate the samples with two washes with 100% ethanol 1 hour each.
5. Clear the specimen with 100% methyl salicylate to observe under the microscope.

Data recording. Chondrogenesis during limb chick development can be observed from stage 22 on. A good exercise would be to stain wild-type embryos at different stages of development (Figures 7.1d and 7.1f).

EXPECTED RESULTS
By using these techniques different results can be obtained:

REMOVAL OF AER
Removal of AER results in limb truncation. The extent of limb truncation depends on the time of AER removal. The skeletal elements can be observed by Alcian Green staining after three days.

Alternatively, a bead soaked in FGF can be placed in the distal mesenchyme after AER removal, and Alcian Green staining can be done after three days to observe the rescue of the phenotype.

CHONDROGENESIS
Treatment in the interdigital space at stage 28 with beads soaked in TGF beta or Activin proteins results in ectopic chondrogenesis in the interdigital area. Again, the skeletal elements can be observed by Alcian Green staining after three days (Figure 7.1e). As a complementary exercise, TGFβ antagonists can be applied at the tip of the digits, leading to inhibition of chondrogenesis. If the implantation is made at stage 28, only the first phalanx will form. Application at more advanced stages will result in the loss of fewer phalanxes.

Apoptosis. Treatment with BMPs at the anterior or posterior margins of the limb bud at stages 20 to 26 induces cell death that can be detected 10 hours after treatment (Figure 7.1g). Application of BMPs in the interdigital spaces, at stages 28–29, will accelerate the onset of cell death in these areas.

DISCUSSION
Coordination of outgrowth during limb development is under the control of organising centres such the AER. The action of the AER is to maintain cells of the underlying mesenchyme in a proliferative state. The classical PZ model suggests that mesenchymal cells at the distal tip of the limb bud receive signals from the AER (namely FGFs) that confer specific fates to these cells. The more time they spend in the PZ receiving the signals, the more distally they are positioned. In support of this model, when the AER is removed, cells in the PZ do not receive the molecules secreted by the AER, and thus cannot proliferate. This results in the truncation of the limb at different levels depending

on the stage when removal is done (Saunders, 1948). As mentioned earlier, FGFs are the key molecules for this action since they are the only molecules able to rescue the truncations induced after AER removal. A new model to explain proximo-distal outgrowth has recently been suggested (Dudley, Ros, and Tabin, 2002; Sun, Mariani, and Martin, 2002). In this new model, the allocation model, the authors suggest that the PZ cells are specified very early in development and that the time they spend in the progress zone is needed only for the growth of the different segments. One of the proofs that sustains this new view is that cell death is triggered 12 hours after AER removal, as observed by staining the embryos with Neutral Red. We encourage students to read about the classic and new PZ models, and do the experiments proposed in the Alternate Experiments section to learn more about these two models.

CHONDROGENESIS

Chondrogenesis is induced *in ovo* by several molecules, among them members of the TGFβ family. Activin A, TGFβ, and BMP, which are expressed physiologically during limb development in the areas where chondrogenic tissue will form, lead to chondrogenic differentiation when applied in the interdigital spaces (Gañán et al., 1996; Merino et al., 1998; Merino et al., 1999*a*). Physiologically, it has been shown that the first secreted molecule detected in the digital rays is Activin A (Merino et al., 1999*a*) while TGFβ2 is detected later during development (Merino et al., 1998); both molecules induce the expression of *bmpR-1b*. Detection of this receptor is observed only in the digital rays. Moreover, misexpression of the constitutively active form of this receptor increases the size of the chondrogenic template, while if the dominant negative form is overexpressed in the autopod, chondrogenesis is inhibited (Zou et al., 1997). Ectopic expression of Activins or TGFβs induces formation of an extra digit when applied in the interdigital space by upregulating the expression of *bmpR-1b*, while treatment with BMPs increases the size of the chondrogenic elements after induction of this receptor. Chondrogenesis can be prevented by treatment with TGFβ antagonists. These antagonists bind TGFβ proteins extracellulary, preventing the TGFβ ligands from acting through their receptors. Several BMP antagonists are expressed during limb development, and each is involved in the control of different actions carried out by TGFβ molecules. *noggin* is expressed in the chondrogenic template and can control the size of the chondrogenic elements (Brunet et al., 1998; Merino et al., 1998), and *chordin* is detected in the forming joints (Francis-West et al., 1999). *follistatin* is detected in the tendons and muscles (Merino et al., 1999*a*) and *gremlin* in the distal part of the limb bud (Capdevila et al., 1999; Merino et al., 1999*b*). Application of all these TGFβ antagonists at the tip of digits inhibits the chondrogenic process.

In the duck, similar experiments have been done but the interdigital mesenchyme has different potency. When TGFβ is applied in the interdigital space, the cells develop only a small distal piece of cartilage (Gañán et al., 1998) although BMPs have the same expression pattern in chick and in duck autopods (Laufer et al., 1997). This is due to the differential expression of *gremlin*, which is lost early in the interdigital mesenchyme of chicken embryos while in the duck, its expression is maintained and enlarged until 10 days of development (Merino et al., 1999*b*). Application of TGFβ in the duck interdigital space has almost no effect because Gremlin prevents the chondrogenic induction mediated by BMPs.

APOPTOSIS

Apoptosis seems to be induced by the dual action of FGFs and BMPs (Gañán et al., 1996; Pizette and Niswander, 1999; Montero et al., 2001). The expression patterns of BMPs overlap with the areas that will undergo programmed cell death and ectopic application of beads soaked in the BMPs induces or accelerates ectopic cell death when applied in the limb bud mesenchyme (Francis et al., 1994; Duprez et al., 1996a; Duprez et al., 1996b; Macías et al., 1997). On the contrary, inhibition of BMP signalling by specific antagonists blocks cell death in physiological areas. Induction of apoptosis can be observed as early as 6 hours after BMP treatment in the anterior or posterior limb bud at stage 20–25 or in the interdigit at stage 28. To observe inhibition of cell death, BMP antagonists can be placed in the interdigital mesenchyme and then cell death inhibition can be observed.

Recently, it has been shown that onset of apoptosis in the limb bud requires the complementary action of FGFs (Montero et al., 2001). When beads soaked in FGF are placed in the anterior or posterior borders of the limb bud or in the interdigital mesenchyme, ectopic cell death is detected after 24 hours. The complementary experiment, application of FGF inhibitor, has been shown to prevent cell death in the interdigit.

As for chondrogenesis, the different BMP antagonists modulate the shape of apoptotic areas preventing cell death in the AER and in mesenchymal areas that will develop chondrogenic tissue (Capdevila et al., 1999; Merino et al., 1998; Merino et al., 1999b). Application of BMP antagonists in the chick interdigit inhibits apoptosis and a webbed space is developed. In duck interdigits, as we mentioned previously, expression of *gremlin* is maintained and BMP treatment only induces a small area of cell death in the distal part. But induction of a free space due to apoptosis can be observed if application of BMPs is combined with the application of FGFs (Gañán et al., 1998).

TIME REQUIRED FOR EXPERIMENTS

Incubation periods vary depending on the desired technique. In one week (without pre-incubation time) any one of the experiments can be done.

POTENTIAL SOURCES OF FAILURE

One important point during manipulation is the positioning of beads. If the bead is not in the proper position, it will not affect the growth of the limb.

TEACHING CONCEPTS

Organising centre: a group of cells that direct the development and outgrowth or spatial orientation of an organ. In this case, the AER is the organising centre for the outgrowth of the proximo-distal axis during limb development.

Chondrogenesis: process by which mesenchymal cells differentiate into cartilage.

Apoptosis: programmed cell death. This process occurs in certain areas and time points during limb development.

Potential or potency: capacity of cells to differentiate into one cell type or another. During limb development, cells of the progress zone retain the potential to differentiate into chondrogenic tissue or undergo cell death.

Differentiation: process characterised by the loss of potential and the onset of expression of tissue specific proteins. Once differentiated, most cells cannot develop into another cell type.

ALTERNATIVE EXERCISES

PROPOSED EXPERIMENTS

To compare the two models for limb growth, students can verify that removing the AER at different stages affects the level of limb truncation, and also that the implantation of FGFs can rescue this truncation. To investigate the allocation model students can apply FGF antagonists to see if cells die after the treatment, and if the expected truncations occur. If cells die by apoptosis and truncation is observed, this would support the allocation model. If the cells do not die but truncation still occurs, the model needs revision.

QUESTIONS FOR FURTHER ANALYSIS

➤ If you place two beads nearby, one soaked with BMPs and the other with BMP antagonists, such as Noggin in the interdigit, what results do you expect?
➤ How could you verify whether the absence of BMPs or the presence of BMP antagonists is responsible for interdigit formation in the duck hindlimb?
➤ If you apply a bead soaked in Activin or TGFβ in the chicken interdigit together with a bead of BMP, what do you expect? If the application of the BMP bead is done 6 hours after application of Activin or TGFβ, what result do you expect?

ADDITIONAL INFORMATION

Evolutionary studies can be carried out by studying the expression patterns of homologous and related genes. The diversity in limb shape throughout the animal kingdom makes limb development an exceptional tool for these evolutionary studies (see Wilkins, 2002).

REFERENCES

Baleman, W., and Van Hul, W. (2002). Extracellular regulation of BMP signalling in vertebrates. A cocktail of modulators. *Dev. Biol.*, 250, 231–50.

Brunet, L. S., McMahon, J. A., McMahon, A. P., and Harland, R. M. (1998). Noggin, cartilage, morphogenesis and joint formation in the mammalian skeleton. *Science*, 280, 1455–7.

Capdevila, J., Tsukui, T., Rodríguez Esteban, C., Zappavigna, V., and Izpisúa Belmonte, J. C. (1999). Control of vertebrate limb outgrowth by the proximal factor Meis2 and distal antagonism of BMPs by Gremlin. *Mol. Cell*, 4, 839–49.

Dudley, A. T., Ros, M. A., and Tabin, C. J. (2002). A re-examination of proximodistal patterning during vertebrate limb development. *Nature*, 418, 539–44.

Duprez, D. M., Bell, E. J., Richardson, M., Archer, C. W., Wolpert, C. W., Wolpert, L., Brickell, P. M., and Francis-West, P. H. (1996a). Overexpression of BMP-2 and BMP-4 alters the size and shape of developing skeletal elements in the chick limb. *Mech. Dev.*, 57, 145–57.

Duprez, D. M., Coltey, M., Amthor, H., Brickell, P. M., and Tickle, C. (1996b). Bone morphogenetic protein-2 (BMP-2) inhibits muscle development and promotes cartilage formation in chick limb bud cultures. *Dev. Biol.*, 174, 448–52.

Fallon, J. F., López, A., Ros, M. A., Savage, M. P., Olwin, B. B., and Simandl, B. K. (1994). FGF-2: Apical ridge growth signal for chick limb development. *Science*, 120, 104–7.

Francis, P. H., Richardson, M. K., Brickell, P. M., and Tickle, C. (1994). Bone morphogenetic proteins and a signalling pathway that controls patterning in the developing chick limb. *Development*, 120, 209–18.

Francis-West, P. H., Robertson, K. E., Ede, D. A., Rodríguez, C. D., Izpisúa-Belmonte, J. C., Houston, B., Burt, D. W., Gribbin, C., Brickell, P. M., and Tickle, C. (1995). Expression of genes encoding Bone morphogenetic proteins and Sonic Hedgehog in Talpid (ta3) limb buds: Their relationships in the signalling cascade involved in limb patterning. *Dev. Dyn.*, 203, 187–97.

Francis-West, P. H., Parish, J., Lee, K., and Archer, C. W. (1999). BMP/GDF-signalling interactions during synovial joint development. *Cell Tissue Res.*, 296, 111–19.

Gañán, Y., Macías, D., Duterque-Coquillaud, M., Ros, M. A., and Hurlé, J. M. (1996). Role of TGFβs and BMPs as signals controlling the position of the digits and the areas on interdigital cell death in the developing chick limb autopod. *Development*, 122, 2349–57.

Gañán, Y., Macías, D., Basco, R., Merino, R., and Hurlé, J. M. (1998). Morphological diversity of the avian foot is related with the pattern of *msx* gene expression in the developing autopod. *Dev. Biol.*, 196, 33–41.

Hamburger, V., and Hamilton, H. L. (1951). A series of normal stages in the development of the chick embryo. *J. Morphol.*, 88, 49–92.

Laufer, E., Pizette, S., Zou, H., Orozco, O. E., and Niswander, L. (1997). BMP expression in duck interdigital webbing: A reanalysis. *Science*, 278, 305.

Macías, D., Gañán, Y., Sampath, T. K., Piedra, M. E., Ros, M. A., and Hurlé, J. M. (1997). Role of BMP-2 and OP-1 (BMP-7) in programmed cell death and skeletogenesis during chick limb development. *Development*, 124, 1109–17.

Merino, R., Gañán, Y., Macías, D., Economides, A. N., Sampath, K. T., and Hurlé, J. M. (1998). Morphogenesis of digits in the avian limb is controlled by FGFs, TGFβs and noggin through BMP signaling. *Dev. Biol.*, 200, 35–45.

Merino, R., Macías, D., Gañán, Y., Rodríguez-León, J., Economides, A. N., Rodríguez-Esteban, C., Izpisúa Belmonte, J. C., and Hurlé, J. M. (1999a). Control of digit formation by activin signaling. *Development*, 126, 2161–70.

Merino, R., Rodríguez-León, J., Macías, D., Gañán, Y., Economides, A. N., and Hurlé, J. M. (1999b). The BMP antagonist gremlin regulates outgrowth, chondrogenesis and programmed cell death in the developing limb. *Development*, 126, 5515–22.

Millan, F. A., Denhez, F., Kondaiah, P., and Akurst, R. J. (1991). Embryonic gene expression pattern of TGFβ1, β2 and β3 suggests different developmental functions in vivo. *Development*, 111, 131–41.

Montero, J. A., Gañán, Y., Macías, D., Rodríguez-León, J., Sanz-Ezquerro, J. J., Merino, R., Chimal-Monroy, J., Nieto, M. A., and Hurlé, J. (2001). Role of FGFs in the control of programmed cell death during limb development. *Development*, 128, 2075–84.

Niswander, L., Tickle, C., Vogel, A., Booth, I., and Martin, G. R. (1993). FGF-4 replaces the apical ectodermal ridge and directs outgrowth and patterning of the limb. *Cell*, 75, 579–87.

Pizette, S., and Niswander, L. (1999). BMPs negatively regulate structure and function of the limb apical ectodermal ridge. *Development,* 126, 883–94.

Saunders, J. W., Jr. (1948). The proximo-distal sequence of origin of the chick wing and the role of ectoderm. *J. Exp. Zool.,* 108, 363–403.

Summerbell, D. (1974). A quantitative analysis of the effect of excision of the AER from the chick limb-bud. *J. Embryol. Exp. Morphol.,* 32, 651–60.

Summerbell, D., and Wolpert, L. (1972). Cell density and cell division in the early morphogenesis of chick wing. *Nature,* 239, 24–6.

Sun, X., Mariani, F. V., and Martin, G. R. (2002). Functions of FGF signalling from the apical ectodermal ridge in limb development. *Nature,* 418, 501–8.

Wilkins, A. S. (ed.) (2002). *The Evolution of Developmental Pathways.* Sunderlan: Sinauer Associates, Inc.

Zou, H., Choe, K. M., Lu, Y., Massague, J., and Niswander, L. (1997). Bmp signaling and vertebrate limb development. *Cold Spring Harbor Symp. Quant. Biol.,* 62, 269–72.

8 Induction of ectopic limb outgrowth in chick with FGF-8

Á. Raya, C. Rodríguez Esteban, and J. C. Izpisúa-Belmonte

OBJECTIVE OF THE EXPERIMENT The chick embryo is extremely amenable to surgical manipulations such as bead implantation. Local delivery of signaling molecules is easily achieved in this experimental model and provides invaluable information as to their effect on a variety of developmental processes. Specifically, our knowledge about limb initiation was significantly increased when members of the Fibroblast Growth Factor (FGF) family were shown to induce ectopic limb formation (Cohn et al., 1995). In this experiment, recombinant FGF-8 is produced, purified, and applied to the flank of chick embryos, where it induces the formation of a fairly normal extra limb.

INTRODUCTION

The analysis of limb outgrowth and patterning during chick embryo development has become an excellent model for understanding the molecular mechanisms underlying general developmental processes. The choice of the chick as an experimental model system has greatly enhanced embryo manipulation and survival in the laboratory.

In most vertebrates, limb appendages originate in the flank of the early embryo where cells group in the so-called limb buds. The first step in limb development, known as limb initiation, comprises the induction of limb budding in specific locations of the embryo's flank. At this stage, most of the flank's extension is competent to give rise to limb buds (Figure 8.1). However, budding will occur only in distinct regions of the embryo's flank (four regions in tetrapods, two on each flank). The limb bud is formed by a mesodermal or mesenchymal core, covered by an ectodermal jacket (Figure 8.2). Specific components of the limb are formed by cell populations originating in the embryo's trunk. For instance, cells in the lateral part of the somites migrate into the limb bud, eventually giving rise to the limb skeletal muscles. The distal-most part of the ectodermal jacket forms the apical ectodermal ridge (AER), an ectodermal thickening that separates the dorsal and ventral sides of the limb bud (Figure 8.2). Maintenance of a functional AER is essential for the proliferation of underlying mesenchymal cells and limb bud outgrowth. Whether these mesenchymal cells acquire positional

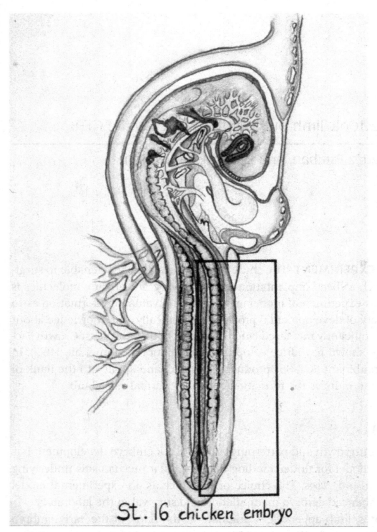

St. 16 chicken embryo

Figure 8.1. Schematic representation of a stage 16 chick embryo. Before limb buds are visible the prospective forelimb, hindlimb and flank regions (rectangle) are competent to induce limb bud outgrowth.

information over time, or possess an initial pre-pattern, is a current subject of controversy. Mesodermal cells in the posterior part of the limb bud, in turn, form the zone of polarizing activity (ZPA), which organizes the antero–posterior pattern during limb bud outgrowth (Figure 8.2).

The molecular mechanisms responsible for limb bud initiation are not fully understood. Our group identified FGF-8 as an early player in this mechanism (Vogel, Rodríguez, and Izpisúa-Belmonte, 1996). The gene encoding FGF-8 is expressed in a transient and dynamic fashion in the intermediate mesoderm of the embryo's flank, at locations corresponding to where the limb buds will appear (Figure 8.3b-d), before any morphological sign of limb bud induction is apparent. This expression pattern

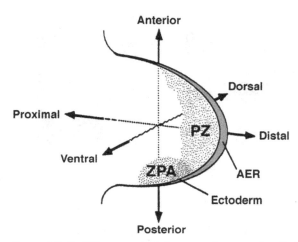

Figure 8.2. Building a complete appendage requires that cells acquire positional information with respect to the three main axes: antero–posterior (thumb to little finger), dorso–ventral (back to palm of the hand), and proximo–distal (shoulder to fingertips). The apical ectodermal ridge (AER) directs limb bud outgrowth by maintaining the immediately underlying mesenchymal cells (progress zone, PZ) in a continuous proliferation state. The 'progress zone' model proposes that, as the limb bud grows, the mesenchymal cells that leave the PZ are determined to form pattern elements according to instructive signals received from organizer regions such as the zone of polarizing activity and the ectoderm.

indicated that *Fgf-8* might have a role during limb initiation. To address this possibility experimentally, we decided to produce recombinant FGF-8 protein and apply it locally in different locations of the flank in early chick embryos (Figure 8.3e).

MATERIALS AND METHODS

PROCEDURES

Expression and purification of recombinant FGF-8

1. An *Fgf-8* cDNA lacking the signal peptide was sub-cloned into the expression vector pET23c (Novagen) in the same way that a six-histidine tag is fused in frame to the C-terminus of the expressed protein.
2. Expression of recombinant protein can be induced in 1-liter cultures at a growth density of $OD_{600} \sim 0.7$ by adding 0.1 mM IPTG and incubating at 30°C for an additional two hours.
3. After cell lysis by sonication in lysis buffer [50 mM HEPES pH 8.0, 150 mM NaCl, 2.5 mM $CaCl_2$, 20 mM imidazole, 10% glycerol, 1 mM Pefablock SC protease inhibitor (Roche)], the recombinant protein was found to be present as inclusion bodies in the insoluble pellet.
4. Recombinant FGF-8 was recovered from the insoluble fraction by solubilization in buffer A (5 M guanidine-HCl, 20 mM Tris-HCl, pH 7.5, 5 mM Pefablock SC protease inhibitor), loaded into a 4 ml Ni-NTA agarose column (Qiagen), and washed sequentially with 10 volumes of buffers A, B, and C. Buffer B contains 8 M urea, 100 mM HEPES pH 8.0. Buffer C contains 6 M urea, 20% glycerol, and 50 mM Tris-HCl pH 7.5.

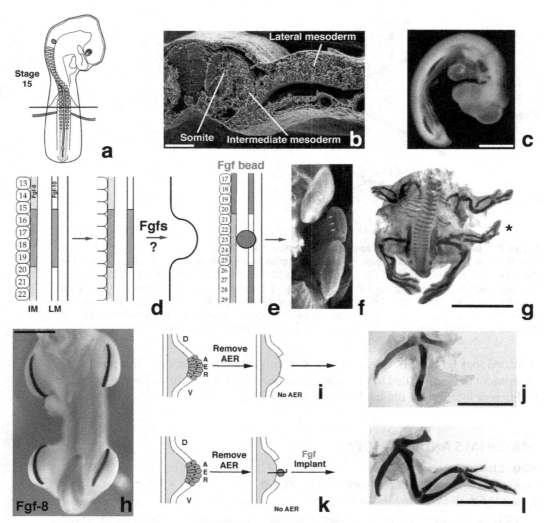

Figure 8.3. Vertebrate appendages are formed from cell populations of the lateral mesoderm (LM) in the flank of the early embryo (a, b), which proliferate under the influence of signals produced by the intermediate mesoderm (IM). *Fgf-8* (c, colored in grey in d and e) is expressed along with other genes in the regions that will give rise to the limb buds (black in d and e). FGF-8–soaked beads, implanted in the flank between the areas where the limb buds will appear (e), are able to induce the growth of an ectopic appendage that develops normally (f, asterik in g). Fgfs expressed in the AER (h) are also implicated in maintaining limb bud outgrowth. Removal of the AER during early limb bud development results in truncated appendages (i, j). Application of FGFs is able to maintain limb bud outgrowth after AER removal, so that a fairly normal appendage is obtained (k, l). Bars represent 50 μm (b), 1 mm (c, h), 1 cm (g) and 5 mm (j, l).

5. Protein refolding was carried out with a 180 ml (2 ml/min) buffer C to buffer D linear gradient. Buffer D contains 1 M urea, 500 mM NaCl, 20% glicerol, 50 mM Tris-HCl pH 7.5.
6. Recombinant FGF-8 was finally eluted with an imidazole gradient (20–250 mM). This procedure yielded >12 mg of recombinant FGF-8 per liter of culture, with a final purity estimated by Coomasie Blue staining greater than 95%.
7. Before implantation, recombinant FGF-8 was dialyzed against 50 mM Tris pH 7.5, 500 mM NaCl, 20% glycerol, and the concentration adjusted to 1 mg/ml.

Staining of skeletal structures in the chick embryo. Embryos are collected and fixed overnight in a solution of 5% trichloracetic acid. The next morning, the embryos are transferred to 0.1% Alcian Green (Sigma) to stain for cartilage, and incubated in this solution for 24 hours. Then, the embryos are dehydrated by washing sequentially with 70% and 100% ethanol. Soft tissues are finally clarified by incubation in methylsalicilate to visualize cartilage and bone structures (see also Chapter 7).

Bead implantation. See Chapter 8.

OUTLINE OF THE EXPERIMENT

EMBRYO MANIPULATION

Fecundated chick eggs are incubated at 37°C until stage 14–15 (around 60 hours). Beads are prepared approximately one hour before embryo manipulation. One μl of acrylic heparin beads (Sigma) is pipetted into a sterile Petri dish. The beads are suspended into an aqueous solution that can be removed by gently applying a small piece of tissue paper. Immediately after drying the beads, they are soaked in 2 μl of recombinant purified FGF-8. We also add several ~10 μl drops of water in the Petri dish to help maintain moisture and prevent evaporation of the FGF-8 solution.

The embryo is windowed by removing the top part of the eggshell and shell membrane. The membrane covering the embryo is removed with fine forceps and india ink (Fount India, Pelikan) is injected under the embryo to help visualization (see Chapter 3). A fine, custom-made tungsten needle is used to make a small incision in the lateral mesoderm of the flank, between somites 20 and 25. One of the FGF-8-soaked beads is taken with fine watchmaker forceps and placed over the embryo, close to the incision. The bead is put in place by dragging it with the tungsten needle. Scotch tape is finally used to seal the window in the eggshell, and the egg is returned to the 37°C incubator until the desired stage. Beads soaked in buffer (50 mM Tris pH 7.5, 500 mM NaCl, 20% glycerol) are used as control. Nine or 10 days after the bead implantation, embryos are collected and fixed.

EXPECTED RESULTS

Four days of incubation will suffice to obtain a small extra growth in the region where the bead was implanted. For a fully formed appendage with skeletal elements, the incubation should proceed for at least 10 days (Figure 8.4).

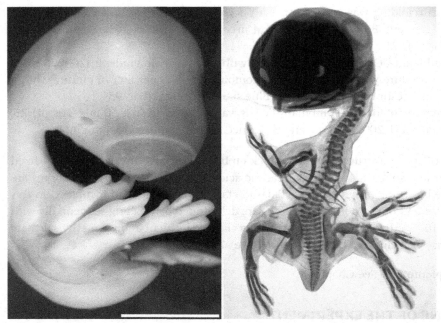

Figure 8.4. Stage 32 chick embryo with an extra leg resulting from the implantation of an FGF-8–soaked bead in a stage 15 embryo at the level of somite 23. Bar = 5 mm.

DISCUSSION

The results obtained by analyzing the expression pattern of *Fgf-8*, and by implanting FGF-8-soaked beads in the flank of the chick embryo, indicate that *Fgf-8* acts from the intermediate mesoderm to maintain cell proliferation in specific locations of the lateral mesoderm, where limb budding will later occur (Vogel et al., 1996).

FGF-8 induces the formation of a complete appendage when ectopically applied to the flank region between fore and hindlimb buds in the chick embryo (Figure 8.3f, g). Thus, *Fgf-8* is sufficient for limb initiation. Moreover, *Fgf-8* is expressed in the AER (Figure 8.3h) and exogeneous FGF-8 application is able to rescue the limb truncation phenotype that results from AER ablation (Figure 8.3i-l). The current model for limb induction postulates that FGF-10, another factor of the same family of secreted proteins, mediates the effect of FGF-8 on lateral mesoderm cells. FGF-8 would, therefore, activate the expression of *Fgf-10* in the lateral mesoderm and FGF-10 protein, in turn, would induce ectodermal cells of the flank to form the AER (Capdevila and Izpisúa-Belmonte, 2001).

ACKNOWLEDGEMENTS

AR is partially supported by a Fellowship from the Ministerio de Educación, Cultura y Deporte, Spain. This work was supported by grants from the March of Dimes, BioCell, NSF, and the G. Harold and Leila Y. Mathers Charitable Foundation.

REFERENCES

Capdevila, J., and Izpisúa-Belmonte, J. C. (2001). Patterning mechanisms controlling vertebrate limb development. *Annu. Rev. Cell. Dev. Biol.*, 17, 87–132.

Cohn, M. J., Izpisúa-Belmonte, J. C., Abud, H., Heath, J. K., and Tickle, C. (1995). Fibroblast growth factors induce additional limb development from the flank of chick embryos. *Cell*, 80, 739–46.

Vogel, A., Rodríguez, C., and Izpisúa-Belmonte, J. C. (1996). Involvement of FGF-8 in initiation, outgrowth and patterning of the vertebrate limb. *Development*, 122, 1737–50.

9 RNAi techniques applied to freshwater planarians (Platyhelminthes) during regeneration

D. Bueno, R. Romero, and E. Saló

OBJECTIVE OF THE EXPERIMENT The mechanisms that silence unwanted gene expression are critical for normal cellular function. Double-stranded RNA (dsRNA) induces gene silencing by the degradation of single-stranded (ssRNA) targets complementary to the dsRNA trigger. RNA interference (RNAi) effects activated by dsRNA have been demonstrated in a number of organisms including plants, protozoa, hydra, planarians, nemertines, annelids, nematodes, insects, and mammals. In this experiment, we introduce a methodology to produce and visualize the loss of function phenotype of dsRNA-injected planarians. In this experiment, two genes will be silenced by post-transcriptional gene silencing: *Gtsix-1*, a gene involved in the determination of the eye field and eye formation, and *Tcen49*, an invertebrate trophic factor involved in regional cell survival.

To achieve *Six/tcen49* loss of function, several procedures will be used:

➤ plasmid linearization
➤ ribosynthesis
➤ dsRNA production
➤ dsRNA delivery by microinjection

DEGREE OF DIFFICULTY This experiment is moderately difficult. It requires some skills in ribosynthesis and microinjection, and the use of appropriate materials.

INTRODUCTION

Platyhelminthes are acoelomate members of the Bilateria, which share with other phyla like Gastrotricha and Acanthocephala a basal position in the Lophotrochozoa clade (Aguinaldo et al., 1997). Planarians (order Tricladida) are free-living platyhelminthes, well known for their ability to regenerate (Baguñà, 1998; Saló and Baguñà, 2002), which can easily be triggered by amputation of body fragments. In addition to the

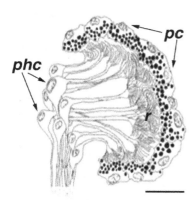

Figure 9.1. Drawing of the planarian eyespot or ocellus. The pigment cells (pc) group to form an eyecup which surrounds the rhabdomeres (r) of the photoreceptors. The photoreceptor cells (phc) are bipolar neurons. Their axons extend towards the cephalic ganglia and their dendritic extremities with rhabdomeric, a regularly ordered microvilli assembly where the opsin accumulates, extend towards the pigment cup. Abbreviations: pc, pigment cells; phc, photoreceptor cells; r, rhabdomeric structures of the photoreceptor cells. Scale bar = 0.1 mm.

regenerative capacity of the planarian body, its morphological simplicity, which nevertheless includes all the basic characteristics of a bilaterian representative – cephalization, dorso–ventral, and anterior–posterior polarity – makes it a useful model for investigating the evolution of molecular developmental processes.

Photosensitivity is an ancient trait, probably preceding multicellularity. Many unicellular organisms possess tiny granules of pigment which, after absorbing light, generate an electrical signal that can alter the behavior of the cell. In the animal kingdom there are eight basic types of eyes, with extensive variation in eye structure within each type. Moreover, morphologically similar eyes can develop using distinctly different mechanisms, such as the camera eyes of vertebrates and cephalopods. These findings suggest that eyes have evolved independently more than 40 times in animals (Salvini-Plawen and Mayr, 1977). However, recent finding that conserved biochemical and developmental pathways generate visual organs in a range of organisms brings the homology of vertebrate and invertebrate visual structures under debate (Gehring and Ikeo, 1999; Fernald, 2000; Pichaud, Treisman, and Desplan, 2001). The common ancestor of primitive bilaterians may have had simple eyes. Planarian eyes, which are composed of only two cell types, photoreceptors, and pigmented cells (Figure 9.1), are considered good representatives of such a primitive visual system (reviewed in Hyman, 1951; Rieger et al., 1991).

The isolation of an increasing number of regulatory genes involved in eye formation in planarians, as well as the possibility of using a functional approach by RNA interference (RNAi) (Pineda et al., 2000), has allowed the exploration of the origin and relationships of molecular building blocks involved in the formation and evolution of the eye. One member of the *Six/sine oculis* gene, *Gtsix-1,* is expressed in both adult and regenerating eyes. *Gtsix-1* is expressed during regeneration in the dorsal anterior region of the cephalic blastema, and later in regions of eye formation. Loss of function by *Gtsix-1* dsRNA caused a non-eye phenotype in planarians regenerating a head (Pineda

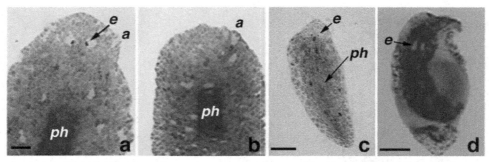

Figure 9.2. Bright-field images of phenotypes produced by RNAi in *Girardia tigrina*. (a) Normal 30-day regenerated head with differentiated eyes visible in control planarian injected with water. (b) Completely regenerated head lacking eyes in a *Gtsix-1* dsRNA-injected planarian. (c) Normal 9-day regenerated planarian injected with water. (d) Central and posterior degeneration of a *tcen-49* dsRNA-injected planarian. Abbreviations: a, auricle; e, eyes; ph, pharynx. Scale bars = a, 0.2 mm; c, d, 0.5 mm.

et al., 2000; Figure 9.2a, b). In intact planarians, *Gtsix-1* RNAi rapidly reduces *Gtsix-1* expression within the first 24 hours, and opsin (*Gtops*) expression decreases progressively in the following week, leading to a blind phenotype (Pineda et al., 2001; Saló et al. 2002). These results are consistent with a key role of *Gtsix-1* in differentiation and maintenance of eye cells which, as in all planarian cell types, are continuously renewed in adults by a neoblast population.

As to the anterior–posterior polarity, in most organisms it is accomplished by the generation of different body regions with different genetic networks. The establishment of these body regions during regeneration and their maintenance in adult organisms are clue processes to understanding the great ability of regeneration exhibited by planarians. Several types of molecules can be involved in such processes, such as master genes and trophic factors. In freshwater planarians, the small secreted protein TCEN49 has been linked to the regional maintenance of the planarian central body, behaving as a trophic factor involved in central body cell survival (Bueno et al., 2002). In planarian tail regenerates, *tcen49* expression inhibition by dsRNA interference causes extensive apoptosis in various cell types, including muscular and nerve cells (Figure 9.2c, d). This phenotype is rescued by the implantation of microbeads soaked in TCEN49 after RNA interference. TCEN49 amino acid sequence has been related to that of EGF and EGF-like repeats of both vertebrates and invertebrates and especially to that of a molluscan trophic factor (Bueno et al., 2002).

In many species, the introduction of double-stranded RNA (dsRNA) exonic sequences of the gene whose expression is to be disrupted induces potent and specific gene silencing, a phenomenon called RNA interference (RNAi; Fire et al., 1998). This phenomenon, based on the targeted degradation of mRNAs (post-transcriptional gene silencing, PTGS; Montgomery, Xu, and Fine, 1997; Zamore et al., 2000; Hannon, 2002), has sparked general interest from both applied and fundamental standpoints. In planarians, the effectiveness of RNAi was first demonstrated through analysis of the specific effects of myosin, β-tubulin, and opsin dsRNAs on regenerating and adult organisms (Sánchez-Alvarado and Newmark, 1999).

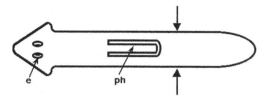

Figure 9.3. Diagrammatic scheme of a planarian. Arrows show where to cut the body for the regeneration experiments. Abbreviations as in Figure 9.2.

MATERIALS AND METHODS

EQUIPMENT AND MATERIALS
Per student
> Razor blade
> Petri dishes
> Mineral water

Per practical group
> Microinjector (soaking method does not require microinjector)
> Dissecting microscope

Biological material
> Planarians
> dsRNA

SPECIES AND CULTURE CONDITION. The freshwater planarians used in this exercise belong to an asexual race of the species *Girardia tigrina*. If other species are to be used, conditions may need to be empirically adjusted to get optimum results. Planarians are maintained in spring water in the dark at 17°C in petri dishes, in 20 ml of water, at 10 organisms/petri dish, and fed once a week with beef liver. Those selected for experiments are starved at 17 ± 1°C for at least 15 days before use. If regenerating planarians are required, organisms are cut behind the pharynx and kept at 17 ± 1°C to allow regeneration as required (see Outline of the Experiments; Figure 9.3).

PREVIOUS TASKS FOR STAFF
Synthesis of plasmid DNA (*GtSix-1/tcen49*-pBluescriptSK) is performed using the QIAprep spin miniprep kit (Qiagen).

PROCEDURES
Plasmid DNA linearization
1. Linearize the appropriate plasmid DNA by incubating 50 μg DNA (*GtSix-1/tcen49*-pBluescriptSK) from a regular plasmid DNA miniprep (Table 9.1). Reagents listed are from Boheringer Manheim.

Table 9.1. Restriction enzyme digestion

	Six-antisense transcription	*Six*-sense transcription	*tcen49*-antisense transcription	*tcen49*-sense transcription
DNA	20 μl	20 μl	20 μl	20 μl
5 × transcription buffer	15 μl	15 μl	15 μl	15 μl
Enzyme	5 μl (*Hind* III)	5 μl (*Xba* I)	5 μl (*Eco*RI)	5 μl (*Xho*I)
H$_2$O	110 μl	110 μl	110 μl	110 μl
Total	150 μl	150 μl	150 μl	150 μl
Temp.	37°C	37°C	37°C	37°C
Timing	2 h to O.N.	2 h to O.N.	2 h to O.N.	2 h to O.N.

2. Add 50 μg/ml Proteinase K and incubate for 30 min at 37°C.
3. Add 100 μl of phenol-chloroform, mix, and spin for 10 min at 4°C.
4. Transfer the aqueous layer to a fresh tube and repeat step 3.
5. Transfer the aqueous layer to a fresh tube, add 0.1 volumes of NaAc 3M pH 5.2 and 2 volumes of 100% EtOH. Spin for 20 min at 4°C.
6. Remove the supernatant and wash the pellet with 80% ice-cold EtOH. Spin for 5 min at 4°C.
7. Remove the supernatant and resuspend the pellet in 10 μl of RNAse-free H$_2$O. Store at −20°C.
8. Run 0.5 μl of linearized DNA on a 0.8–1% TBE agarose electrophoresis gel.

Production of dsRNA

1. *Six* and *tcen49* sense and antisense RNAs. This generates a 348 bp long-*tcen49* riboprobe corresponding to most of the coding sequence, and a 250 bp long-*Gtsix-1* riboprobe corresponding to the 5' end coding sequence (Gen Bank accession no. AJ251661) (Table 9.2).
   (a) DNA from *Gtsix-1*-pBluescript clone has been previously linearized by digestion with *Hind*III for antisense T3 transcription, and with *Xba*I for sense T7 transcription.
   (b) DNA from *tcen49*-pBluescriptSK has been previously linearized by digestion with *Xho*I for sense RNA T3 transcription, and with *Eco*RI for antisense RNA T7 transcription.
2. Digest DNA template by adding 1 μl of DNAse I to each transcription reaction, and incubate for 15 min at 37°C.
3. Remove 1 μl from each reaction and store at −20°C.
4. Combine sense and antisense reactions.
5. Add 380 μl of 1 × STOP buffer and 0.3 μl of glycogen. Stop buffer (1 ×): 1 M NH$_4$OAc + 10 mM EDTA + 0.2% SDS.
6. Add 200 μl of phenol-chloroform, and spin for 10 min at 4°C.

Table 9.2. Transcription

	T3 polymerase	T7 polymerase
DNA	4 to 8 μl	4 to 8 μl
5 × transcription buffer	16 μl	16 μl
5 × rNTPs DIG or FITC UTP-labelled	16 μl	16 μl
RNAse inhibitor	4 μl	4 μl
DTT	8 μl	8 μl
T3 or T7 polymerase	4 μl	4 μl
H_2O	24 to 28 μl	24 to 28 μl
Total	80 μl	80 μl
Temp.	25°C	37°C
Timing	1.5 h	1.5 h

7. Transfer aqueous layer to a fresh tube, add 200 μl of chloroform, and spin for 10 min at 4°C.
8. Transfer aqueous layer to a fresh tube, and incubate for 10 min at 68°C. Then incubate for 30 min at 37°C.
9. Add 1 ml of 100% EtOH, mix, and spin for 15 min at 4°C.
10. Discard supernatant, wash with 0.5 ml of 80% EtOH, and spin for 5–10 min at 4°C.
11. Discard supernatant and resuspend in 10 μl of RNAse-free H_2O. Final dilution should be ≈ 1 μg/μl. Store at −20°C.
12. Run 1 μl of ssRNA from step 3 and 0.5 μl dsRNA on a 1% TBE agarose electrophoresis gel at 120 V to prevent RNA degradation (ssRNA should migrate faster than dsRNA).

dsRNA injection. *tcen49* double-stranded RNA should be injected 30 minutes after amputation, when *tcen49* is not yet expressed in these regions, and re-injected every 3–4 days. A similar protocol can be followed for the *Gtsix-1* double-stranded RNA, but the injection frequency can be once per week. Each injection delivers a total volume of 23 nl and is administered twice per day in each animal. Microinjections are performed under a binocular (× 20) with a Drummond Scientific Nanoject injector and micromanipulator. The planarians are anesthetized by cold. To this end, the organisms are placed in a slide surrounded by ice. After injection, the planarians are kept in spring water with 0.1 mg/l of gentamicin sulphate at 17°C and the media are replaced daily to avoid contamination.

dsRNA soaking. A simple method was recently developed for RNA interference (RNAi) (Orii et al., 2003). After transverse cutting, the regenerate piece is soaked in a drop of water (100 μl) containing dsRNA (0.5 μg/μl) for 5 hours, and then allowed to regenerate in spring water as described previously. This method avoids the use of a microinjector, but requires extra dsRNA synthesis.

OUTLINE OF THE EXPERIMENTS

Once the dsRNA is produced, the injections can be initiated. To this end, the population of planarians should be divided in two groups: a control group injected with sterilized water or β-*gal* dsRNA (obtained following the same protocol but using a β-*gal* containing pBluescript plasmid); and the *Gtsix-1* and *tcen49* minus groups that will be injected with dsRNA. To properly compare both groups, all the animals should be cut at the same level and at the same time, and the injections should be performed with the same periodicity.

PLANARIANS GROUP 1 – CONTROL GROUP
Procedure
1. Cut organisms at the appropriate level.
2. Microinject control organisms with sterilized water or β-*gal* dsRNA 30 minutes after amputation (2×23 μl per regenerate). Alternatively, you can soak the cut planarian in a drop of 100 μl of water or water containing β-*gal* dsRNA solution (0.5 μg/μl) in a piece of Parafilm for 5 hours.
3. Reinject or resoak the regenerates every 3–7 days as stated above.

Data recording
- *Gtsix-1* control regenerates: monitor the regeneration of the eyes. The eyes should regenerate as in classical regeneration experiments, i.e., they start to differentiate at 5 days of regeneration and are easily visible at 7 days of regeneration as two brown dots at the base of the new unpigmented regenerated tissue or blastema.
- *tcen49* control regenerates: monitor the formation of the new central region. The formation of the new central region should be monitored by the formation of the pharynx, which is clearly visible at 7 days of regeneration as a dark cylinder under a dissecting microscope.

PLANARIANS GROUP 2 – *Gtsix-1* AND *Tcen49* MINUS GROUP
Procedure
1. Cut organisms at the appropriate level.
2. Microinject control organisms with *Gtsix1* or *tcen49* dsRNA after amputation (2×23 μl per regenerate). Alternatively, you can soak the cut planarian in a drop of 100 μl of water containing dsRNA solution (0.5 μg/μl) in a piece of Parafilm for 5 hours.
3. Reinject or resoak the regenerates as stated above.

Data recording
- *Gtsix-1* control regenerates: monitor the regeneration of the eyes. The eye spots do not form, and a blind planaria is observed.
- *tcen49* control regenerates: monitor the formation of the new central region. After pharynx formation, at 9–12 days of regeneration, cells from the central region enter apoptosis and the regenerate dies.

EXPECTED RESULTS

Gtsix-1 RNAi EXPERIMENT

Gtsix-1 RNAi rapidly reduces *Gtsix-1* expression within 24 hours after the first injection or soaking (Pineda et al., 2000; Figure 9.2a, b). *Gtsix-1* dsRNA-injected fragments regenerating the head show a normal speed of blastema formation and brain regeneration. At 6 days of regeneration at 17°C, the control organisms injected with water or β-*gal* dsRNA show normal differentiation of the eyes, i.e., two brown spots at the edge of the blastemal and postblastema region, while the head regenerating organisms without functional *Gtsix-1* show a non-eye phenotype (Pineda et al., 2000). Such phenotype can be maintained by regular injection or soaking of the same dsRNA once per week (Pineda et al., 2001). After one month of regeneration, normal and well-proportioned heads without eyes can be observed. If the same organisms are maintained for two further weeks without re-injection, normal eyes are regenerated without further decapitation (Pineda et al., 2000). The injection or soaking of *Gtsix-1* dsRNA in intact organisms induces a similar rapid decrease in *Gtsix-1* expression and progressively lowers the expression of the planarian opsin gene, which leads to the loss of phototactic behavior in the animal (blind phenotype). At the same time, the size and number of eye cells slowly and continuously decrease (Pineda et al., 2001).

Tcen-49 RNAi EXPERIMENT

In *tcen49* RNAi experiments performed on regenerating tails, regeneration proceeds normally until day 9–12 of regeneration, i.e., a complete new pharynx is formed. However, at day 9–12 of regeneration, when the pharynx usually recovers functionality and TCEN49 is first secreted from cyanophilic secretory cells (Bueno, Baguna, and Romero, 1996), the organisms microinjected with *tcen49* dsRNA lysed (Figure 9.2c, d). Just before lysis, the pharynx is ventrally and posteriorly displaced, and finally expelled from the planarian body.

At this stage, cells and tissues from the putative central (around the pharynx) and posterior (tail) regions degenerate. The process of degeneration does not affect the anterior region (head) or the pharynx itself. To visualize this process, the TUNEL assay and a molecular marker specific for muscle cells (phalloidin) can be used. Muscle fibers from the affected regions are completely disorganized and disrupted. Moreover, the TUNEL assay reveals extensive cell apoptosis in the putative central and posterior regions, including muscle and nerve cells, among other cell types from the planarian body. Apoptotic cells in the putative anterior region and in the pharynx are detected as in controls.

To perform the TUNEL assay, organisms are fixed in paraformaldehyde 4% on an ice-cold slide, dehydrated in an ethanol series, and embedded in paraffin. The 10 μm-sagittal paraffin sections are deparaffinated and rehydrated through xylene and a reversal ethanol series, washed in PBS, and permeabilised with 20 μg/ml Proteinase K in PBS at pH 7.4 for 15 minutes at room temperature. After permeabilisation, sections are microwave irradiated at 600 W for 1 minute ($\times$ 5) with 0.01 M citrate buffer pH 3. The 3'-OH ends of DNA are labelled for two hours at 37°C by addition of digoxigenin

11-UTPs using the enzyme TdT and then detected with a fluorescein-conjugated anti-digoxigenin antibody (the TUNEL kit is from Boheringer Manheim).

To perform phalloidin staining, organisms are fixed for 15 minutes at room temperature (RT) in 3.7% formaldehyde (Merck). After washing for 10 minutes in PBS they are permeabilised in 0.3% Triton X-100 (Merck) in PBS for 10 minutes at RT. They are then washed for 15 minutes in PBS and incubated with fluorescein isothiocianate (FITC)-phalloidin (Molecular Probes) diluted 1:100 in PBS overnight at 4°C. After washing in PBS three times, fluorescence is detected with confocal microscopy.

DISCUSSION

Gtsix-1 RNAi EXPERIMENT

In planarians, the *sine oculis* gene *Gtsix-1* is continuously expressed in adult eyes. Initial expression also coincides with the first signs of eye differentiation during cephalic regeneration. The expression of developmental regulatory genes in adults is normal, because planarians show great morphological plasticity in the continuous growth and regression or regeneration processes. Because heads that fail to regenerate eyes after *Gtsix-1* dsRNA injection contain normal differentiated cephalic ganglia and auricles, we can assume that this loss of function affects only the process of eye formation. The maintenance of such eye inhibition in the head blastemas requires dsRNA re-injection at weekly intervals, thus indicating that the head is always competent for eye regeneration. The loss of *opsin* expression in the photoreceptor cells and the production of a blind phenotype in the differentiated eyes can be interpreted as the disruption of this conserved eye network. *Gtsix-1* RNAi–induced loss of function points to a crucial function of *Gtsix-1* in early eye determination and in the maintenance of the differentiated state. These data also provide additional support for the evolutionary conservation of the initial genetic pathway in eye determination of triploblastic metazoans.

Tcen-49 RNAi EXPERIMENT

In metazoans, trophic factor molecules are associated with cell survival. This highly heterogeneous group of molecules also patterns structures and organs during embryogenesis and regulates neuronal plasticity. The loss of function experiments on planarian regenerating tails presented in this practical, in which the lack of TCEN49 leads to planarian death by extensive cell apoptosis, indicate that, at least during a specific period of planarian regeneration, TCEN49 behaves as a trophic factor involved in regional cell survival. However, this molecule is not involved in the determination of its most representative organ, the pharynx, since it forms as in control regenerates, and thus is not involved in the determination of planarian body regions.

When considering TCEN49 as a planarian trophic factor involved in regional cell survival, several features should be highlighted. TCEN49 is a trophic factor for cells from the central and posterior regions of the body, but not for those from the anterior region, i.e. head and cephalic ganglia, or the pharynx. This may explain why the pharynx becomes detached from the rest of the body before the lysis of the whole organism, as the neuromuscular cells from the central body region that are located at the implantation zone of the pharynx, those holding the pharynx, enter apoptosis and die. We would also like to stress that when TCEN49 is absent from the central region, cells from the posterior region enter apoptosis. This may affect other unknown molecules that are trophic factors for the corresponding cells of the posterior body region or disrupting the ventral nerve chords, which run from the cephalic ganglia to the tip of the tail.

TIME REQUIRED FOR THE EXPERIMENTS

Miniprep DNA preparation, linearization, ribosynthesis, and production of dsRNA, including gel electrophoresis for testing the quantity and quality, take 3 days. The observation of the phenotype requires 9 to 12 days of periodic injections every 3–4 days of *tcen49* dsRNA, or one month of periodic injections every week of *Gtsix-1* dsRNA.

POTENTIAL SOURCES OF FAILURE

ss-RNA is very sensitive to degradation by RNAses, but once annealed into ds-RNA, it becomes resistant to the same RNAses. RNA ribosynthesis should thus be performed very carefully to avoid RNA degradation. We also strongly recommend starting with a large number of organisms per group (12 animals each group minimum), because some will die after several rounds of injection.

TEACHING CONCEPTS

Developmental network: a network of related genes involved in a specific morphogenetic process, e.g. the eyes.

Loss of function phenotype: the phenotype of organisms in which we specifically inhibited a precise gene.

Regeneration and anterior-posterior body pattern maintenance: restitution and maintenance of a complete organism from an adult fragment.

ALTERNATIVE EXERCISES

QUESTIONS FOR FURTHER ANALYSIS

➤ Why does the tail of planaria degenerate when *Tcen-49* dsRNA is injected? What experiment would you do to prove your hypothesis?

➤ Provided that you can graft a head to pharynx and the grafted head induces the formation of an ectopic "neck" tissue, how can you prove whether *Gtsix-1* is involved in eye or eye region specification?

REFERENCES

Aguinaldo, A., Turbeville, J., Linford, L., Rivera, M., Garey, J., Raff, R., and Lake, J. (1997). Evidence for a clade of nematodes, arthropods and other moulting animals. *Nature*, 387, 489–93.

Baguñà, J. (1998). Planarians. In *Cellular and Molecular Basis of Regeneration: From Invertebrates to Humans*, eds. P. Ferretti and J. Gèraudie, pp. 135–65. Chichester: John Wiley and Sons Ltd.

Bueno, D., Baguñà, J., and Romero, R. (1996). A central body region defined by a position-specific molecule in the planaria *Dugesia (Girardia) tigrina*. Spatial and temporal variations during regeneration. *Dev. Biol.*, 178, 446–58.

Bueno, D., Fernàndez-Rodríguez, J., Cardona, A., Hernàndez-Hernàndez, V., and Romero, R. (2002). A novel invertebrate trophic factor related to invertebrate neurotrophins is involved in planarian body regional survival and asexual reproduction. *Dev. Biol.*, 252, 188–201.

Fernald, R. (2000). Evolution of eyes. *Curr. Opin. Neurobiol.*, 10, 444–50.

Fire, A., Xu, S., Montgomery, M. K., Kostas, S. A., Driver, S. E., and Mello, C. C. (1998). Potent and specific genetic interference by double-stranded RNA in *Caenorhabditis elegans*. *Nature*, 391, 806–11.

Gehring, W. J., and Ikeo, K. (1999). Pax 6 mastering eye morphogenesis and eye evolution. *Trends Genet.*, 15, 371–7.

Hannon, G. J. (2002). RNA interference. *Nature*, 418, 244–51.

Hyman, L. H. (1951). *The Invertebrates. II. Platyhelminthes and Rhynchocoela. The Acoelomate Bilateria*. New York: McGraw-Hill.

Montgomery, M. K., Xu, S., and Fire, A. (1997). RNA as a target of double-stranded RNA-mediated genetic interference in *Caenorhabditis elegans*. *Proc. Natl. Acad. Sci. USA*, 95, 15502–7.

Orii, H., Mochii, M., and Watanabe, K. (2003). A simple "soaking method" for RNA interference in the planarian *Dugesia japonica*. *Dev. Genes Evol.*, 213, 138–41.

Pichaud, F., Treisman, J., and Desplan, C. (2001). Reinventing a common strategy for patterning the eye. *Cell*, 105, 9–12.

Pineda, D., González, J., Callaerts, P., Ikeo, K., Gehring, W. J., and Saló, E. (2000). Searching for the prototypic eye genetic network: *sine oculis* is essential for eye regeneration in planarians. *Proc. Natl. Acad. Sci. USA*, 97, 4525–9.

Pineda, D., González, J., Marsal, M., and Saló, E. (2001). Evolutionary conservation of the initial eye genetic pathway in planarians. *Belg. J. Zool.*, 131, 85–90.

Rieger, R., Tyler, M. S., Smith III, J. P. S., and Rieger, G. R. (1991). Platyhelminthes: Turbellaria. In *Microscopic Anatomy of Invertebrates*, eds. F. W. Harrison and B. J. Bogitsch, pp. 7–140. New York: Wiley-Liss.

Sánchez-Alvarado, A., and Newmark, P. A. (1999). Double-stranded RNA specifically disrupts gene expression during planarian regeneration. *Proc. Natl. Acad. Sci. USA*, 96, 5049–54.

Saló, E., Pineda, D., Marsal, M., González, J., Gremigni, V., and Batistoni, R. (2002). Eye genetic network in Platyhelminthes: Expression and functional analysis of some players during planarian regeneration. *Gene*, 287, 67–74.

Saló, E., and Baguñà, J. (2002). Regeneration in planarians and other worms. New findings, new tools and the steps ahead. *J. Exp. Zool.*, 292, 528–39.

Salvini-Plawen, L. V., and Mayr. E. (1977). On the evolution of photoreceptors and eyes. *Evol. Biol.*, 10, 207–63.

Zamore, P. D., Tuschl, T., Sharp, P. A., and Bertel, D. P. (2000). RNAi: double-stranded RNA directs the ATP-dependent cleavage of mRNA at 21 to 23 nucleotide intervals. *Cell*, 101, 25–33.

10 Microinjection of *Xenopus* embryos

R. J. Garriock and P. A. Krieg

INTRODUCTION

Microinjection of *Xenopus* embryos is an important technique with multiple applications in the fields of Cell Biology and Developmental Biology. Literally thousands of publications have resulted from use of these microinjection approaches. Fortunately, the equipment required for microinjection is inexpensive and compact and, due to the extremely large size of the *Xenopus* eggs and embryos, little practice is needed before the researcher becomes proficient with the technique. The purpose of this chapter is to provide a straightforward guide to microinjection methods. We will emphasize the most important factors when considering the equipment and materials required and we will describe procedures known to be reliable and efficient.

EQUIPMENT AND MATERIALS

INJECTION EQUIPMENT

Microscope. The technical specifications of a stereomicroscope suitable for microinjection are rather simple because the *Xenopus* embryos are large (about 1.2 mm across) and easily viewed under low magnification. The first concern is that the microscope has sufficiently good optics to be used for several hours (the length of a typical injection session) without causing eye strain. Second, the microscope must have a large working distance between the objective and the bench (at least 8–10 cm) to allow room for the microinjection apparatus, the injection dish, and the operator's hands. To maximize the working distance the stereomicroscope should be supported by a boom stand rather than a conventional raised base (Figure 10.1a). Use of a boom stand provides a large working distance and also facilitates movement of the injection apparatus and dishes of embryos. A halogen light source with flexible fiber-optic arms is nonintrusive and provides a cold light that does not heat up the media in the culture dish.

Micromanipulators and micromanipulator stand. The controls on the micromanipulator allow for precise positioning of the injection needle and also for delicate movements

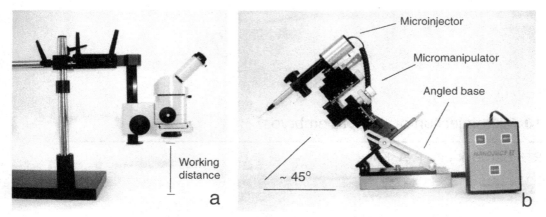

Figure 10.1. Microinjection apparatus. (a) Stereomicroscope and stand suitable for microinjection experiments. Note that the standard base beneath the objective lens has been replaced by a weighted base with an extension arm. (b) Microinjector supported on a micromanipulator and a movable weighted base. The angle of the micromanipulator is adjusted to approximately 45°. The control box for the micromanipulator is visible at the right of the weighted base.

during the injection procedure. For greatest control, the micromanipulator should position the injector at approximately a 45-degree angle to the injection dish (Figure 10.1b). The micromanipulator stand is a heavy base that holds the micromanipulator. Suitable micromanipulators and stands are manufactured by Narishige, Stoelting, and several other providers. An alternative to the locked angle models is the Singer micromanipulator which allows great flexibility of movement, but which is more difficult to use for the inexperienced worker.

Microinjector. *Xenopus* embryos are typically injected with volumes ranging from about 5 nl to 50 nl. Drummond manufactures an affordable and accurate microinjector suitable for *Xenopus* microinjection sold under the name Nanoject II. This injector uses a positive displacement mechanism and injects volumes ranging from 2.3 nl up to 69 nl in discrete increments. We find this range of volumes to be suitable for all of our microinjection requirements. Because the injection volume is controlled mechanically, no calibration of injection volumes for individual needles is required. For certain applications, however (for example injection of single cells in late-stage embryos), it may be desirable to inject volumes smaller than 2.3 nl. In this case, it will be necessary to use one of the gas pressure–regulated microinjectors that allow injection of extremely small volumes and permit continuous variation in injection volume. With these devices, it is necessary to calibrate the injection volumes for different needles and for different pressure conditions. Suitable devices are provided by Narashige, Picospritzer, Harvard Apparatus, Sutter, and others. Whichever type of microinjector you choose, it is convenient to use a foot-pedal injection control because this keeps your hands free and allows you to pay constant attention to the embryos.

Needle puller. The injection procedure requires fine glass needles prepared from glass capillary tubes using a needle puller. Needle pullers are available in either horizontal or vertical configurations. In general the horizontal needle pullers are used for preparation of exceptionally fine needles, for example those for mouse transgenesis methods or for patch clamping probes. These machines are expensive and the precision provided is much greater than that required for *Xenopus* microinjection needles. However, if such a machine is available, it can certainly be programmed to pull needles suitable for *Xenopus* microinjection. Although not as versatile as the horizontal models, vertical needle pullers are relatively inexpensive and produce needles perfectly suitable for microinjection of *Xenopus* embryos. We routinely use a vertical needle puller made by Narishige (Model #PC-10). Details of the needles themselves will be discussed in the next section.

INJECTION MATERIALS

Injection media. For injection, embryos are maintained in 0.4 × MMR solution supplemented with 4% Ficoll (Sigma). (1 × MMR is 0.1 M NaCl, 2 mM KCl, 1 mM $MgSO_4$, 2 mM KCl, 5 mM Hepes-OH pH 7.8.) Alternatively, several other Tris-based Amphibian culture media (all supplemented with 4% Ficoll) are also suitable for incubating embryos during injection (Sive, Grainger, and Harland, 2000). Ficoll is included in the injection buffer because it increases the viscosity of the medium and holds the fertilization envelope tightly against the membrane of the embryo. This helps to prevent leaking of the injected solution and blebbing of the embryos at the site of injection. Although many workers use injection media containing much higher salt concentrations, we find that these high-salt solutions do not offer any obvious advantages. On the other hand, use of low-salt concentrations effectively avoids the possibility of gastrulation defects (especially exogastrulation), which may occur with high salt media. Approximately 8–16 hours after injection (usually after overnight incubation) the embryos are transferred into 0.2 × MMR solution, without Ficoll, for the duration of development.

Injection dish. We microinject Xenopus embryos in a 60-mm plastic petri dish, the bottom of which has been lined with a circle of Whatmann 3MM filter paper. The paper has a soft but slightly rough surface that provides sufficient friction to prevent the embryos from sliding around. After filling the injection dish with injection media plus Ficoll, the filter paper should be pushed down to the bottom of the injection dish to secure it in place and any bubbles should be removed. An alternative to the filter paper is to use a disk of fine nylon mesh (e.g. Nitex 500 micrometer mesh, Tetko) at the bottom of the dish. In this case, the embryos settle into the holes in the mesh and this also prevents them from moving around freely.

Injection needles. For the Nanoject injector, all needles must be prepared from borosilicate glass capillary tubing provided by Drummond (90 mm in length, 1.5 mm outside diameter × 0.9 mm inside diameter). To make an injection needle, the glass capillary tube is heated until molten in the middle and is stretched by weights to create a very thin hourglass-shaped taper about 30 mm in length. We use a two-step protocol on a vertical needle puller to prepare injection needles. First, the capillary tube is fastened on the needle puller, the heating element is adjusted to a setting of 90.6 and the needle is pulled through a distance of about 10 mm. This produces a thin tapered region in

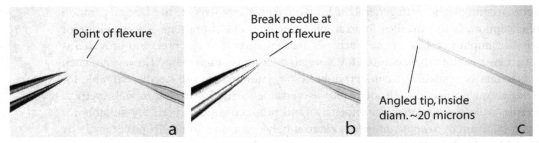

Figure 10.2. Preparation of an injection needle. (a) The finely drawn-out capillary is bent using fine forceps to identify the point of flexure. (b) The forceps are used to very gently break the capillary at an angle at approximately the point of flexure. (c) Magnified view of needle tip, showing the angled break. This needle has an inside diameter of approximately 20 microns.

the middle of the capillary. Second, the heating element moves to the middle of the taper and the second pull occurs (also at 90.6) to create a very finely tapered needle. In general the bottom needle has a better shape and the top needle is discarded. We make quite large numbers of needles at one time and store the extras on plasticine (modeling clay) supports in any convenient container. Just prior to use, the beveled end of the needle is generated by breaking the needle at an angle using a pair of very fine forceps (e.g., Dumont number 5 – Figure 10.2). We determine the approximate position where the needle resists bending (the point of flexure) and break it at this point. The inside diameter at the tip of the needle should be about 10–30 microns (measured using a stage micrometer) although quite a range of diameters will still yield acceptable results.

PROCEDURES
The Microinjection protocol
LOADING AND TESTING THE INJECTION NEEDLES

1. For the Nanoject injector, the injection needle must first be completely filled with mineral oil (e.g. Sigma mineral oil) to act as a noncompressible displacement medium. Mineral oil is loaded into the blunt end of the injection needle using a 26 gauge Hamilton syringe, making sure not to introduce any air bubbles.
2. The oil-filled needle is now attached to the Nanoject injector following the manufacturer's instructions.
3. To test the needle, place a few microliters of water (plus any visible dye) onto the sterile side of a piece of Parafilm and then draw the liquid into the needle using the "fill" setting on the injector. It is quite easy to see the water moving into the needle by observing the oil/water interface. If the water does not draw steadily into the needle the tip of the needle is probably too fine and will need to be retrimmed.
4. The Drummond Nanoject injectors use manual controls that adjust the volume of liquid to be expelled. To test for injection, raise the tip of the needle into the air, press the injection button, and observe the small bolus of liquid that is expelled from the end of the needle.
5. Alternatively, the needle tip can be placed below the surface of a drop of oil before pressing the inject button. In this case, a tiny droplet of water will form in the oil and detach from the needle tip. By moving the needle slightly between injections,

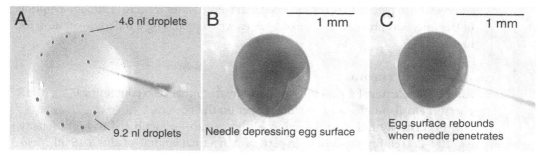

Figure 10.3. Testing and use of an injection needle. (a) Injection of dye solution beneath the surface of a drop of mineral oil, illustrating consistent injection volumes. (b) Injection needle contacting a one-cell embryo at approximately a right angle and depressing the surface. (c) Once the needle has penetrated, the embryo and vitelline membrane revert to their normal shape.

a series of identical water droplets will form under the oil, indicating that constant volumes of liquid are being expelled (Figure 10.3a).

6. Solutions to be injected into embryos should be deposited onto clean Parafilm and taken up into the injection needle exactly as described for water in step 3 above.

7. This should only be done immediately before the embryos are ready to be injected (otherwise the solution may dry out and block the tip of the needle).

INJECTION OF EMBRYOS. *Xenopus* embryos can be injected at any time from the one-cell stage, after cortical rotation, until about the 64-cell stage.

1. After this time, injection of a volume of 2.3 nl (the minimum volume of the Nanoject II injector) into a single cell will damage the cell. In the case of the smaller *Xenopus tropicalis* embryos it is not possible to inject 2.3 nl volumes later than the 16-cell stage.

2. After loading the needle with the solution to be injected, place a group of embryos (about 25–50 embryos) in injection medium under the microscope and position the microinjector (angled at about 45°) ready for injection.

3. The needle should be positioned just above the group of embryos so that the tip of the needle and the embryos are in the same focal plane under the microscope.

4. Move the dish to position the needle over the first embryo and now use the micromanipulator to push the needle directly into the embryo. The embryo usually compresses a little under the pressure of the needle before the needle suddenly penetrates (Figure 10.3b, c). It is important that the needle should contact the embryo more or less perpendicularly, otherwise the embryo will roll in response to the pressure and the needle will not cleanly pierce the membrane.

5. As soon as the needle penetrates the embryo, press the "Inject" button (or use the foot pedal). The needle does not need to remain inside the embryo for more than a second. The time it takes to inject the embryos is limited only by the time required to align the next embryo and inject it. After some practice it should be possible to comfortably inject up to one hundred embryos in 5 minutes.

6. Remember that the tip of the needle is extremely fine and the contents can rapidly dry out and block the needle if it is left exposed to the air. Therefore, we find it

convenient to place the tip of the needle in sterile water if there is any interruption to the injection procedure, for example, as when preparing a different solution for injection.

CARE OF INJECTED EMBRYOS

1. After injection, transfer the embryos into a fresh culture dish containing 0.4 × MMR plus 4% Ficoll, but without the filter paper on the bottom.

2. For the first day following injection, embryos can be incubated at high density, e.g. 100–200 embryos per 60-mm petri dish in a total volume of about 15 ml. *Xenopus laevis* embryos are normally maintained at 22°C (Nieuwkoop and Faber, 1994) and *tropicalis* embryos at 25°C. We recommend that injected *laevis* embryos be cultured at a temperature between 15 and 18°C and *tropicalis* at 21–23°C. These slightly lower temperatures slow the rate of development and also the decomposition of any embryos that may be dead or dying. Antibiotics may also be added to the culture medium as an extra precaution against bacterial contamination (e.g. Gentamycin at 50 micrograms/ml and Streptomycin at 100 micrograms/ml), but the major factor effecting embryo health is good husbandry and thus the embryos should receive daily attention.

3. After approximately stage 8–9, embryos may be transferred into a 0.2 × MMR solution, without Ficoll, for further culture.

4. At this time it is also important to decrease the number of embryos per dish or to transfer them to larger culturing dishes. A 60-mm dish should contain no more than 50 embryos and a 90-mm dish no more than about 200 embryos.

5. As the embryos develop to neurula stages it is necessary to further reduce the density of embryos (we suggest no more than 100 embryos in a 90-mm dish).

6. Injected embryos frequently show the presence of blebs of tissue attached to their bodies at the original site of injection. These blebs are not a viability concern, but they detract from the appearance of the embryo. If desired, the blebs can be removed using fine forceps. *Xenopus* embryos are extremely robust and small wounds caused during removal of blebs will quickly heal.

Injection of dyes, mRNA, and oligonucleotides

GENERAL CONSIDERATIONS. When multiple different experimental solutions are to be used in an injection session it is necessary to rinse the needle with water between each injection solution. We usually expel all of the old solution into a drop of water being careful not to expel the mineral oil. Rinse the needle by drawing up about 2 mm of sterile water and then expelling it again. This should be repeated several times with changes of water to ensure that there is no cross contamination of experimental solutions.

The cleavage pattern of the *Xenopus* embryo is highly reproducible and fate maps are available showing the development of each of the different blastomeres (up to the 32 cell stage) in the embryo (Moody, 1987; Dale and Slack, 1987; and available at www. xenbase.org). By targeting specific blastomeres for injection it is possible to direct the injected material preferentially to a particular tissue, or region within the embryo.

INJECTION OF DYES FOR LINEAGE TRACING. Fluorescent dextrans are routinely used as lineage tracing dyes in the *Xenopus* embryo. We prefer to use 10,000 MW Lysine fixable dyes because these can be visualized following in situ hybridization and immunocytochemistry detection protocols. Lysine-fixable Rhodamine dextran (Molecular Probes) and Lysine-fixable Texas Red dextran (Molecular Probes) are amongst the most useful since the wavelength of emission of the fluor is distinct from the yolk autofluorescence. Injection of 4.6 or 9.2 nl of a 1% (w/w) dextran solution into *Xenopus* embryos is sufficient to provide a bright signal that can be easily visualized throughout embryonic development. Although dextrans make excellent lineage tracers, there are two reasons why they are not ideal for lineage tracing of injected mRNA. First, fluorescent dextrans are highly diffusible and will rapidly distribute throughout the injected blastomere, and sometimes into the adjacent blastomere if cleavage is not complete. On the other hand, injected mRNAs tend to be rather nondiffusible and they stay localized fairly near the site of injection. Therefore, the distribution of the fluorescent dextran may not accurately represent the distribution of the injected mRNA (see the next section for the use of mRNAs encoding green fluorescent protein or beta-galactosidase as lineage tracers). Second, fluorescent dextrans are not necessarily free of ribonucleases and we have seen cases where the dyes caused rapid degradation of the synthetic mRNA.

INJECTION OF SYNTHETIC mRNA. Injection of synthetic mRNAs is one of the most commonly used techniques in *Xenopus* developmental biology. High efficiency transcription kits are commercially available (e.g. Ambion, Roche, Epicentre) but the reaction can also be carried out using standard laboratory reagents (Krieg and Melton, 1987). Synthetic mRNAs may be stored as ethanol precipitates at $-20°C$ or dissolved in water and stored at $-80°C$.

The amount of mRNA to be injected depends on the specific experimental objectives and also on the nature and bioactivity of the encoded protein.

1. Typically mRNA is injected in amounts ranging from about 20 pg to 2 ng per embryo.
2. The concentration and integrity of the mRNA should be checked by electrophoresis of an aliquot of the mRNA on a denaturing agarose gel followed by staining with ethidium bromide (Sambrook and Russell, 2000).
3. Although spectrophotometric analysis is useful to quantify mRNA yield, it is important to examine the quality of the mRNA on a gel to ensure that it is not degraded.
4. Since the Nanoject apparatus injects only certain specific volumes, injection of a specific amount of mRNA is achieved by altering the concentration of the mRNA solution. We find a convenient way to calculate the concentration of mRNA to inject is to dissolve 1000 times more mRNA in 1000 times more volume of water than will be introduced per injection. For example, if you wish to inject 100 pg of mRNA per embryo in a volume of 4.6 nl, you will dissolve 100 ng of mRNA (1000 × amount) in 4.6 microliters of water (1000 × volume). For injection of 200 pg of mRNA it is only necessary to double the injection volume to 9.2 nl, however for

larger amounts it is probably wise to prepare another mRNA solution at higher concentration.

5. The mRNA solutions can be stored on ice for a short period (several hours) but should be stored at $-20°C$ for longer periods of time in order to reduce the possibility of degradation.

For most purposes it is useful to inject a mixture of the experimental mRNA plus a second mRNA encoding a reporter molecule (e.g. green fluorescent protein (GFP) or beta-galactosidase). This serves two purposes. First, detection of the reporter protein serves as a control that the co-injected mRNAs did not degrade during the injection procedure. Second the reporter protein acts as a lineage tracer that accurately indicates the location of the injected experimental mRNA. GFP is a very sensitive reporter that has the advantage of being visible in the living embryo when viewed by UV fluorescence. Injection of 100–500 pg of GFP mRNA will indicate the location of the cells receiving the injected mRNA at least until stage 45 of development. Alternatively, GFP can also be detected by immunocytochemistry in fixed embryos using specific antibodies. Beta-galactosidase protein must be detected in fixed embryos using a color reaction. Several different colorimetric substrates are available and these produce a number of different colored stains (e.g., Gibco, Roche, Biosynth). Most of these stains remain stable during in situ hybridization or immunohistological assays and so act as permanent markers of the location of the injected material.

INJECTION OF ANTISENSE OLIGONUCLEOTIDES. Injection of antisense oligonucleotides or morpholino modified nucleotides (Gene Tools, LLC) is now commonly used to interfere with gene expression in developing *Xenopus* embryos (Dagle et al., 2000; Heasman et al., 2000). In general, the oligonucleotides used for embryo injections are chemically modified to increase their stability. In the case of oligonucleotides, formation of sequence specific hybrids with endogenous mRNAs causes the mRNA strand to become a target for RNase H–mediated endonuclease activity and the target mRNA is degraded. On the other hand, morpholino oligonucleotides bind specifically to target transcripts and block translation of the mRNA but the transcript is not degraded (Nasevicius and Ekker, 2000; Heasman et al., 2000). Among the advantages of morpholinos are that very large molar quantities of the antisense oligonucleotide can be injected with very low toxicity and that the oligonucleotides are very stable in the embryo. Oligonucleotides or morpholinos are injected exactly as described previously for synthetic mRNA.

1. In the case of morpholinos, it is useful to resuspend the oligonucleotide stock in a solution of 50 mM Hepes-OH, at pH 8.0, because the solution may become highly basic.
2. For morpholinos it is usually necessary to carry out a dose curve (about 5 to 50 ng) and then select the dose that gives a clear phenotype without nonspecific toxicity.
3. As with injection of synthetic mRNAs it is helpful to coinject a lineage tracer together with the antisense oligo or morpholino in order to determine the location of the injected material. Finally, it is important to be aware that antisense reagents, including morpholinos, have the potential to produce both false positive and false negative results, and so careful attention to controls is essential (Heasman, 2002).

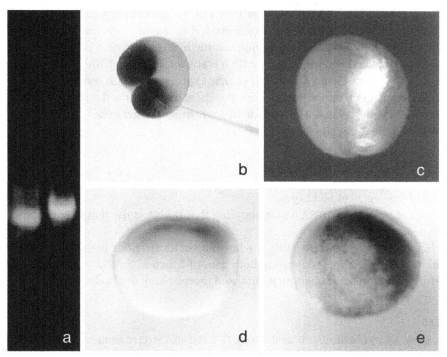

Figure 10.4. Overexpression of *myocardin* mRNA in the *Xenopus* embryo. (a) Agarose gel showing full-length synthetic mRNAs encoding GFP (left) and myocardin (right). (b) A mixture of mRNAs encoding GFP (125 pg) and myocardin (125 pg) is injected into one cell of the embryo at the 2-cell stage. (c) Dorsal view of a GFP-injected neurula stage embryo visualized under UV light. Fluorescence indicates that the right side of the embryo received the mRNA injection. (d) Lateral view of neurula embryo (anterior at right) injected with *GFP* mRNA alone and assayed by in situ hybridization for the muscle marker *cardiac α-actin*. Expression of *cardiac α-actin* is limited to the developing somites. (e) Lateral view of neurula embryo (anterior at right) injected with both *GFP* and *myocardin* mRNAs. Note ectopic expression of the *cardiac α-actin* muscle differentiation marker. Data provided by Eric Small (unpublished).

ALTERNATIVE EXERCISE

To illustrate the use of the microinjection method, we will present an experiment in which the potent cardiac and smooth muscle transcription factor, myocardin (Wang et al. 2001), is overexpressed in the *Xenopus* embryo. Figure 10.4a shows in vitro synthesized mRNAs encoding *GFP* and myocardin fractionated on an agarose gel and visualized using ethidium bromide. Note that both of the mRNA bands are clean and show no incomplete transcripts or degradation products. A mixture of both *GFP* mRNA (lineage tracer) and *myocardin* mRNA (experimental transcript) was injected into a single cell of the two-stage embryo as illustrated in Figure 10.4b. During subsequent development, the injected side of the embryo can easily be distinguished by the presence of GFP fluorescence when viewed under UV illumination (Figure 10.4c). As stated above, myocardin is a potent transcription factor expressed in both cardiac muscle and smooth muscle tissues. In this experiment, we have assayed the injected *Xenopus* embryos to determine whether expression of *myocardin* is sufficient to activate the ectopic expression of the muscle marker, *cardiac α-actin*. Figure 10.4d shows a neurula stage control embryo that has been injected with *GFP* mRNA, but no *myocardin* mRNA, and then assayed for *cardiac α-actin* expression using in situ hybridization. Note that expression of *cardiac α-actin* is limited to the developing somitic muscle at this stage of development. An embryo injected with *myocardin* mRNA is shown in Figure 10.4e. In this case, overexpression of *myocardin* has resulted in ectopic expression of high levels of *cardiac α-actin* transcript in lateral regions of the embryo. These experiments demonstrate that myocardin is sufficient to activate expression of downstream target genes precociously and ectopically in the *Xenopus* embryo.

TEACHING CONCEPTS

➤ Handling of frog eggs and embryos.
➤ Preparation of synthetic mRNA. Understanding that mRNA is the template for protein synthesis.
➤ Simple lineage tracing demonstrating the fate map of a vertebrate embryo.
➤ Manipulation of gene expression using both gain-of-function (mRNA injection) and loss-of-function (antisense oligonucleotide and morpholino) methods.

REFERENCES

Dagle, J. M., Littig, J. L., Sutherland, L. B., and Weeks, D. L. (2000). Targeted elimination of zygotic messages in *Xenopus laevis* embryos by modified oligonucleotides possessing terminal cationic linkages. *Nucleic Acids Res.*, 28, 2153–7.

Dale, L., and Slack, J. M. (1987). Fate map for the 32-cell stage of *Xenopus laevis*. *Development*, 99, 527–51.

Heasman, J. (2002). Morpholino oligos: Making sense of antisense? *Dev. Biol.*, 243, 209–14.

Heasman, J., Kofron, M., and Wylie, C. (2000). Beta-catenin signaling activity dissected in the early *Xenopus* embryo: A novel antisense approach. *Dev. Biol.*, 222, 124–34.

Krieg, P. A., and Melton, D. A. (1987). In vitro RNA synthesis with SP6 RNA polymerase. *Methods Enzymol.*, 155, 397–415.

Moody, S. A. (1987). Fates of the blastomeres of the 32-cell-stage *Xenopus* embryo. *Dev. Biol.*, 122, 300–19.

Nasevicius, A., and Ekker, S. C. (2000). Effective targeted gene "knockdown" in zebrafish. *Nat. Genet.*, 26, 216–20.

Nieuwkoop, P. D., and Faber, J. (1994). *Normal Table of Xenopus laevis (Daudin)*, 2nd ed. New York: Garland Publishing.

Sambrook, J., and Russell, D. (2000). *Molecular Cloning: A Laboratory Manual.* 3rd ed. New York: Cold Spring Harbor Laboratory Press.

Sive, H., Grainger, R. M., and Harland, R. M. (2000). *Early Development of Xenopus Laevis: A Laboratory Manual.* New York: Cold Spring Harbor Laboratory Press.

Wang, D.-Z., Chang, P. S., Wang, Z., Sutherland, L., Richardson, J. A., Small, E., Krieg, P. A., and Olson, E. N. (2001). Activation of cardiac gene expression by myocardin, a transcriptional cofactor for serum response factor. *Cell*, 105, 851–62.

SECTION V. GENETIC ANALYSIS

11 Segmental specification in *Drosophila melanogaster*

L. De Navas, M. Suzanne, D. Foronda,
and E. Sánchez-Herrero

OBJECTIVE OF THE EXPERIMENT *Drosophila melanogaster*, the fruitfly, is a model organism for the study of multiple biological problems. Likewise, it has served as a teaching instrument to illustrate basic concepts of genetics. In fact, many fundamental genetic principles were discovered in *Drosophila*.

As is true for all insects, *Drosophila* is a segmented organism, with well-differentiated cephalic, thoracic, and abdominal regions. The segmental specification of these different domains depends on the activity of a group of genes called homeotic or Hox genes, which are distinctively deployed along the antero–posterior axis. These genes also determine the antero–posterior axis in most animal species, including humans. In *Drosophila*, mutations in the Hox genes cause one segment (or part of it) to substitute for another one. Many of these transformations are easily observed in the embryonic or adult cuticle.

The objective of the experiments detailed in this chapter is to study the segmental transformations caused by homeotic mutations in *Drosophila* and to carry out a complementation analysis between several homeotic mutations. To this aim, two types of experiments are described: (1) A phenotypical study of different homeotic mutations in the adult cuticle and a complementation analysis. (2) A study of the phenotype of homeotic mutations in the embryonic cuticle.

DEGREE OF DIFFICULTY Experiment 1 requires some experience in the handling of the flies, distinguishing males from females, recognising the dominant mutations of the balancer chromosomes, and identifying the homeotic transformations, but this is easily acquired. Experiment 2 is more demanding technically, because it requires the preparation of the embryonic cuticle, more material, and a microscope equipped with dark-field or phase contrast optics.

INTRODUCTION

In the second decade of the past century, T. H. Morgan and his collaborators began to establish the mechanisms of Mendelian heredity using *Drosophila melanogaster* as a

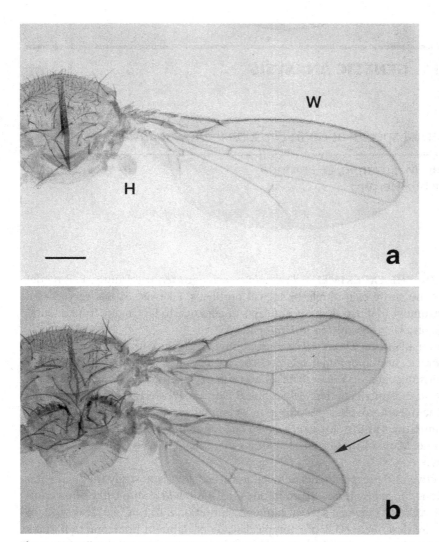

Figure 11.1. Effect in the adult cuticle of mutations in the *Ubx* regulatory region. (a) A wild-type thorax, showing the right wing (W; in the second thoracic segment) and the right haltere (H; in the third thoracic segment). (b) Transformation of the third thoracic segment into the second one. Note the partial duplication of the notum (central part of the thorax) and the transformation of the haltere into wing (arrow). The mutant shown is a triple mutant combination in *Ubx* regulatory sequences that almost eliminates *Ubx* function in the third thoracic segment. The bar in (a) indicates 0.5 mm. Adapted from Investigación y Ciencia (1995), 222, 48–56.

model organism. Among the mutations discovered by his group was one, called *bithorax* (*bx*), that partially transformed one segment into another. The study of this homeotic mutation, and of similar mutations discovered later on, has led to crucial discoveries in the field of developmental biology (McGinnis and Kuziora, 1994).

Homeotic mutations in *Drosophila* transform, for instance, antennae into legs, or halteres into wings (halteres are little structures that develop, instead of wings, in the third thoracic segment; Figure 11.1). In the 1940s Ed Lewis began his studies on mutations in

Drosophila Hox genes, showing that many of them were clustered in a genetic complex (which he called the Bithorax complex). He also proposed a model of how Hox genes work to specify thorax and abdomen. In this model, beginning with the third thoracic segment and finishing with the eighth abdominal one, different genes of the complex were sequentially activated in each segment. Thus, in the third thoracic segment the first Hox gene was activated, while in the first abdominal segment this same gene and the second one were activated, and so on, down to the eighth abdominal segment, in which all the genes of the complex were transcribed (Lewis, 1978). Ed Lewis received the Nobel Prize in 1995 for his work on the Bithorax complex.

Lewis' model was modified in 1985, after a detailed genetic and developmental analysis of new mutations induced in the Bithorax complex demonstrated that there are only three genes in this complex. How could only three genes determine the morphology of nine segments (one thoracic and eight abdominal ones)? Three features account for the distinct morphology of each segment (see also Chapter 20): the different levels of protein products of each of the three genes, their combinatorial activity, and their particular spatial and temporal distribution in the different metameres. The three genes are called *Ultrabithorax* (*Ubx*), *abdominal-A* (*abd-A*), and *Abdominal-B* (*Abd-B*). *Ubx* specifies the third thoracic segment (T3) and the first abdominal one (A1), *abd-A*, specifies abdominal segments A2–A4, and *Abd-B* specifies abdominal segments A5–A8 and the genitalia (Sánchez-Herrero et al., 1985; Tiong, Bone, and Whittle, 1985; Figure 11.2)

In 1984 it was discovered, also in *Drosophila*, that homeotic genes (and, as was later found, many other genes involved in development) shared a conserved sequence of 180 bp. This sequence was called the "homeobox" (Gehring, 1985). At the same time that genetic analyses showed the existence of three genes in the Bithorax complex, each gene was also found to contain a homeobox. The homeobox is conserved in the Hox genes of all the metazoans, and has allowed the isolation of these genes from other organisms, such as mouse and human. As in *Drosophila*, Hox genes determine the antero–posterior axis in vertebrates and are clustered in genetic complexes. Moreover, mutations in mouse Hox genes also cause transformations in this axis (De Robertis, Oliver, and Wright, 1990). *Drosophila*, therefore, serve as a model for the study of Hox genes in different organisms.

MATERIALS AND METHODS

EQUIPMENT AND MATERIALS

The equipment necessary for handling the flies is described in Chapter 14.

In addition, the preparation of the embryonic cuticles requires the following:

Per Student

Slides: Standard size.

Coverslips: Different sizes may be used, preferably 24 × 32 mm.

Scintillation counter vials or similar glass vials.

Egg-laying cage: This apparatus can be made with a 52-mm diameter cylindrical plastic tube. Cover one end with a mesh fine enough to prevent flies from

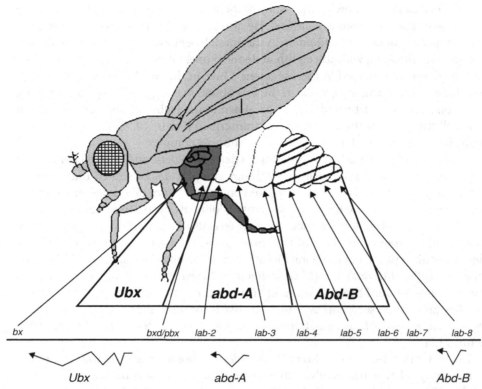

Figure 11.2. Scheme of the Bithorax complex showing the three genes (*Ubx, abd-A* and *Abd-B*) and the regulatory sequences (above the line representing the DNA). The region expressing *Ubx* is shown in dark grey, that expressing *abd-A* is shown in white, and the region expressing *Abd-B* is hatched.

escaping. After transferring the flies to the tube, the open end is covered with a plastic petri dish filled with juice–agar food (see Appendix), which precisely fits the open end and is fixed to the tube with tape (Figure 11.3a). The flies will lay their eggs on this plate, and the tape and juice-agar plate can then be easily removed to allow transferring the flies to a bottle or replacing the plate with a new one for continued egg collection.

Funnel: To help in transferring flies from the bottles or vials to the egg-laying cage and vice versa.

Bottle with water: Plastic bottle with an opening to squirt water over the petri dish and drag the eggs to a little basket.

A small, fine-haired paint brush

A small beaker

Pasteur pipettes

Microcentrifuge tubes (Eppendorf)

Baskets: To collect the embryos after they are removed from the food. The baskets can be made as follows: (a) cut off 50-ml Falcon tubes (or any similar plastic tube)

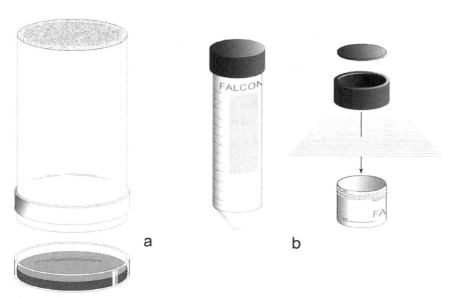

Figure 11.3. Drawing showing how to make (a) an egg-laying cage, and (b) a basket to collect embryos.

about one third of their length from the top; (b) cut off the centre of the screw cap, forming an open ring; (c) cover the open top of the tube with a thin mesh; (d) screw back the cap holding the mesh in place, thus forming a basket (Figure 11.3b).

Disposable tips: Two sizes: 0 to 200 μl or 200 μl to 1 ml.

Paper towels

Gilson pipettes

Small petri dishes

Microscope equipped with phase contrast and dark field optics.

Per practical group. To prepare the larval cuticle an incubator that can be set at 65°C is required. TO WORK WITH METHANOL AND HEPTANE THE USE OF A FUME HOOD IS RECOMMENDED!

Biological material

FLY STRAINS. The *Drosophila melanogaster* stocks used are (the slash separates homologous chromosomes and the semicolon non-homologous chromosomes):

a   For the phenotypical and complementation analysis of mutations:

$Ubx^1/TM6B$ $AbdB^{M5}/TM3$

$Ubx^{9.22}/TM6B$ $iab-7^{Sz}/TM3$

$abdA^{M1}/TM6B$ $bx^3/TM6B$

b   For the analysis of the embryonic cuticle:

$Ubx^1/TM6B$

If experience in working with flies is lacking, it may be necessary to have a wild-type stock (Oregon-R, for example) to compare with the mutants and to identify the mutations present in the balancer chromosomes.

The homeotic mutations Ubx^1, $Ubx^{9.22}$, $abdA^{M1}$, $AbdB^{M5}$, bx^3, and $iab-7^{Sz}$ are kept over balancer chromosomes (*TM6B* or *TM3*). Each balancer chromosome is homozygous lethal and carries several inversions that prevent recombination, thus maintaining the homeotic mutation. The balancer chromosomes are recognised by one or more dominant mutations (see below).

STOCK REQUEST. Fly stocks can be requested from the authors or from international stock centers.

REAGENTS

Bleach (undiluted)	Strawberry juice
Agar (Merck)	Chloral hydrate (Merck)
Glucose (Merck)	Methanol (Merck)
Arabic gum (Merck)	Heptane (Merck)

The rest of reagents are shown in Chapter 14.

PREVIOUS TASKS FOR STAFF

Maintenance of fly strains and preparation of fly food. Described in Chapter 14.

Solutions. The solutions needed to preserve and mount the halteres are identical to those used for mounting wings as described in Chapter 14.

For the analysis of the embryonic cuticle:

Bleach (undiluted)

Hoyer's medium: This is used to observe the embryonic cuticle (it can also be used for the adult cuticle). The procedure to make Hoyer's medium is the following:

(a) Add 30 g of Arabic gum to 50 ml of distilled water, and stir the solution overnight until the Arabic gum is completely dissolved.

(b) Understirring, add 200 g of chloral hydrate in small amounts, to prevent the formation of lumps (see Chapter 22 for hazards prevention).

(c) Add 20 g of glycerol.

(d) Centrifuge the mixture until it clarifies and has no particles on it (3 h to overnight at 12,000 g).

PROCEDURES

Distinguishing males from females. Basic to the experiments proposed here is distinguishing *Drosophila* males from females. Males have six abdominal segments, of which the fifth and sixth show black pigmentation in the whole segment. By contrast, females have seven abdominal segments and none has complete black pigmentation. Newly eclosed males are barely pigmented, so care must be taken not to mistake them for females. Male and female genitalia (observed ventrally at the end of the abdomen) are also notably different: male genitalia have many more structures whereas female genitalia have only two vaginal plates with a row of small modified bristles each. Finally, males have a sex comb, a small structure formed by a row of black thick bristles in the distal region (about one-fourth of the length of the leg from the tip) of the first pair of legs.

Virgin collection. To collect virgins: the stocks are kept in vials with fly food (at least two vials for each stock), and the flies transferred from one vial to a new one every 5–7 days. It is advisable to write on the vial the genotype of the flies and the date. When flies from one of the vials are eclosing, hatched flies are discarded the first thing in the morning and females are then collected from this vial in the evening (not more than 8–10 hours later). In this way, the females are guaranteed to be virgins since they do not mate until 8–10 hours after eclosion. If necessary, virgins can be collected from the same vials the following day, early in the morning. In this case, only females with a light body colour should be selected, since they are more likely to be virgins.

Protocol for the preparation of the embryonic cuticle

1. Place 80–100 flies from the $Ubx^1/TM6B$ stock in an egg-laying cage with a juice–agar plate for egg laying. Tape the plate to the tube so that flies cannot escape. Leave the egg-laying cage at 25°C.
2. After 6–8 hours at 25°C remove the petri dish, transfer the flies into a bottle or vial with food (use the funnel), put a new plate with food in the egg-laying cage, and leave the old plate with the eggs at 25°C for about 11–14 h.
3. Squirt some water in the plate to remove the eggs, dragging them with a paintbrush, and dropping them into a basket. The eggs are cleaned by adding some more water to the basket.
4. Put the basket into a small beaker with bleach covering about half of the basket, for 2 minutes. Shake the basket gently and rinse the embryos with bleach using a Pasteur pipette to remove the chorion (external cover) from the eggs. The chorion has been removed when the embryos float together on the surface.
5. Take the basket out of the beaker and wash the embryos with water from the squirt bottle to remove the bleach left. Dry the bottom of the basket briefly with a paper towel.
6. Transfer the embryos with a brush from the basket to a glass vial (scintillation vial or similar) filled with 3 ml of heptane and 3 ml of methanol (they form two layers, with methanol at the bottom). The embryos will remain at the interphase. Methanol and heptane are toxic. Use plastic or rubber gloves when working with them and be careful with methanol and heptane fumes. WORK IN THE FUME HOOD!
7. Once the embryos are set in the interphase of the two liquids, close the vial and shake it vigorously for 30 seconds. This removes the vitelline membrane (internal cover) from the embryos and most of them settle at the bottom of the vial (methanol layer).

8. Take the embryos from the bottom with a Pasteur pipette and transfer them to a microcentrifuge tube. Be careful not to take heptane (upper layer) with the embryos.

9. Wash the embryos four times with methanol.

10. Take the embryos with a disposable 0 to 200 μl tip (cut off at the tip) and put them gently over a cleaned glass slide.

11. Tip the slide against a paper towel (without touching the embryos) to remove the excess methanol, and allow most of the methanol evaporate. Be careful not to let the embryos dry completely. Add 1 or 2 drops of Hoyer's medium, enough to cover the embryos without spreading out to the edges of the slide.

12. Put a coverslip gently over the embryos and leave the slide at 65°C overnight. The cuticle of the embryos can be observed the following day in a microscope equipped with dark field or phase contrast.

OUTLINE OF THE EXPERIMENTS

EXPERIMENT 1A. STUDY OF THE PHENOTYPE OF DIFFERENT MUTATIONS AND COMPLEMENTATION ANALYSIS

The aim of the experiment is to carry out a complementation analysis between different homeotic mutations of *Drosophila* and to analyse the phenotype caused by mutations in regulatory regions of these same Hox genes. If two mutations *in trans* (that is, one on each homologous chromosome) do not show any visible phenotype they are said to complement, whereas if they show a phenotype they fail to complement. Mutations in the coding regions of each of the three Hox genes of the Bithorax complex, *Ubx, abd-A,* and *Abd-B*, complement each other. Mutations in the regulatory regions of any of the three genes fail to complement mutations in the coding region of its corresponding gene. These regulatory mutations control the expression of the homeotic gene in a certain segment or in part of it. For instance, mutations in a regulatory region of the *Ubx* gene, such as the *bx* mutations, fail to complement *Ubx* mutations. Flies of the genotype *bx/Ubx* show a transformation of the anterior part of the third thoracic segment into the homologous region of the second one (see below). The genetic combination "mutation in a regulatory region/mutation in the coding region" gives a phenotype in the segment controlled by the regulatory region, since one of the chromosomes does not synthesise the product of the gene (mutation in the coding region), and the other does not express the protein in a certain segment (mutation in the regulatory region). Some of the mutations in these regulatory regions do not affect the whole segment but its anterior or posterior part (compartment; see Chapter 14). The *bx* mutations are part of this group.

To carry out the experiment, females (or males) from stocks carrying mutations in coding regions of homeotic genes are crossed to males (or females) carrying mutations in regulatory regions of these same genes, or to mutations in other homeotic genes. In the complementation analysis, mutations in the three Hox genes of the bithorax complex, *Ubx, abd-A,* and *Abd-B*, are used. To study the transformation caused by mutations in regulatory regions we use a *bx* mutation, which regulates *Ubx* expression in part of the third thoracic segment, and an *infraabdominal-7* (*iab-7*) mutation, which controls *Abd-B* expression in the seventh abdominal segment.

The mutations are "balanced," that is, the mutant chromosome is kept over a balancer chromosome (see above). The dominant mutations of the balancer chromosomes allow one to distinguish, in the progeny of the cross, the flies that carry the balancer chromosomes from those *trans*-heterozygous for the homeotic mutations under study. Therefore, the general scheme of the crosses is the following:

$$\text{mutation A/Balancer 1} \times \text{mutation B/Balancer 2}$$

From these crosses, the progeny could be of the following genotypes:

1. *mutation A/Balancer 2*
2. *mutation B/Balancer 1*
3. *Balancer 1/Balancer 2*
4. *mutation A/mutation B.*

For this experiment, the phenotype of the fourth class of progeny is the one of interest.

EXPERIMENT 1B. PROTOCOL FOR THE COMPLEMENTATION ANALYSIS

1. Cross 5–7 males of the *mutation A/Balancer* genotype to 7–10 virgin females of the *mutation B/Balancer* genotype in vials with fly food. It is important to be sure that the females are virgins (see above). Otherwise, females can store sperm from their brothers in the stock and lay eggs with a genotype different from that intended, even after being mated with males of the correct genotype.
2. Maintain the cross at 25°C, transferring the parents to a new vial after 2–3 days.
3. Observe the phenotype of the progeny after 10–13 days for the presence of *mutation A/mutation B* individuals and, if they appear, study their phenotype.

The crosses and the progeny expected for the different crosses are as follows (either males or females from each stock can be selected for the crosses):

➤ *Ubx^1/TM6B* × *Ubx$^{9.22}$/TM6B*. The *TM6B* balancer chromosome can be recognised (in one copy; two copies are lethal) because the flies are somewhat broader ("tubby;" best observed in the thorax). *TM6B* larvae and pupae are also broader. Besides, *TM6B* shows, in the upper left and right parts of the thorax, called the humerus (as if they were the "shoulders" of the fly), a group of long bristles (3–5) instead of only two in the wild-type (Figure 11.4a). The progeny of interest are the flies that do not carry the *TM6B* chromosome; they will be of the *Ubx^1/Ubx$^{9.22}$* genotype.

➤ *Ubx^1/TM6B* × *abdA^{M1}/TM6B*. Look for progeny that are not *TM6B* (flies of normal size, without extra bristles in the humerus). These flies are of the *Ubx^1/abdA^{M1}* genotype.

➤ *abdA^{M1}/TM6B* × *AbdB^{M5}/TM3*. The *TM3* balancer has the mutation *Stubble*, which results in short bristles. Look for progeny that are neither *TM3* nor *TM6B*, that is, of normal size, without extra bristles in the humerus, and with bristles of normal size. These flies are of the *abdA^{M1}/AbdB^{M5}* genotype.

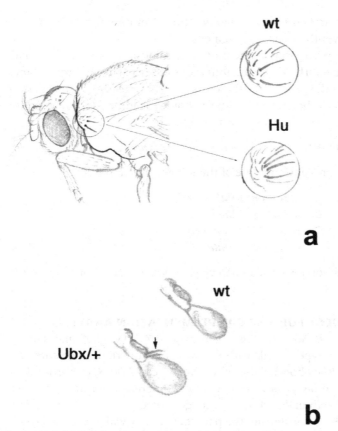

Figure 11.4. (a) Drawing showing the mutation *Humeral* (*Hu*), a dominant marker of the *TM6B* balancer chromosome (more bristles in the humerus, dorsal lateral region of the thorax). Compare it to the wild-type (wt). (b) Drawing showing the phenotype of the *Ubx/+* flies: one or a few bristles at the base of the haltere, which is also slightly enlarged compared to wild-type (wt).

➤ *bx³/TM6B* × *Ubx¹/TM6B*. Look for progeny that are not *TM6B* (flies of normal size, without extra bristles in the humerus). The genotype of these flies is *bx³/Ubx¹* and their phenotype is described in "Expected Results" (Experiment 1).

➤ *iab-7ˢᶻ/TM3* × *Abd-Bᴹ⁵/TM3*. Look for progeny that are not *TM3*, that is, that have normal-sized bristles. The genotype of these flies is *iab-7ˢᶻ/Abd-Bᴹ⁵* and their phenotype is described in "Expected Results" (Experiment 1).

Data recording. The results of the crosses will be annotated according to the different phenotypic classes. At least 15–20 individuals of each genotype should be observed. If the two mutations are in the coding regions of different genes, for example *Ubx* and *abd-A*, *Ubx/abd-A* flies will show no phenotype and will constitute one-third of the progeny (*TM6B/TM6B* flies die). If the two mutations correspond, one to a mutation in the coding region of a gene (i.e. *Ubx*), and the other to a mutation in a regulatory region of the same gene (for instance, *bx*), the *bx/Ubx* flies (also one-third of the progeny) will show a phenotype. The transformations can be observed in more detail by mounting

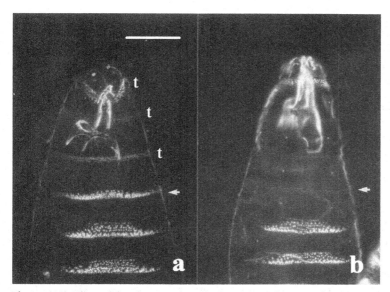

Figure 11.5. Effect of the absence of *Ubx* in the embryonic cuticle. (a) Anterior part of a normal embryo, with three thoracic segments (t) with fine denticles and some abdominal segments, the first one having no trapezoidal form (arrow). (b) Embryo mutant for a *Ubx* mutation: the first abdominal segment has been transformed into a thoracic segment (arrow). The bar in (a) indicates 0.3 mm. Adapted from Investigación y Ciencia (1995), 222, 48–56.

the flies on a slide and studying them using a microscope. The procedure is described in Chapter 14.

EXPERIMENT 2. STUDY OF THE PHENOTYPE OF LETHAL MUTATIONS IN THE EMBRYONIC CUTICLE

Mutations in the coding region of each of the three genes of the Bithorax complex are homozygous lethal and, therefore, their phenotype cannot be observed in adults. However, the transformations caused by the mutations can be studied in the embryonic cuticle (Wieschaus and Nüsslein-Volhard, 1986). *Drosophila* embryos develop a cuticle in which the thoracic and abdominal segments can be distinguished in the ventral region by their characteristic denticle belts. Mutations in the Bithorax complex genes result in changes in the pattern of these belts corresponding to each segmental transformation.

The embryonic phenotype of *Ubx* lethal mutations is studied by analysing the cuticle of embryos from a *Ubx¹/TM6B* stock. One-fourth of the embryos are homozygous for the *Ubx* mutation and show a characteristic phenotype in the embryonic cuticle described in "Expected Results" (Experiment 2).

Data recording. The three thoracic and eight abdominal segments of *Drosophila* embryos can be distinguished in the microscope (in dark field or phase contrast optics) because the thoracic segments have thin ventral belts of fine denticles whereas the abdominal ones bear trapezoidal ventral belts of thick denticles (Figure 11.5a).

EXPECTED RESULTS

PHENOTYPE OF DIFFERENT MUTATIONS AND COMPLEMENTATION ANALYSIS (EXPERIMENT 1)

The expected results are:

➤ Ubx^1 and $Ubx^{9.22}$ mutations fail to complement ($Ubx^1/Ubx^{9.22}$ flies die).

➤ Ubx^1 and $abd\text{-}A^{M1}$ mutations complement ($Ubx/abd\text{-}A$ flies show no phenotype).

➤ $abd\text{-}A^{M1}$ and $Abd\text{-}B^{M5}$ mutations complement.

➤ bx^3 and Ubx^1 mutations fail to complement; bx^3/Ubx^1 flies show a partial transformation of the anterior region of the metathorax (third thoracic segment, including the halteres) into the anterior part of the mesothorax, or second thoracic segment. The transformation is detected by the bristles and cuticle that appear in the central part of the metathorax (between the mesothorax, which bears bristles and occupies most of the thorax, and the abdomen), and by the appearance of small wings instead of halteres. The phenotype is similar to that of Figure 11.1b (Lewis, 1998), except that only the anterior part of the halteres is transformed.

➤ $iab\text{-}7^{Sz}$ and $Abd\text{-}B^{M5}$ mutations fail to complement; males of the $iab\text{-}7^{Sz}/Abd\text{-}B^{M5}$ genotype have seven abdominal segments instead of six (wild-type condition). The seventh segment, which normally does not develop, is transformed into the sixth one. The transformation also takes place in females, although it is more difficult to see.

➤ The bx^3 and $iab\text{-}7^{Sz}$ mutations transform just one segment (or part of it), whereas Ubx and $Abd\text{-}B$ mutations affect several segments. This indicates that bx^3 and $iab\text{-}7^{Sz}$ are mutations in regulatory regions of the Ubx and $Abd\text{-}B$ genes, respectively. The *bithorax* regulatory region controls the expression of Ubx in part of the third thoracic segment and the *infraabdominal-7* regulatory region does the same with respect to $Abd\text{-}B$ in the seventh abdominal segment.

STUDY OF THE PHENOTYPE OF LETHAL MUTATIONS IN THE EMBRYONIC CUTICLE (EXPERIMENT 2)

The Ubx homozygous mutants show a transformation of the third thoracic segment and the first abdominal one into the second thoracic segment. The transformation of the third thoracic segment is very difficult to observe, but that of the first abdominal segment is easily detected because the embryo now has 4 segments of the thoracic type (thin belts of fine denticles) instead of three (Figure 11.5b) and only 7 abdominal segments. One-fourth of the embryos from a $Ubx^1/TM6B$ stock will show this phenotype.

DISCUSSION

The results allow the following conclusions:

1. Mutations in Ubx, $abd\text{-}A$, or $Abd\text{-}B$ complement each other, but individuals homozygous for Ubx mutations die (the same occurs with flies homozygous for $abd\text{-}A$ or $Abd\text{-}B$ mutations). That is, the Ubx, $abd\text{-}A$, and $Abd\text{-}B$ genes form different complementation groups. The phenotype of these mutations affects several segments,

dividing the region between the third thoracic segment and the eighth abdominal one into three subregions, corresponding to the domains of the three genes (see Figure 11.2). We have observed that the *Ubx* mutations affect more than one segment because they transform the third thoracic segment (Experiment 1, complementation analysis with mutations in regulatory regions) and the first abdominal segment (Experiment 2, analysis of transformations in the embryonic cuticle).

2. Mutations in regulatory regions (*bx³*, *iab-7^{Sz}*) affect only part of the domains of each of the three genes. *bx³/Ubx¹* flies transform part of the third thoracic segment but not the first abdominal one, and *iab-7^{Sz}/Abd-B^{M5}* flies show a transformation of just the seventh abdominal segment. The *bx* regulatory region controls the expression of the *Ubx* gene in the third thoracic segment, and the *iab-7* regulatory region does the same with respect to the *Abd-B* gene in the seventh abdominal segment.

TIME REQUIRED FOR THE EXPERIMENTS

The selection of virgin females and males and the setting up of the crosses should only take 10–15 minutes for each experiment. However, the preparation of the stocks for these crosses, the maintenance of the crosses, etc., require a small amount of work during more than two weeks before and after setting up the crosses.

The analysis of the larval cuticle also requires previous stock keeping and raising as well as the preparation of media and equipment. The preparation of the cuticle, once the eggs have been collected, takes about 30–35 minutes. The embryos are analysed the following day.

If these experiments are previously carried out by the teachers, the observation of the embryonic cuticle and the study of the F1 progeny obtained in the complementation analyses could take only one or two days.

POTENTIAL SOURCES OF FAILURE

The main potential pitfall in Experiment 1 is selecting females that are not virgins. It is important to empty completely the vials or tubes from which virgins are to be subsequently picked (see above). Another problem could stem from the wrong identification of the dominant markers of the balancer chromosomes when scoring the progeny. It is advisable to check that these are easily recognised in the parental stocks. Experiment 2 has many potential pitfalls because it requires many steps in the preparation of the embryonic cuticle. For instance, if most embryos do not settle to the bottom of the methanol phase after shaking (step 7), it is likely that the chorion has not been completely removed previously, which may be rectified by putting the embryos in bleach for a longer time. If the final preparation is cloudy, it is likely that the methanol did not evaporate completely before adding Hoyer's medium.

TEACHING CONCEPTS

The gene as a unit of complementation. If two mutations located on homologous chromosomes do not show a phenotype they are said to complement and if they do, they are said to fail to complement. Mutations that complement belong to different genes, if the gene is defined as a unit of complementation. In Experiment 1, there are examples of mutations that fail to complement.

Hox (homeotic genes). The Hox genes specify the antero–posterior axis of all the animals studied. They are expressed and required in different regions along this axis. In the experiments described we observe that the mutations under study affect either the thorax (*Ubx*) or the abdomen (*Abd-B*).

Activity of Hox genes in embryonic, larval, and pupal stages. The Hox genes are active throughout development and the proteins encoded by these genes are detected at embryonic, larval, and pupal stages (when metamorphosis is taking place). These products are also required at all the stages, since the embryonic and adult cuticle is affected in mutations for these genes.

ALTERNATIVE EXERCISES

QUESTIONS FOR FURTHER ANALYSIS

Will mutations in different regulatory regions of the same Hox gene complement? Why? You can test your prediction by crossing flies mutant for a *bx* mutation to flies carrying a *pbx* allele. Both *bx* and *pbx* are mutations in regulatory regions of *Ubx*; *bx* mutations affect the anterior compartment of the third thoracic segment and *pbx* mutations the posterior one.

ADDITIONAL INFORMATION

THE HOX COMPLEX IN OTHER ORGANISMS

As stated in the "Objective of the Experiment," the Hox genes specify the antero–posterior axis in all animals studied. The Hox genes are grouped in genetic complexes. In *Drosophila*, there is one Hox complex, divided in two clusters, the Bithorax complex (analyzed in this work) and the Antennapedia complex, required to specify head and anterior thorax. Apart from *Drosophila*, the organism in which the Hox complex has been best studied is the mouse. In this animal (as well as in other vertebrates including human), there are 4 different Hox complexes, with 39 Hox genes forming 13 groups (none of the complexes has a gene for each of the 13 groups, but there is at least one gene for each group among the 4 complexes). The genes of these 13 groups are homologous to the *Drosophila* Hox genes. In addition, in both *Drosophila* and mouse Hox complexes there is a phenomenon known as "colinearity." This means that the genes located at one end of the complex are expressed at the anterior region of the animal, whereas those located at the other end are expressed at the posterior end of the organism (Krumlauf, 1994).

In spite of this similarity, the functioning of the *Drosophila* and mouse Hox complexes differs in several respects: The mouse Hox genes are transcribed following a temporal order, so that genes from one end of the complex are transcribed before (and more anteriorly in the animal) than those located at the other end (which are also transcribed at the back of the organism). This temporal expression pattern does not exist in *Drosophila*. In addition, some mouse Hox genes are also required to specify the proximo–distal axis of the limbs, something that does not happen in the appendages of *Drosophila*. Finally, the regulation of homeotic genes in the fly and the mouse differs in some significant respects. For example, two mouse Hox genes can share a regulatory sequence, and in the mouse there are regulatory sequences outside the homeotic complex that can regulate several Hox genes simultaneously.

ACKNOWLEDGEMENTS

Work in the laboratory is supported by grants from the Dirección General de Investigación Científica y Técnica (N⁰ PB98-0510), the Comunidad Autónoma de Madrid (N⁰ 08.9/0003/98 and N⁰ 08.1/0031/2001.1), and an Institutional Grant from the Fundación Ramón Areces to the Centro de Biología Molecular "Severo Ochoa."

REFERENCES

De Robertis, E. M., Oliver, G., and Wright, C. V. (1990). Homeobox genes and the vertebrate body plan. *Sci. Am.*, 263, 46–52.

Gehring, W. J. (1985). The molecular basis of development. *Sci. Am.*, 253, 153–62.

Krumlauf, R. (1994). Hox genes in vertebrate development. *Cell*, 78, 191–201.

Lewis, E. B. (1978). A gene complex controlling segmentation in *Drosophila. Nature*, 276, 565–70.

Lewis, E. B. (1998). The bithorax complex: The first fifty years. *Int. J. Dev. Biol.*, 42, 403–15.

McGinnis, W., and Kuziora, M. (1994). The molecular architects of body design. *Sci. Am.*, 270, 58–61.

Sánchez-Herrero, E. Vernós, I., Marco, R., and Morata, G. (1985). Genetic organization of the *Drosophila* Bithorax complex. *Nature*, 313, 108.

Tiong, S., Bone, L. M., and Whittle, J. R. (1985). Recessive lethal mutations within the bithorax complex in *Drosophila. Mol. Gen. Genet.*, 200, 335–42.

Wieschaus, E., and Nüsslein-Volhard, C. (1986). In *Drosophila: A Practical Approach*, ed. D. B. Roberts, pp. 199–227. Oxford: IRL Press.

APPENDIX

ADDITIONAL PROTOCOLS

Juice–agar plates. To make juice–agar plates for the egg-laying cage, do as follows:

(a) In one receptacle mix:

H₂O	500 ml
Strawberry juice	500 ml
Agar	23 g
Glucose	75 g

(b) Boil the mixture for 20 minutes.

(c) Allow the mixture to cool to at least 65°C and then add 4 ml of propionic acid.

(d) Distribute the mixture in the plates (small petri dishes) and wait for the medium to solidify.

(e) Store plates upside-down at 4°C.

Ubx and Abd-B phenotypes. The *Ubx* and *Abd-B* mutations have, in heterozygosis, a subtle mutant phenotype. In *Ubx* heterozygous flies the halteres are slightly enlarged with respect to wild-type flies and frequently bear one or two bristles at the base (Figure 11.4b). *Abd-B* heterozygous males have some bristles in the ventral part of the sixth abdominal segment, while wild-type flies have none, and have a very small seventh abdominal segment, barely visible in the dorsal abdomen. These phenotypes are difficult to observe and do not alter the results of the experiments proposed.

12 Genetic analysis of flower development in *Arabidopsis thaliana*. The ABC model of floral organ identity determination

J. L. Riechmann

OBJECTIVE OF THE EXPERIMENT *Arabidopsis thaliana* is a model organism for the study of many aspects of plant biology. The acceptance of Arabidopsis as a model system since the 1980s has resulted in part from its advantageous characteristics for genetics and molecular biology research: ease of growth in the laboratory, short life cycle, self-fertilization and ample seed production, ease of mutagenesis and of transformation by exogenous DNA, and a small genome size (Somerville and Koornneef, 2002). The complete genome sequence of Arabidopsis was reported in 2000 (Arabidopsis Genome Initiative, 2000), and many genetic and genomic research tools have been developed in this species. Our understanding of the control of floral organ identity in particular has been greatly advanced by the use of genetic analysis in Arabidopsis.

The objective of the experiment described in this chapter is to study the alterations in floral organ identity that are caused by mutations in the homeotic genes that form the basis of the ABC model of flower development (Bowman, Smyth, and Meyerowitz, 1991; Coen and Meyerowitz, 1991).

DEGREE OF DIFFICULTY The experiment described in this chapter is easy to perform, because it consists of studying the phenotype of wild-type and mutant flowers under a stereo microscope. The alternative exercises that are proposed require some experience in crossing Arabidopsis, but this is easily acquired.

INTRODUCTION

In animals, determination of the segmental identities along the antero–posterior body axis depends on the activity of a group of genes identified as homeotic (see Chapter 11), a term coined by William Bateson in 1894 to describe variations that resulted in normal body parts or organs developing at abnormal positions. Homeotic variations and mutations have long been recognized in plants, and they have formed the basis for the studies that have elucidated the genetic mechanisms responsible for establishing the patterns of floral organs (Meyerowitz et al., 1991).

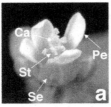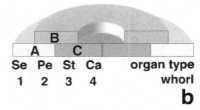

Figure 12.1. The basics of floral organ identity determination. (a) Wild-type Arabidopsis flower (Ler accession). Organs are four sepals (Se; first whorl), four petals (Pe; second whorl), six stamens (St; third whorl), and two fused carpels (Ca) that occupy the center of the flower. The size of open Arabidopsis flowers is ~2–3 mm (apical–basal length). (b) Schematic representation of the ABC model of floral organ identity determination, as originally proposed, showing how the three organ-identity functions (A, B, and C) combinatorially specify the identity of the different types of organs in the four whorls of a wild-type flower.

The majority of angiosperms bear flowers that consist of four different organ types: sepals, petals, stamens, and carpels. These organs are usually arranged in concentric rings or whorls, in that order: sepals in the outside whorl (whorl 1), petals (whorl 2), stamens (whorl 3), and carpels (in the central, fourth whorl) (Figure 12.1a). The homeotic mutations that led to the formulation of the ABC model of floral organ identity determination (by E. M. Meyerowitz, E. Coen, and colleagues in 1991) usually fall into one of three different classes, each affecting two adjacent whorls of the flower and defining one of three organ-identity functions, A, B, or C (Bowman et al., 1991; Coen and Meyerowitz, 1991; Meyerowitz et al., 1991). Function A affects whorls 1 and 2, function B whorls 2 and 3, and function C whorls 3 and 4. According to the ABC model (Figure 12.1b), each whorl of a flower primordium has either a single or a unique combination of organ identity activities that is responsible for specifying the developmental fates of the organ primordia that form in that whorl. The specification of sepal identity (whorl 1) is dependent on class A gene activities, whereas the combination of class A and class B gene activities determines petal identity (whorl 2). Stamens (whorl 3) are specified by the combined activities of classes B and C, and specification of the carpels (whorl 4) is achieved by class C activity alone (Figure 12.1b). In addition to its combinatorial nature, the model has two other basic tenets. First, A, B, and C organ identity activities function in a manner that is independent of the position they occupy in the developing floral primordium (i.e., a combination of, for example, A and B activities specifies petals, regardless of the whorl in which that combination may occur). Second, the A and C functions are mutually antagonistic: in the absence of either one of them, the domain in which the other is active expands to occupy the entire floral primordium. This simple genetic model successfully explains the homeotic changes that take place in flowers that are singly, doubly, or triply mutant in the different organ identity loci (Bowman et al., 1993; Bowman et al., 1991).

Arabidopsis class A genes include *APETALA1* (*AP1*) and *APETALA2* (*AP2*), class B genes are *APETALA3* (*AP3*) and *PISTILLATA* (*PI*), and the prototypic class C gene is *AGAMOUS* (*AG*). *AP1, AP2, AP3, PI,* and *AG* have all been cloned, and found to encode transcription factors belonging to the MADS (AP1, AP3, PI, and AG) and AP2/ERF (AP2) protein families (reviewed in Lohmann and Weigel, 2002; Riechmann and Meyerowitz, 1997). The determination of the expression patterns of *AP1*, *AP3*, *PI*, and *AG* revealed that their activity is regulated primarily at the transcriptional level, because, for the most part, each gene is expressed in the region of the flower that exhibits homeotic conversions when the gene is mutated (reviewed in Jack, 2001; Lohmann and Weigel, 2002; Weigel and Meyerowitz, 1994).

Subsequent genetic and molecular studies have uncovered additional genetic functions that are required for the determination of floral organ identity. Three other members of the Arabidopsis MADS gene family, the *SEPALLATA* genes (*SEP1*, *SEP2*, and *SEP3*), are highly related in sequence and are expressed in the developing floral whorls. None of the single nor the different doubly loss-of-function *sep* mutants shows a dramatic floral phenotype; however, in the triple *sep1 sep2 sep3* mutant, all floral organs are transformed to sepal-like organs (Pelaz et al., 2000). Thus, the *SEP* genes are largely redundant in function and are required for the development of petals, stamens, and carpels (reviewed in Jack, 2001; Theiben, 2001). Further experiments have shown that the *SEP* genes act in concert with the ABC genes through direct interactions between their respective protein products (Theiben, 2001).

It should be noted, however, that the ABC model does not explain all the phenotypic changes that are seen in some of the different mutant flowers, in part because these genes have other functions in addition to specifying organ identity (see below).

MATERIALS AND METHODS

EQUIPMENT AND MATERIALS

A stereo microscope is required for these experiments. Fine-point tweezers can be obtained from general laboratory suppliers, such as VWR or Fisher (for example, Tweezers SuperFine Point Uni-Fit, VWR).

Supplies for growing Arabidopsis (soil mix, perlite, vermiculite, pots, and fertilizer) can be obtained at nurseries or gardening stores.

Biological material. Seed stocks can be obtained from the Arabidopsis Biological Resource Center (ABRC; http://www.biosci.ohio-state.edu/~plantbio/Facilities/abrc/abrchome.htm), or from the Nottingham Arabidopsis Resource Center (NASC; http://nasc.nott.ac.uk/). The ABRC database and ordering system are incorporated into The Arabidopsis Information Resource (TAIR; http://arabidopsis.org/). Required seed stocks are as follows:

Allele	Germplasm/Stock #	Genotype	Inheritance
ag-1	CS25	Heterozygous	Recessive
ap1-1	CS28	Homozygous	Recessive
ap2-8	CS3083	Heterozygous	Recessive
pi-1	CS77	Heterozygous	Recessive
ap3-3	CS3086	Heterozygous	Recessive

The number of seeds for these stocks in the vials shipped by ABRC is 50–60 (30 in the case of *ap1-1*, since it is homozygous). All these alleles are in the Landsberg *erecta* (L*er*) background, which can also be obtained from the ABRC (Germplasm/Stock No. CS20). Heterozygous stocks will segregate plants of wild-type phenotype, which can serve as control in the experiments.

PROCEDURES

Soil, planting, and growth conditions. Arabidopsis plants can be grown in a variety of settings, including growth rooms, growth chambers, and greenhouses. With good growth conditions (continuous light, ~25°C, and adequate watering and fertilization) Arabidopsis seeds will germinate 3–5 days after planting, forming rosette leaves, bolt, and flowers within a month. Seeds can usually be harvested within two months. However, the rate of development, and the time to flowering and seed set, depend on environmental variables (as well as on the genetic background). Among these variables are light, temperature, and water and nutrient availability. Cool white fluorescent lamps supplemented with incandescent lamps are used in growth chambers (optimum light is ~150 μE/m^2 per second). A regime of continuous light accelerates the transition to flowering. Temperatures lower than 25°C can be used, but higher temperatures are not recommended, especially for germination and early development. Although not very demanding in growth conditions, Arabidopsis does not tolerate stress very well, and responds to it by flowering early, and becoming unhealthy. Stress can result from crowding, high temperature, too low or too high humidity, serious insect infestation, fungus infestation, too low or too high nutrient level, and over- or under-watering.

In addition to the information provided below, further details on the growth and care of Arabidopsis plants and seeds can be found in books of Arabidopsis methods (Scholl, Rivero-Lepinckas, and Crist, 1998). Protocols for growing Arabidopsis, pest control, and seed harvest and storage are also provided by the Arabidopsis seed stock centers, ABRC (http://www.biosci.ohio-state.edu/~plantbio/Facilities/abrc/handling.htm) and NASC (http://nasc.nott.ac.uk/protocols/newgrow.html).

A. SOIL AND POT PREPARATION

1. Put together potting soil (such as a commercial peat moss–based mix), perlite, and vermiculite in a 4:3:2 ratio in a large tray or pan. Add tap water and mix well (water will prevent dust formation during mixing). Perlite and vermiculite facilitate aeration of soil and allow better flow of water.
2. Fill small pots (for example, 10-cm square plastic pots) to about 1 cm from the top with the soil mix. It is convenient to put pots on a plastic tray, large enough to accommodate ~10 pots.
3. Add water to the tray (up to approximately 2–3 cm depth around the base of pots). Spray the surface of the soil with water before planting. This ensures that the seeds will fall on a wet surface.

B. PLANTING

1. A reduced number of plants per pot allow for better growth. In pots of the dimensions indicated above, up to 10–16 plants will grow well. When planting, try to space seeds evenly, as far apart as possible.
2. Seeds can be deposited onto the soil by pushing them off a folded piece of paper with a brush, shaking them off of a folded piece of paper, or planting each seed with tweezers.
3. After planting, cover the tray with clear plastic food wrap. This keeps the humidity high, encouraging germination. Put the tray at 4°C (refrigerator or cold room) for at least 3 days (5–7 days is optimal).

In some instances it may be useful to imbibe the seeds on Whatman filter paper in Petri dishes, which are placed at 4°C for 3–7 days. Seeds are then transferred to soil with tweezers, one at a time. Though labor intensive, this leads to high germination rates. In this case, trays can go directly into the growth room after planting, covered with plastic wrap.

C. GROWING PLANTS

1. After trays have been in the cold room for the appropriate period of time, transfer them to the growth area. The plastic wrap covering the plants should be well secured, to keep humidity high; replace if it is torn (but be careful that the pots do not overheat if using a greenhouse).
2. There should be standing water in the tray during the germination phase (~2 cm depth; add some water if necessary). Plants should germinate 3–4 days after trays have been put in the growth room.
3. After germination is complete, remove the plastic wrap (the plastic wrap may be removed gradually, first opening several slits in it, and a day later removing it completely). While the plants are young (up to the 4–5 leaf stage), the soil should be kept moist. After that the soil should be allowed to partially dry out periodically.

Initial watering should be with very dilute fertilizer (for example, 1/2 tablespoon per 4–5 liters of water). Be careful not to over-fertilize (signs of over-fertilization include leaves and inflorescences that turn yellow, or discolored and weak). Over-fertilized plants are unrecoverable. Signs of under-fertilization include the rosette leaves or sepals of young flowers turning purple. Fertilization in later growth phases increases plant health and seed set.

D. SEEDS AND SEED COLLECTION. As plants mature, they will need less water, but watering should continue while seeds are setting. It is always best to collect seeds promptly after all siliques are dry (when the siliques break with little or no applied pressure). To collect seeds, the entire inflorescence is cut off at its base. Gently rubbing the fingers around the siliques will release the seeds, which are collected on a piece of paper placed below. The seeds are then filtered through a section of cheese cloth onto another piece of paper,

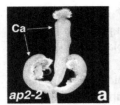  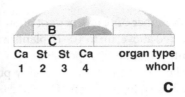

Figure 12.2. Organ identity genes of the A class: *APETALA1* and *APETALA2*. (a) *ap2-2* homozygous flower. The two medial first whorl organs are transformed to carpels (lateral first whorl organs are missing). No organs have developed in the second and third whorls. (b) *ap1-1* homozygous flower. Sepals are converted to bract-like organs, in the axils of which secondary flowers develop, whereas second whorl petals are absent. (c) Schematic representation of the ABC model reflecting the pattern of gene activities in A function mutant flowers (*ap2*), indicating the organ types formed on each whorl.

to get rid of the siliques (if siliques are stored along with the seeds, fungus growth can result). For certain stocks, especially in the Landsberg *erecta* (L*er*) background, siliques will often break open prematurely.

Collected seeds can be stored in small plastic or glass tubes. The tubes should be left at room temperature for approximately one week, then a hole should be poked on the top or cap of the tube, and the tube should then be placed in a sealed container with Drierite (desiccant). Alternatively, if seeds are well dried, they could be stored in tightly sealed cryovials. After another week or so the seeds may be planted.

PREVIOUS TASKS FOR STAFF
Growth conditions should be tested well in advance of the planned experiment. Seeds should be planted ~3–6 weeks in advance of the scheduled date for phenotypic analysis, to ensure availability of plants with flowers.

OUTLINE OF THE EXPERIMENTS

EXPERIMENT: STUDY OF THE PHENOTYPE OF DIFFERENT MUTATIONS
The aim of the experiment is to characterize how mutations in the floral homeotic genes cause organ identity changes in the Arabidopsis flower. All the mutations used in this study are strong or null alleles. The resulting phenotypes are explained in the light of the ABC model of flower development outlined earlier. Phenotypes are easily observable under the stereo microscope, particularly in older, open flowers.

EXPECTED RESULTS

PHENOTYPE OF DIFFERENT MUTATIONS
ap2-8: Flowers of plants homozygous for the *ap2-8* allele consist mostly of staminoid and carpelloid organs. Organ identities in the outer two whorls are altered and, in addition, the numbers and positions of organs in all four whorls can be affected (Bowman et al., 1991). In the flower shown in Figure 12.2a (which corresponds to an *AP2* allele of

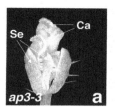
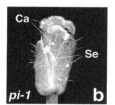
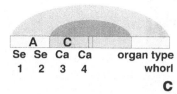

Figure 12.3. Organ identity genes of the B class: *APETALA3* and *PISTILLATA*. (a) *ap3-3* homozygous flower. One first-whorl sepal has been removed to reveal the organs of the inner whorls. (b) *pi-1* homozygous flower. One first-whorl sepal has been removed. (c) Schematic representation of the ABC model reflecting the pattern of gene activities in B function mutant flowers (*ap3* or *pi*), indicating the organ types formed on each whorl.

similar phenotype, *ap2-2*), the first whorl is occupied by two medial solitary carpels topped with stigmatic tissue and with ovules along their margins; no organs have developed in the second and third whorls, and two fused carpels occupy the fourth whorl. This floral phenotype results from the absence of *AP2* function leading to ectopic expression of *AG* (C function) in all floral whorls (Figure 12.2c).

ap1-1: In flowers of plants homozygous for the *ap1-1* allele, sepals are converted to bract-like organs, in the axils of which secondary flowers develop, whereas second whorl petals are often absent (Figure 12.2b). The complexity of this phenotype derives from the dual roles that *AP1* plays in flower development: it acts early in the process to generally specify floral identity, and later contributes to A function (Bowman et al., 1993).

ap3-3 and *pi-1*: Flowers of plants homozygous for either the *pi-1* or *ap3-3* alleles are essentially indistinguishable: four additional sepals are present in the positions normally occupied by petals in whorl two, and whorl three is occupied by filamentous or carpelloid organs that fuse to the central gynoecium (Figure 12.3a and b) (Bowman et al., 1991). This floral phenotype is explained because, in the absence of B activity, for which both *AP3* and *PI* are required, the identity of second whorl organs is determined by A activity alone, whereas C activity alone will lead to carpel identity in the third whorl (Figure 12.3c).

ag-1: Flowers of plants homozygous for the *ag-1* allele are characterized by organ identity changes in whorls three and four. In whorl three, the positions that in the wild-type are occupied by stamens are occupied by petals, and the gynoecium in whorl four is replaced by a new flower, consisting of an outer whorl of sepals and two whorls of petals. The process of flowers within flowers continues, resulting in a reiterated pattern of (sepals-petals-petals)n (Figure 12.4a; the *ag-1* and *ag-3* mutations are phenotypically similar) (Bowman, Smyth, and Meyerowitz, 1989; Bowman et al., 1991). In the absence of C function, A function extends to the inner whorls of the flower, thus explaining the organ identity changes (Figure 12.4b). The *ag* mutant phenotype also reveals that *AG* has another function in addition to specifying organ identity: suppression of the indeterminate growth of the floral meristem (Bowman et al., 1989; Bowman et al., 1991). Thus, the loss of *AG* activity leads to indeterminacy, causing the repeating pattern of organ types.

 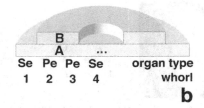

Figure 12.4. Organ identity gene of the C class: *AGAMOUS*. (a) *ag-3* homozygous flower, showing petals and sepals in place of stamens and carpels. (b) Schematic representation of the ABC model reflecting the pattern of gene activities in C function mutant flowers (*ag*), indicating the organ types formed on each whorl.

DISCUSSION

The phenotypes of these single mutants illustrate that the activities that regulate organ identity in flower development are homeotic in nature. They also provide for some of the conclusions on which the ABC model is based. These are:

➤ Each organ identity activity affects the development of two adjacent whorls,
➤ Organ identity activity is independent of organ position,
➤ A and C functions are antagonistic.

TIME REQUIRED FOR THE EXPERIMENTS

Observation of the different floral mutant phenotypes under the stereo microscope and data recording can take 1 hour. Soil and plant preparation and seed planting may take 2 hours, and should take place ~6 weeks before the desired observation date (depending on the growth conditions).

POTENTIAL SOURCES OF FAILURE

The main (and almost only) potential problem in the experiment described above is inadequate seed and plant handling and growth conditions.

TEACHING CONCEPTS

➤ Homeotic gene activity in plants.
➤ Combinatorial genetic interactions in floral organ identity determination

ALTERNATIVE EXERCISES

PROPOSED EXPERIMENTS

Several alternative exercises can be easily performed with the mutants described earlier. These include complementation tests (such as those described in Chapter 11), double mutant analyses, as well as teaching the concept of an allelic series (for this experiment, additional seed stocks have to be obtained). Complementation analysis and the construction of doubly mutant strains require crossing Arabidopsis, for which a protocol is provided below. If these experiments are planned, the generation time of Arabidopsis (2–3 months) should be taken into consideration.

Complementation test. Strong *ap3* and *pi* alleles cause similar floral phenotypes. That *AP3* and *PI* are indeed different genes can be shown by a simple complementation test using the corresponding mutants. The phenotype of the *ap3 pi* double mutant is indistinguishable from the single-mutant phenotypes.

Double mutant analysis. The elucidation of the ABC mechanism of organ identity determination was possible in part because of the phenotypic analysis of double and triple organ identity mutants (Bowman et al., 1991). Such double or triple mutants can be constructed by cross-pollination using plants homozygous for the individual mutations, except in the case of *ag* plants, which require the use of heterozygotes because homozygous plants do not bear any reproductive organs; F_1 plants are allowed to self-pollinate and double and triple mutants are selected from the F_3 plants.

Allelic series. Different mutations in the same gene can cause phenotypic changes of varying severity. Allelic series are available for the Arabidopsis organ identity genes, and this concept can be studied using *ap2* mutants, for example. Additional seed stocks to obtain from the stock centers are *ap2-1* (CS29), *ap2-2* (CS3082), and *ap2-9* (CS3084). The *ap2* allelic series has been described in detail (Bowman et al., 1991).

Crossing Arabidopsis
1. Select the appropriate female and male parent plants for the cross.
2. With a pair of fine-point tweezers and under the stereomicroscope, remove from the female parent plant flowers and floral buds not to be used in the cross (young floral buds as well as flowers with white petals easily visible), leaving a few large floral buds on the inflorescence.
3. Open the buds carefully and, with tweezers, remove the six anthers, without damaging the carpels.
4. From the male parent plant, pick up stamens from fully open flowers and, with tweezers, rub the anthers with the stigma of the female parent flowers, making sure that pollen (yellow color) is released from the anthers.
5. The floral buds that have been used for the cross can be marked with a color thread, or with a label attached to the stem below them.
6. Success of the cross can be determined a few days later (~4) by inspecting the flowers: female parent pistils should be developing as young siliques. Seeds can be collected when the siliques are dry, usually ~3 weeks after the cross.

The mutant flowers described in this chapter show changes in organ identity that can facilitate or complicate the crosses between them. For example, *ap3-3* and *pi-1* homozygous flowers do not bear stamens, obviating the need for emasculation, whereas homozygous *ag-3* mutant flowers do not have any reproductive organs, making necessary the use of heterozygous plants.

ACKNOWLEDGEMENTS

I am grateful to Dr. Elliot Meyerowitz and former members of his laboratory at the California Institute of Technology, who developed the ABC model of organ identity determination in Arabidopsis, and introduced me to Arabidopsis research.

REFERENCES

Arabidopsis Genome Initiative (2000). Analysis of the genome sequence of the flowering plant *Arabidopsis thaliana*. *Nature*, 408, 796–815.

Bowman, J. L., Smyth, D. R., and Meyerowitz, E. M. (1989). Genes directing flower development in *Arabidopsis thaliana*. *Plant Cell*, 1, 37–52.

Bowman, J. L., Smyth, D. R., and Meyerowitz, E. M. (1991). Genetic interactions among floral homeotic genes of *Arabidopsis*. *Development*, 112, 1–20.

Bowman, J. L., Alvarez, J., Weigel, D., Meyerowitz, E. M., and Smyth, D. (1993). Control of flower development in *Arabidopsis thaliana* by *APETALA1* and interacting genes. *Development*, 119, 721–43.

Coen, E. S., and Meyerowitz, E. M. (1991). The war of the whorls: Genetic interactions controlling flower development. *Nature*, 353, 31–7.

Jack, T. (2001). Relearning our ABCs: New twists on an old model. *Trends Plant Sci.*, 6, 310–6.

Lohmann, J. U., and Weigel, D. (2002). Building beauty: The genetic control of floral patterning. *Dev. Cell*, 2, 135–42.

Meyerowitz, E. M., Bowman, J. L., Brockman, L. L., Drews, G. N., Jack, T., Sieburth, L. E., and Weigel, D. (1991). A genetic and molecular model for flower development in *Arabidopsis thaliana*. *Dev. Suppl.*, 1, 157–67.

Pelaz, S., Ditta, G. S., Baumann, E., Wisman, E., and Yanofsky, M. F. (2000). B and C floral organ identity functions require *SEPALLATA* MADS-box genes. *Nature*, 405, 200–3.

Riechmann, J. L., and Meyerowitz, E. M. (1997). MADS domain proteins in plant development. *Biol. Chem.*, 378, 1079–1101.

Scholl, R., Rivero-Lepinckas, L., and Crist, D. (1998). Growth of plants and preservation of seeds. In *Arabidopsis Protocols*, eds. J. M. Martínez-Zapater and J. Salinas, pp. 1–12. Totowa: Humana Press.

Somerville, C., and Koornneef, M. (2002). A fortunate choice: The history of Arabidopsis as a model plant. *Nat. Rev. Genet.*, 3, 883–9.

Theiben, G. (2001). Development of floral organ identity: stories from the MADS house. *Curr. Opin. Plant. Biol.*, 4, 75–85.

Weigel, D., and Meyerowitz, E. M. (1994). The ABCs of floral homeotic genes. *Cell*, 78, 203–9.

13 Genetic analysis of vulva development in *C. elegans*

S. Canevascini

OBJECTIVE OF THE EXPERIMENT In this chapter, I will introduce the experimental model organism *Caenorhabditis elegans*. Particular emphasis will be given to the advantages that this nematode offers for genetic studies. Several genes of ancient origin have been conserved during evolution in nematodes and vertebrates. They control essential aspects of cell viability, proliferation, signalling, and differentiation in these organisms. Thus, characterisation of the function of a new gene (and consequently, of the biological pathway that it modulates) in *C. elegans* will allow better understanding of the corresponding pathway in vertebrates.

The characterisation of the genetic factors involved in inter- and intracellular signal transduction during vulva development in *C. elegans* will be presented as a paradigm to demonstrate the potential that this organism offers to the developmental biologist.

DEGREE OF DIFFICULTY Easy (Experiment 1a) to difficult (Experiments 1b and 2). Lab instructors and preferably students as well need to be experienced in basic molecular biology techniques and handling of *C. elegans*.

INTRODUCTION

C. elegans is a small worm (1 mm as adult), that is easy and cheap to grow under laboratory conditions. It has two sexes, hermaphrodite (two X chromosomes, XX) (Figure 13.1a) and male (one X chromosome, XO) (Figure 13.1b), allowing both self- and cross-fertilisations. It has a short reproductive cycle (three days at 20°C) and it can give rise to abundant progeny (up to 300 animals). In spite of its small size, it has a nervous system, muscles, and gut that are miniaturised and simplified versions of their vertebrate counterparts. Its fully sequenced genome, together with proficient genetic tools to introduce mutations in the genome, RNA interference to specifically knock down a given gene, and a large collection of mutant strains, have made this organism particularly suited for genetic studies.

153

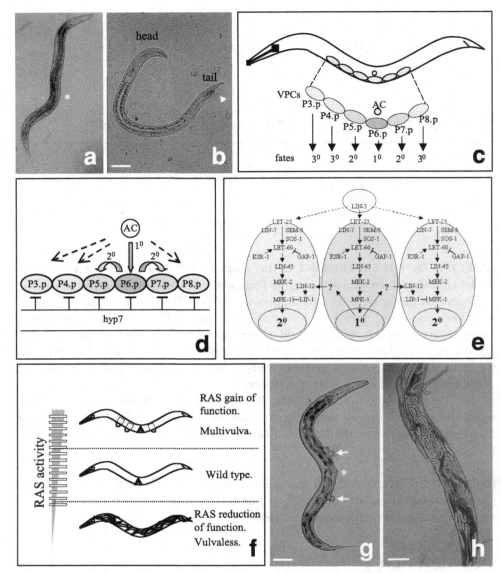

Figure 13.1. Vulva induction and mutant vulval phenotypes in *C. elegans*. (a) Hermaphrodite. (b) Male. (c) Schematic representation of the nematode with the six VPCs lying on its ventral site. The VPCs are named P3.p to P6.p. The anchor cell is located centrally in close proximity to P6.p. The head is on the left. (d) Three signals modulate vulval induction. A signal from the AC induces P6.p to acquire a 1° fate, a second signal from P6.p induces P5.p and P6.p to adopt a 2° fate. P3.p, P4.p, and P8.p do not receive enough of either inducing signals and are not able to overcome the inhibitory signal from the hypodermis. (e) Schematic representation of the EGF/RAS/MAPK and NOTCH pathways controlling induction in P5.p, P6.p, and P7.p. Only the core components of the pathways and the factors investigated in this practice are shown. (f) Schematic representation of the phenotypes associated with gain and reduction of function mutations in the RAS pathway. (g), Multivulva phenotype. (h), Vulvaless ("bag of worms") phenotype. AC, anchor cell; VPC, vulva precursor cell; arrowhead, fan; asterisk, vulva; arrow, pseudovulva. Bars = 0.1 mm.

VULVA DEVELOPMENT IN *C. ELEGANS*

The vulva in *C. elegans* is the organ that connects the uterus to the outside and allows the gravid hermaphrodite to lay eggs and receive sperm from the males. Vulva differentiation starts in the third of four larval stages that precede adulthood. Pioneer studies disclosed the genetic basis of vulva development (Ferguson and Horvitz, 1985). At the beginning of the third larval stage, a cell in the gonad (the anchor cell, AC) secretes a protein, LIN-3, that is homologous to epidermal growth factor (EGF) of vertebrates. LIN-3 is a ligand for the LET-23 (EGFR) receptor that is expressed by six cells on the ventral side of the animal. These cells, the vulval precursor cells (VPCs), have an equivalent potential to respond to the LIN-3 signal and to differentiate into the cells that form a functional vulva. The activated LET-23 receptor transduces the signal intracellularly through SEM-5 (GRB2) and SOS-1 (GNEF) to the LET-60 (RAS) protein. From LET-60, the signal is then transduced further through LET-45 (RAF), MEK-2 (MAPKK), and MPK-1 (MAPK) to the nucleus where it induces changes in gene expression (Figure 13.1c–e).

P6.p, the VPC closest to the AC, receives most of the LIN-3 signal and acquires a 1° fate. In turn P6.p sends a signal through LIN-12 (NOTCH receptor) that will prevent the neighbouring cells (P5.p and P7.p) from acquiring a 1° fate; they acquire a 2° fate instead. The remaining VPCs (P3.p, P4.p, and P8.p) do not receive enough LIN-3 signal and they adopt a 3° fate. The cells acquiring the 1° and 2° fate go through three rounds of cell divisions, generating 8 and 7 cells, respectively. The 22 cells resulting from the induction of P5.p, P6.p, and P7.p together form a functional vulva. The VPCs that have acquired the 3° fate do not contribute to vulva formation. They divide once and fuse with the hypodermis (Figure 13.1c–e). See Wang and Sternberg (2001) for a review.

In order to induce three out of the six VPCs to differentiate into a functional vulva, the extent of activation of the RAS pathway must be tightly controlled. Nonetheless, mutations in the core elements of the RAS pathway or in factors responsible for its control can lead to its over-activation. In this case, additional VPCs (P3.p, P4.p, and P8.p) can be induced to acquire vulval fates. As a result, ectopic pseudovulvae will form on the ventral side of the animal (a multivulva "Muv" phenotype, Figure 13.1f and Figure 13.1g). Mutations that lead to an inactivation of the RAS pathway can cause fewer than three VPCs to be induced. As a consequence, the animal will not have a functional vulva and it will not be able to lay eggs (a vulvaless "Vul" phenotype). The larvae will hatch inside the hermaphrodite resulting in a characteristic "bag of worms" phenotype (Figure 13.1f and Figure 13.1h).

The Muv and Vul phenotypes have been used in genetic screens as markers for the selection of mutations in genes that are candidates for playing a role in the RAS pathway. This method has been extremely efficient in *C. elegans* and has allowed the characterisation of most of the factors known to modulate the RAS pathway in the nematode. In the last years, the use of classical mutagens in genetic screens has been complemented by the application of the RNAi technique in *C. elegans*. This technique exploits the ability of a dsRNA sequence to target for degradation gene transcripts with which it shares sequence complementarity (see also Chapter 9). In *C. elegans*, the use of this technique is improved by the fact that the dsRNA can be expressed in bacteria that are then fed to the worms. By a not-yet-fully-understood mechanism, the dsRNA released from the bacteria upon digestion diffuses into the worm compromising the stability of the mRNA of the corresponding gene in the entire animal, resulting in a transient knock down (Fire et al., 1998, Kamath et al., 2003).

In the experiments proposed in this chapter, we will learn how to recognise the Muv and Vul phenotypes and will perform a simple RNAi test to characterise the genetic interaction between different factors of the RAS pathway. Furthermore, a mutation causing a Vul phenotype will be mapped to a single chromosome using SNP (single nucleotide polymorphism) mapping.

MATERIALS AND METHODS

EQUIPMENT AND MATERIALS

Per student

Dissecting microscope

Tool for picking the worms

Alcohol lamp to sterilize the platinum wire

96-well PCR plate

96-well PCR machine

Electrophoresis apparatus

Pipettor set (1 to 1,000 μl) and sterile tips

Ice

Liquid nitrogen

Sterile dH$_2$O (1 ml)

Bleach (1 ml)

Mineral oil (5 ml) (Sigma)

Worm lysis buffer (500 μl)

LB+Amp+Tet (20 ml)

NGM+IPTG plates (15)

NGM+OP50 plates (25)

The worm strains growing on a NGM+OP50 plate

Plates with Hawaii strain males

Per practical group

Microscope with Nomarski optics

15°C, 20°C, 30°C, and 37°C incubators

Apparatus for UV imaging

Primer mixes, 10 × PCR Buffer, dNTPs, 15 mM MgCl$_2$

AvaII, DraI, and EcoRI restriction enzymes

Agarose (Invitrogen)

Ethidium bromide (EthBr) stock solution (10 mg/ml)

1 × TBE buffer

Biological material

C. ELEGANS STRAINS. MT2124 *let-60(n1046)*, cB1413 *lin-7(e1413)*, AH12 *gap-1(ga133)*, PS1123 *lin-3(syIs1)*, cB1417 *lin-3(e1417)*, cB4856 Hawaii, N2 Bristol, RU51 *xyz-1; unc-5(e53)*.

BACTERIA. The bacterial strains OP50, ksr1RNAi, gap1RNAi, lin7RNAi, and him6RNAi can be ordered from the author. The RNAi strains must be grown on LB+Amp+Tet.

STOCK REQUEST. The worm strains can be ordered from the Caenorhabditis Genetic Center, University of Minnesota (http://biosci.umn.edu/CGC/Strains/request.htm).

REAGENTS
Agar (BD)
Tryptone (BD)
Yeast extract (BD)
Bromophenol Blue (BIO-RAD)
Reagents for PCR (Roche)
Agarose (Invitrogen)
Proteinase K (Roche)
Mineral oil (Sigma)
IPTG (Roche)
Restriction enzymes (Roche)
Glycerol (Merck)
Ethidium bromide (Merck)

PREVIOUS TASKS FOR STAFF
Maintenance of *C. elegans* strains. The strains are grown at 20°C on NGM+OP50 plates. Avoid starving the worms. Before the bacteria have been depleted, transfer some worms (either with a pick, or by chunking a small square of agarose containing worms) to a new NGM+OP50 plate. Starving can affect vulva development.

A simple tool for picking the worms. Attach a 3-cm long platinum wire (0.3 mm diameter, Sigma) to a metal or glass holder. Flatten the tip of the free wire with a hammer or a pincer. Smooth the edges of the wire with fine sandpaper. Bend the wire as shown in Figure 13.2a.

To generate *C. elegans* males. Put 20 young adult hermaphrodites on a NGM+OP50 plate. Incubate for 4 hours at 30°C and then shift the worms back to 20°C for three days. 1–2% of the progeny will be males. Prepare a mating plate by putting a small amount of OP50 bacteria in the middle of a NGM plate. Transfer 6–8 males and 5 young hermaphrodites to the bacteria on the mating plate (use Figure 13.1a–b as reference). Incubate the plate for three days at 20°C. Up to 50% of the progeny from this mating will be males.

Preparation of *C. elegans* food
NGM+OP50. Grow a culture of OP50 bacteria in LB overnight. Pipette a small amount of OP50 onto the NGM plates leaving a ring of 5 mm without bacteria near the walls of the plate. This will keep the worms in the middle of the plate. Allow the plates to dry for at least one day before seeding them with worms.

NGM+IPTG. Add 100 μl of 1 M IPTG to each plate. Distribute the solution over the entire surface of the plate. Let the solution soak into the medium.

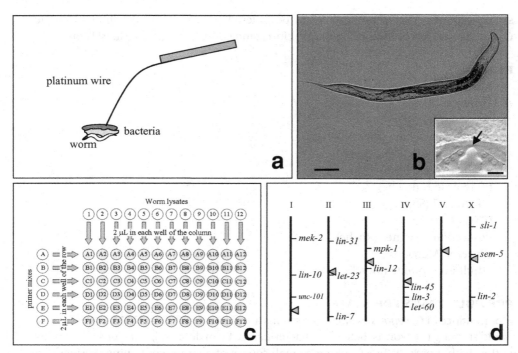

Figure 13.2. (a) Drawing of a tool used to pick the worms. (b) Hermaphrodite at the L4 larval stage. Inset: Morphology of a differentiating vulva (arrow) at the L4 stage. (c) Schematic representation for the preparation of the samples for the PCR reactions. (d) Schematic representation of the six *C. elegans* chromosomes. The approximate position of the SNPs (arrowheads) used in the experiment and of some of the genes involved in the RAS pathway are shown. Bars represent 0.1 mm (b) and 0.01 mm (b inset).

Solutions

WORM LYSIS BUFFER. 10 mM Tris-HCl pH 8.3, 50 mM KCl, 2.5 mM MgCl$_2$, 0.45% NP-40, 0.45% Tween20, 0.01% gelatin, 100 μg/ml Proteinase K.

PCR REACTION MIX. For one reaction: 2 μl 10 × PCR Buffer, 2 μl 15 mM MgCl$_2$, 2 μl 2 mM dNTPs mix, 0.5 U DNA Polymerase, add dH$_2$O to a final volume of 16 μl. Mix the components on ice.

5 × DNA LOADING BUFFER. 30% v/v glycerol, 0.25% Bromophenol Blue.

1 × TBE. For 1 l: 10.8 g Tris-Base (Fluka), 5.5 g boric acid, 4 ml 0.5 M EDTA pH 8.

3% AGAROSE GEL. 3% Agarose, 0.5 μg/ml EthBr in 1 × TBE Buffer.

NGM. For 1 l: To 800 ml H$_2$O add: 3 g NaCl, 18 g Agar, 2.5 g Peptone, 1 ml cholesterol stock solution (stock: 5 mg/ml in 100% EtOH). Add H$_2$O to a final volume of 973 ml. Sterilise by autoclaving. Let the solution cool down to 50°C with continuous stirring.

Table 13.1. Primer pairs and restriction enzymes for SNP mapping

| Chrom. | Primer mix | Primers | RE | DNA fragments after digestion (bp) | |
				Bristol	Hawaii
I	mix-A	CTCATGCATGATTTCGAGGG& AAATCCAACAGGAGCAGGAC	EcoRI	285 241	526
II	mix-B	TCCACACTATTTCCCTCGTG& GAGCAATCAAGAACCGGATC	DraI	299 125 70	369 125
III	mix-C	CATTAGGAAGTGATGCAAGTGG& TGGATTTGAGAGGTGTCCATAG	AvaII	300 198	498
IV	mix-D	CCAAACAACCTACAGAAAATGC& AAGATATTCATGCGTCGTAGTG	DraI	455 71 57	283 172 71 57
V	mix-E	GTGCTAATTCCAGAAATGATCC& TAGTGTTCATAGCATCCCATTG	DraI	432 79	297 135 79
X	mix-F	TTTCTTGACACCTCCGGTAG& CTCACTCTGGTCTTTTTCCG	EcoRI	517 259	776

Add 25 ml of 1 M KPO$_4$ pH 6, 1 ml 1 M CaCl$_2$, 1 ml 1 M MgSO$_4$. These salt solutions must be filter sterilised or autoclaved separately to avoid precipitation. Pour the medium into 60 × 10 mm plates (10 ml per plate).

LB. For 1 l: 10 g tryptone, 5 g yeast extract, 10 g NaCl. Adjust to pH 7.0 with NaOH. Sterilise by autoclaving. *LBA:* Add 15 g/l Agar to LB solution. *LB+Amp+Tet:* Add ampicillin and tetracyclin to 50 μg/ml final concentration. Before adding the antibiotic, cool the medium to 50°C.

PROCEDURES

How to pick the worms. To pick the worms, collect some bacteria underneath the flattened surface of the wire. Touch the worm(s) gently with the bacteria on the wire. The worm(s) will adhere to the bacteria. Gently lay the worm(s) on a new plate.

SNP mapping. The primers and the restriction enzymes described in Table 13.1 must be ordered by the organisers of the course. Mix the two primers to 2 μM final concentration. Alternative primers and restriction enzymes can be chosen according to Wicks et al. (2001).

OUTLINE OF THE EXPERIMENTS

Two short experiments are described. In the first, we will investigate how genes in the RAS pathway interact, and what types of vulval phenotypes mutations in these genes produce. In the second experiment we will take the first step in identifying an unknown mutation: mapping the mutation to a single chromosome.

EXPERIMENT 1A. VULVAL PHENOTYPES IN MUTANTS OF THE RAS PATHWAY

Naturally occurring or artificially induced mutations can influence the activity of a gene. Mutations in the promoter region of a gene can enhance its transcription. Alternatively, a mutation in the coding region can increase the affinity of the encoded protein for its substrate. We call this class of mutations "gain-of-function (gf)" mutations. Mutations that reduce the activity of a gene or completely inactivate it are called "reduction-of-function (rf)" and "loss-of-function (lf)" mutations, respectively. Studying the phenotypes associated with these mutations not only gives insights into the function of the genes, but also allows testing of whether they are components of a given pathway.

To become familiar with the vulval phenotypes resulting from alterations in activity of the RAS pathway, one must first characterise the phenotypes of the following strains under the dissecting microscope:

➤ Wild-type (Bristol).
➤ *lin-3(e1417)*, an rf allele of the EGF.
➤ *lin-7(e1413)* is an rf mutation. The LIN-7 protein is necessary for the correct localisation of the LET-23 receptor (EGFR). In the *e1413* mutants the receptor is mislocalised and the VPCs receive less EGF signal.
➤ *lin-3(syIs1)* is a gf mutation. This strain contains additional copies of the *lin-3* gene, resulting in increased synthesis of the EGF ligand.
➤ *let-60(n1046)* is a gf mutation in the RAS protein.
➤ *gap-1(ga133)* is an lf mutation of an inhibitor of the RAS protein. In the *ga133* mutants, the RAS protein, once activated, cannot be efficiently down regulated. The resulting over-activation of the RAS pathway is not extreme enough to result in a Muv phenotype. The worms look wt.

Day 1. To synchronise the development of the worms, transfer twenty gravid hermaphrodites from each of the strains in the preceding list to a drop of bleach (10–20 μl) placed on a new NGM+OP50 plate. Collect the worms first, then quickly place the drop of bleach on the plate, and transfer the worms into the bleach before it is absorbed by the medium. Keep the plates at 20°C for 3 days. By this procedure the hermaphrodite is killed, but the fertilised eggs survive and give rise to developmentally synchronised progeny. This will facilitate the analysis of the phenotypes at day 4.

Data recording. At day 4, determine for each strain the percent of normal, multivulva, and vulvaless animals (use Figure 13.1f–h as reference).

Table 13.2. Worm and bacterial strains for the study of genetic interactions

Strain	Number of worms per plate	RNAi plate	Number of plates
let-60(n1046)	2	ksr-1RNAi	3
let-60(n1046)	2	him-6RNAi	3
lin-3(e1417)	5	gap-1RNAi	3
wild type	2	lin-7RNAi	3
gap-1(ga133)	2	lin-7RNAi	3

EXPERIMENT 1B. CHARACTERISATION OF THE GENETIC INTERACTIONS BETWEEN FACTORS OF THE RAS PATHWAY DURING VULVAL DEVELOPMENT

In this experiment we will exploit the RNAi (RNA interference) technique to selectively inactivate some genes of the RAS pathway in *C. elegans*. Bacteria expressing dsRNA complementary to these genes will be fed to the worms (see Introduction). These artificially generated mutants will allow us to investigate the genetic interactions between different genes of the RAS pathway.

Day 1. Inoculate 5 ml LB+Amp+Tet with each of the four RNAi bacterial clones. Grow the four bacterial cultures overnight at 37°C.

Day 2. Plate 100 μl of each bacterial culture onto NGM+IPTG plates. The IPTG will induce dsRNA expression in the bacteria. For ksr-1RNAi, him-6RNAi, and gap-1RNAi, make three plates of each bacteria. For the lin-7RNAi bacterial culture, make six plates. Leave the plates overnight at room temperature.

Day 3. Following the instructions given in Table 13.2, put L4-stage worms on each plate. Wild-type animals can be used to practice recognition of the L4 larval stage. In this stage the differentiating vulval cells form an invagination easily recognisable under the microscope as an ovoid clear area (Figure 13.2b). Alternatively, L4-stage animals can be collected from the synchronised populations of Experiment 1a 36 hours after bleaching.

Day 5. Remove the parent worms from the plates.

Data recording. At days 7 and 8, count the percent of wild-type, vulvaless, and multivulva animals on each plate. Compare these data with those from Experiment 1a.

EXPERIMENT 2. MAPPING OF A MUTATION OF THE RAS PATHWAY

To discover new factors involved in a given pathway, mutations can be induced at random in the genome of the model organism with chemicals or gamma radiation and a specific phenotype used as marker to select those of interest. In our case, mutations causing a vulval phenotype are good candidates for genes able to modulate the RAS pathway.

To identify the gene carrying the mutation, the position of the lesion must be narrowed down to a genomic region as small as possible. The genes within this region are then tested for the presence of the mutation.

To quickly map a mutation, single nucleotide polymorphisms (SNP) between two intercrossing *C. elegans* varieties – isolated in Bristol and Hawaii, respectively – will be used as markers. SNPs are the result of point mutations that have accumulated in the genomes of the two worm populations after they have become geographically isolated. The mutations used here have resulted in a restriction site in one of the two strains. Thus, to visualize the markers, it will be enough to perform a restriction analysis of a DNA fragment amplified from this genomic region. Our mutant Bristol strain will be crossed into the Hawaii strain to generate heterozygous animals. In the next generation, we will then investigate the segregation of the mutation and of one selected SNP for each chromosome. The position of the mutation will be inferred in relation to the position of these known genomic markers. If the mutation and the SNP lie on different chromosomes (or far away on the same chromosome), they will segregate independently from each other and 75% of the worms showing the vulval mutant phenotype will carry the Hawaii SNP marker (25% homozygous and 50% heterozygous for the Hawaii SNP). On the other hand, if the mutation and the SNP lie near to each other on the same chromosome, less than 75% of the progeny showing the vulval mutant phenotype will carry the Hawaii SNP. In other words, the nearer the mutation and the marker, the lower the probability of a recombination event that will cause the vulval mutation and the Hawaii SNP marker to cosegregate.

The mutation that we want to map is recessive and temperature sensitive. We arbitrarily name it *xyz-1*. At the permissive temperature of 20°C the animals will look normal. At the nonpermissive temperature of 15°C a large proportion of the worms will be vulvaless. The strain carries a second mutation, *unc-5(e53)*, that causes an "uncoordinated movement" phenotype. The *unc-5* marker and the *xyz-1* mutation are on different chromosomes and segregate independently. The *unc-5* marker is used only to facilitate the recognition of cross progeny on day 3.

Day 1. Transfer a dozen Hawaii males (Figure 13.1b) on NGM+OP50 plates with 40 *unc-5; xyz-1* hermaphrodites. (The cross may be difficult because not all the hermaphrodites will have a functional vulva. Therefore more than one cross should be set.) Make sure that you transfer only males from the Hawaii strain! Incubate the plates at 20°C.

Day 3. From the mating plates, transfer twelve F1 cross progeny (cross progeny will move normally because they are heterozygous for the *unc-5* mutation) each to a new NGM+OP50 plate. At this time, it may not be possible to distinguish between males and hermaphrodites in the progeny. Nonetheless, it is worth transferring single worms before they mature sexually, to avoid mating. Incubate the plates at 20°C.

Day 4. Incubate the plates at 15°C.

Day 9–10

1. Discard the plates from day three in which a male was transferred. On the other plates, 25% of the progeny should be vulvaless.
2. Pipette 12 μl lysis buffer in each of the 12 wells of the first row of the 96-well PCR plate. Transfer 1 vulvaless animal to each of the first 10 wells. As controls, transfer one *unc-5; xyz-1* and one Hawaii worm, respectively, in the last two wells. Cover the buffer with one drop of mineral oil to avoid evaporation.
3. Freeze and thaw the probes in liquid nitrogen to permeabilise the worms. Incubate the probes for one hour at 60°C and subsequently for 15 min at 94°C (these incubation steps can be performed in the PCR machine).
4. Prepare the primer mixes as described in Table 13.1 (2 μM final concentration for each primer in the mix).
5. Aliquote each lysate into six PCR wells (2 μl each, Figure 13.2c). Transfer the probes on ice. Add 2 μl of primer mix and 16 μl of reaction mix to each probe as described in Figure 13.2c. Cover the probes with one drop of mineral oil.
6. Perform the PCR reaction: Step1: 2 min at 94°C; Step2: 30 s at 94°C, 30 s at 58°C, 1 min at 72°C; repeat 35 times Step2; Step3: 3 min at 72°C.
7. Remove the samples from the PCR machine and let them cool down to RT for 5 min.
8. Add 10 μl of the corresponding restriction mix (3 × concentrated) to each probe according to Table 13.1 (mix the solution by pipetting up and down without disturbing the oil layer on top). Incubate the samples for two hours at 37°C.
9. With the help of an assistant, prepare a 3% Agarose gel. When preparing the gel, slowly heat up the solution. The high agarose concentration will cause foaming.
   ETHER IS CARCINOGENIC. USE ADEQUATE PROTECTION WHEN HANDLING IT.
10. Let the samples cool down to room temperature for 10 min. Add 7 μl of 5 × DNA loading buffer to each sample and mix without disturbing the oil layer.
11. Load the probes on the 3% Agarose Gel.
12. Run the gel at 70 V in 1 × TBE until the Bromophenol Blue band has reached 2/3 of the length of the gel.

Data recording. With the help of an assistant, take a picture of the gel under UV light (protect eyes and skin from UV light). Compare the pattern of the bands of each probe with the corresponding Bristol and Hawaii controls. Determine for each set of probes what percentage have the Hawaii pattern. Look at the genetic map shown in Figure 13.2d, which indicates the position of the SNPs used in this experiment and some of the genes involved in the RAS pathway. To which one of these genes could *xyz-1* be allelic?

EXPECTED RESULTS

EXPERIMENT 1

In this experiment, we investigate the ability of different genes to modulate the same phenotype. In *let-60(n1046)* mutants the RAS pathway is over-activated. Inhibition of *ksr-1*, another positive regulator of RAS signalling, by RNAi counteracts the effects of the *let-60(n1046)* gf mutation and attenuates the strength of the Muv phenotype. The same effect cannot be achieved by inactivating *him-6*, a gene required for proper meiotic chromosome segregation (used in this experiment as negative control).

Both the *lin-3(e1417)* mutation and inactivation of *lin-7* by RNAi inactivate the RAS pathway leading to a Vul phenotype. In these genetic backgrounds, knocking out *gap-1*, an inhibitor of RAS, either by a genomic lesion or by RNAi, increases the strength of the RAS pathway. This attenuates the Vul phenotypes, and even induces a Muv phenotype in the *lin-7* RNAi treated animals (see Discussion).

EXPERIMENT 2

The goal of this experiment is to map the *xyz-1* mutation relative to the position of known genetic markers, the SNPs. If the *xyz-1* mutation is linked to one of these Bristol SNP markers, they will co-segregate in the F2 offspring. In the F2 Vul animals examined, the Bristol specific sequence will be overrepresented (present in more than 75% of the animals) whereas the Hawaii-specific sequence will be underrepresented (present in less than 75% of the animals). In our experiment, the SNP for chromosome II behaves in that way. Figure 13.2d shows that it maps next to the *let-23* gene. The *xyz-1* mutation is indeed *let-23(n1045)*.

DISCUSSION

EXPERIMENT 1A

The phenotypes caused by mutations in the RAS pathway controlling vulva development in *C. elegans* are easily recognisable. Most of the animals showing these phenotypes are still fertile, giving progenies that are essential for the characterisation of the genetic lesion.

EXPERIMENT 1B

A large number of mutant strains and efficient techniques to inactivate genes in the nematode are powerful tools to investigate the interactions between different genes. More than 86% of the predicted genes of *C. elegans* are represented in RNAi libraries. These libraries have also been successfully used for genome-wide genetic screens (Kamath et al., 2003).

In our experiment, the double mutant *gap-1(ga133)*; lin-7RNAi shows a Muv phenotype. This can be explained by the dual role played by the LET-23 receptor during induction. LET-23 plays both a positive role by transducing the EGF signal into P6.p as well as a negative role by binding most of the LIN-3 ligand and preventing it from reaching the distally located P3.p, P4.p, and P8.p cells. In the *gap-1(ga133)*; lin-7RNAi double mutants, mislocalisation of the LET-23 receptor results in a less efficient binding of LIN-3 by P6.p, so that LIN-3 can diffuse more distally. The increased LIN-3 signalling in P3.p, P4.p, and P8.p adds to the positively sensitised *gap-1(ga133)* background and leads to the activation of these distally located VPCs.

EXPERIMENT 2

For a long time, fine mapping a mutation has been laborious and time consuming. With the adoption of the SNPs as molecular markers, the mutations can be mapped to a small region of a chromosome in a very short time. Several SNPs are known for each chromosome and more can be identified by sequencing DNA fragments from the Hawaii strain and comparing the sequences to those from the Bristol strain.

TIME REQUIRED FOR THE EXPERIMENTS

All the experiments can be performed in less than two weeks. Allow flexible time planning, depending on how fast the worms grow and reach the desired developmental stages.

POTENTIAL SOURCES OF FAILURE

Generating Hawaii males starting from a hermaphrodite plate is not trivial. The most difficult step is getting males after the heat shock and finding them on the plate.

The Vul phenotype is easily recognisable only in mature adults. The eggs accumulate inside the animal and only at a later stage do the larvae hatch and form "bags of worms" within the parent hermaphrodite. Young Vul animals can be incorrectly scored as wild type in Experiment 1. Before collecting data, learn how to recognise young and old animals on a synchronised plate.

In Experiment 2, the DNA digestion (step 8) must be complete. To be certain of complete digestion, the digestion time can be increased, and the absence of undigested bands in the Bristol and Hawaii controls should be verified. Partial digestion will confuse the interpretation of the data.

TEACHING CONCEPTS

SNP mapping. Inferring the position of a gene using small (single nucleotide) DNA sequence differences as molecular markers.

Multivulva phenotype. Formation of ventral blips resulting from induction of more than three vulva precursor cells in the third larval stage of *C. elegans*.

Vulvaless phenotype. Failure to form a functional vulva due to an induction of less than three vulva precursor cells in the third larval stage of *C. elegans*.

Equivalence group. A set of cells with the same developmental potential.

Lateral inhibition. Signalling from one cell that prevents its neighbouring cells from adopting the same developmental fate.

Gain of function mutation. A genomic lesion enhancing the activity of a gene.

Loss of function mutation. A genomic lesion that fully inactivates a gene.

Reduction of function mutation. A genomic lesion reducing the activity of a gene.

RNAi (RNA interference). see Chapter 9.

ALTERNATIVE EXERCISES

QUESTIONS FOR FURTHER ANALYSIS

➤ What phenotype do you expect from a gain-of-function mutation in *gap-1*?

➤ After reading the article by Timmons and Fire, 1998, could you describe how you would build a bacterial strain expressing dsRNA for genes not (yet) represented in the RNAi libraries?

➤ Is it possible to measure relative distances within the chromosome between mutations and the SNP markers by using SNP mapping?

REFERENCES

Ferguson, E. L., and Horvitz, H. R. (1985). Identification and characterization of 22 genes that affect the vulval cell lineages of the nematode *Caenorhabditis elegans*. Genetics, 110, 17–72.

Fire, A., Xu, S., Montgomery, M. K., Kostas, S. A., Driver, S. E., and Mello, C. C. (1998). Potent and specific genetic interference by double-stranded RNA in *Caenorhabditis elegans*. *Nature*, 391, 806–11.

Kamath, R. S., Fraser, A. G., Dong, Y., Poulin, G., Durbin, R., Gotta, M., Kanapin, A., et al. (2003). Systematic functional analysis of the *Caenorhabditis elegans* genome using RNAi. *Nature*, 421, 231–7.

Timmons, L., and Fire, A. (1998). Specific interference by ingested dsRNA. *Nature*, 395, 854.

Wang, M., and Sternberg, P. W. (2001). Pattern formation during *C. elegans* vulval induction. *Curr. Top. Dev. Biol.*, 51, 189–220.

Wicks, S. R., Yeh, R. T., Gish, W. R., Waterston, R. H., and Plasterk, R. H. A. (2001). Rapid gene mapping in *Caenorhabditis elegans* using a high density polymorphism map. *Nature Genetics*, 28, 160–4.

SECTION VI. CLONAL ANALYSIS

14 The role of the gene *apterous* in the development of the *Drosophila* wing

F. J. Díaz-Benjumea

OBJECTIVE OF THE EXPERIMENT The objective of the experiment detailed in this chapter is to study the function of the *Drosophila* gene *apterous* (*ap*) in the development of the wing in clones of mutant cells (Díaz-Benjumea and Cohen, 1993). The analysis of this experiment is a way to introduce the concept of an *organizer center* in development.

DEGREE OF DIFFICULTY Moderate. The experiments require some experience in the handling of flies and in the preparation of microscope slides with fly wings. The original technique of mitotic recombination clones requires the availability of an X-ray source, which is the greatest restriction for this exercise.

INTRODUCTION

Vertebrate and invertebrate appendages have been used as model systems to study the mechanisms that underlie pattern formation in multicellular organisms. Experimental embryologists addressed these studies by microsurgical analysis (grafting, cauterization, etc.). *Drosophila* offers the advantage of a genetic system in which to decipher the roles of specific genes. Many genes involved in patterning *Drosophila* appendages have been identified and characterized. One of the most striking discoveries was the identification of "developmental compartments" (García-Bellido, Ripoll, and Morata, 1973). Compartments were first identified in *Drosophila* as developmental restrictions detected by cell lineage analysis. These restrictions subdivide the wing into four different compartments: anterior (A), posterior (P), dorsal (D), and ventral (V) (Figure 14.1). The A/P restriction is established during early embryonic development, and the D/V restriction is established during larval development (60 hours after egg laying [AEL]). Cell lineage analysis has also indicated that compartments have a polyclonal origin. However, after the specification of a compartment, proliferating cells at both sides of a compartment boundary do not mix and maintain minimum contacts.

There are mutations in *Drosophila* that transform one structure into another. These mutations cause what is called a homeotic transformation, and the genes responsible

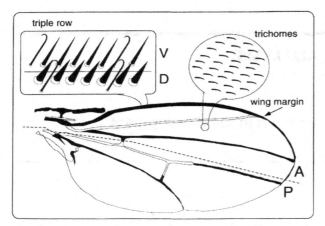

Figure 14.1. A sketch of the *Drosophila* wing. The wing is formed by a single flat epithelium folded over itself at the wing margin (WM). This gives rise to the D and V surfaces that are closely attached in the adult wing. Each cell differentiates a small hair or trichome that can be easily seen under the compound microscope. The WM differentiates a precise pattern of bristles that varies along the WM and is different in each surface. The A WM differentiates a triple row of bristles and the P WM differentiates a double row. The wing is lined by the veins which are linear folds in either the D (full veins) or the V (empty veins) surfaces. The D/V compartment boundary coincides with the WM. The A/P compartment boundary runs along an imaginary line that crosses the wing (dashed line).

for such phenotypes are members of the Hox cluster of genes (described in Chapter 11). Some examples of homeotic transformations are seen in *engrailed* (*en*) mutants, where P wings are transformed into A, and in *Ultrabithorax* (*Ubx*) mutants, where the halteres (dorsal metathoraxic appendages) are transformed into wings (García-Bellido and Santamaría, 1972; Morata and Lawrence, 1975). The characterization of these genes has led to the proposal that compartments are determined by the function of specific genes called "selectors" (García-Bellido, 1975). [For example, in *Drosophila* appendages, the expression of the gene *en* is restricted to P cells. In *en* mutant flies P cells acquire A identity. Once *en* expression is activated in early embryo development, it activates an autoregulatory mechanism that maintains its expression when cells divide, establishing the posterior lineage. Together these results indicate that *en* is the selector gene that specifies P identity.] From these results it was proposed that *Drosophila* appendages develop through the reiterative subdivision of a presumptive field into different compartments by the expression of distinct selector genes (García-Bellido, 1975).

The *apterous* (*ap*) gene encodes a transcription factor required for the development of the wing (Cohen et al., 1992). *ap* expression is restricted to the D wing cells and homozygous (viable) *ap⁻* adult flies have no wings. If *ap* is the selector gene that specifies D identity, the expected *ap* mutant phenotype should be wings with two V sides. Instead, surprisingly, both D and V surfaces of the wing are lost, even though the V surface does not express *ap*. This result suggests that the development of the V wing requires the expression of *ap* in D wing cells. The experiment described in this chapter

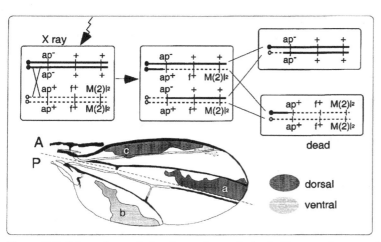

Figure 14.2. Scheme for producing clones of *ap* mutant cells. X rays produce a random DNA breakage that, when repaired, can give rise to an exchange of DNA between chromatids in homologous chromosomes. If this exchange happens proximally to the *ap* gene, it will yield, after a cell division and in 50% of the cases, a cell *ap/ap* and another cell +/+. The *ap* mutant cells will also be *f*⁻ and *M*⁺, which will mark the clone cells and will give a proliferative advantage. The other cell will be homozygous *M*⁻, which is cell lethal. Clones can be independently scored in each wing surface by focusing up and down under the compound microscope. Clone size depends on the time of irradiation, which determines the number of cell divisions that remain until the end of development. Clones induced after 60 hours AEL will be either D (a) or V (b). Clones induced before 60 hours AEL will mostly be D+V (c). Clones will never cross the A/P compartment boundary (dashed line) that is established in early embryo development.

is based on this paradoxical result and addresses the question of how *ap* functions in wing development. In this experiment, we analyze flies that are mosaic for the *ap* mutation, allowing the examination of how *ap* mutant cells behave in the context of wild-type cells.

Wing imaginal discs develop from a group of about 50 cells in the first instar larva. The cells within this group that are destined to be P cells express *en*, whereas those destined to be A cells do not. These 50 cells proliferate to give rise to 50,000 cells at the end of larval development. Thirty thousand of these cells give rise to both dorsal and ventral wing surfaces, whereas the rest give rise to the body wall (dorsal notum and pleura) (García-Bellido, 1971) (Figure 14.1). By X-ray irradiation of heterozygous larvae, it is possible to induce recombination in mitotic cells (Figure 14.2), thus generating homozygous mutant clones of cells from a background of heterozygous mutant cells at any developmental time. To recognize the clones of mutant cells, the chromosome that carries the *ap* mutant allele is also mutant for a gene required for the formation of trichomes, small hairs, on every cell of the cuticle (García-Bellido and Dapena, 1974). Thus, cells mutant for this gene have altered the trichome morphology, and are easily distinguished from wild-type cells (Figure 14.1). To produce larger clones of mutant cells, flies mutant for a *Minute* (*M*) gene, which encodes ribosomal proteins, are used. *Minute* mutant flies are homozygous lethal, but viable when heterozygous, and show a developmental delay. By generating *ap*⁻ *M*⁺ homozygous clones from a background of *ap*⁻ *M*⁺/*ap*⁺

M^- cells (Figure 14.2), the ap^- mutant cells will have a proliferative advantage over the surrounding M^-/M^+ cells, thus generating larger clones (Morata and Ripoll, 1975).

MATERIALS AND METHODS

EQUIPMENT AND MATERIALS

Per student

Dissecting and compound microscopes

Hair brush

Fly aspirator

50-ml etherizing bottle (see Roberts, 1986, p. 21 for a description)

To use CO_2: CO_2 tank, a plastic bottle and a sintered glass plate are required (see Roberts, 1986, p. 22 for a description).

Morgue (a bottle with 70% ethanol for the discarded flies)

2-ml tubes to keep flies in SH (70% ethanol, 30% glycerol)

2 fine-tipped forceps for dissecting flies (watchmaker's forceps)

Micro slides

Coverslips (24 × 40)

Glass multiwell

Small foam rubber mat

Per practical group

25°C incubator

100-ml vials with cotton caps for fly culture

X-ray source (or γ-rays; usually available in hospitals)

Biological material

FLY STRAINS

ap^{UGO35}/CyO

$f^{36a}/FM6; Dp(2R)f^+_{44} M(2)58F/CyO$

Flies of this genotype can be easily sorted out from anaesthetized flies under the dissecting scope (males *Bar*$^+$ (not carrying *FM6*) *Cy*$^+$ (not carrying *CyO*) which means normal eyes and straight wings).

STOCK REQUEST. Fly stocks can be requested from the Bloomington, Drosophila Stock Center (http://fly.bio.indiana.edu.)

REAGENTS

Diethyl ether (Merck)

Ethanol (Merck)

Glycerol (Merck)

Acetone (Merck)

Araldite (Fluka)

Fly food components: agar, yeast, soy and corn flour, malt extract, syrup, propionic acid, and nipagin.

PREVIOUS TASKS FOR STAFF

Maintenance of fly strains. The *Drosophila* life cycle at 25°C lasts 10 days from egg laying to the eclosion of the adult fly. Mutant strains usually show some developmental delay. This delay is more apparent in *Minute* stocks, which require special care. Fly strains should be maintained in individual vials and transferred into new vials with fresh food every four days. Overcrowding in the cultures should be avoided. Discarded vials containing larvae and adult flies should be sterilized in an oven or frozen in a −20°C fridge.

Preparation of fly food. Enough fly food must be prepared to maintain the stocks. Staff can follow this recipe:

Water	10 l
Agar	120 g
Yeast	180 g
Soy flour	100 g
Corn flour	800 g
Malt extracts	800 g
Syrup	220 g

Mix all the components and boil the mix for one hour. Add water to compensate for the loss of liquid during boiling. When the temperature is lower than 60°C add 62.5 ml of propionic acid and 24 g of nipagin and pour the mix into the vials (25 ml each). Let cool before tapping, cover with a gauze to avoid contamination and then keep in the fridge. Fly food should keep in good conditions for a couple of weeks.

Distributors of *Drosophila* research supplies: Labscientific, Inc. (USA). Sci-Mart, Inc. (USA, UK, or Germany).

Solutions
SH: 70 glycerol/30 ethanol
Ethanol: acetone (1:1)

PROCEDURES

Fly crosses. To collect males or females from a fly stock, vials containing flies should be tapped on the rubber mat, then the cotton cap removed and, with a fast movement, flies should be transfered into the etherizing bottle (USE ETHER IN VENTILATED AREAS). Once flies stop moving (a few seconds) they can be transfered on a small piece of hard paper for their analysis under the dissecting microscope. Experience will tell you how much time you have before flies recover from the anaesthesia.

To insure the virginity of the female in the crosses, these must be no more than 8 hours old. Thus, vials giving new flies should be cleared of flies at 8-hour intervals; newly-hatched females should be collected (for more information on handling flies see Strickberger (1962) and Roberts (1986)).

Mounting wings. To mount wings, separate them from the fly body with fine-tipped forceps in SH, in a multi-well glass slide. Wash with water, dehydrate (70% ethanol, 100% ethanol, 1/1 ethanol/acetone, 100% acetone), and then mount the wings in araldite. These slides can be maintained indefinitely.

OUTLINE OF THE EXPERIMENTS

PROPOSED EXPERIMENT

1. Cross around 40 virgin females from the stock f^{36a}/FM6; Dp(2R)f_{44}^+ M(2)58F/CyO with the same number of males from the stock ap^{UGO35}/CyO. The appropriate number of females required for the cross depends on the size of the vials. Too many females will produce overcrowding in the culture, which will mainly affect the survival of the *Minute* larvae.
2. Leave flies mating in a 25°C incubator and after two days transfer them into a new vial.
3. Transfer flies into a new vial once every day during the development of the experiment. To keep the flies laying eggs, it is a good idea to replace flies that stick on the food with new females each day.
4. Irradiate 80 ± 12 hours old larvae at 1,000 R (approximately 1,000 cGy) under an X-ray source. Because of the developmental delay of the *Minute* stock (19 hours), this developmental time corresponds to late second instar larvae (Ferrús, 1975).
5. Leave the larvae to develop until adulthood. Once the flies hatch, sort males with the genotype f^{36a}; Dp(2R)f_{44}^+ M(2)58F/ap^{UGO35} out and place them in a vial with SH solution.
6. Mount the wings according to the protocol in the previous section.
7. To find mutant clones, individual wings may be scored under the compound microscope looking for patches of *forked* (f) trichomes. Both wing surfaces must be scored independently focusing up and down to distinguish the trichomes of each surface.

Data recording. f mutation affects the morphology of bristles and trichomes that are short, bent, and with split ends in this mutant.

➤ Determine the location of the clone (dorsal or ventral).
➤ Determine the clone location within the anterior–posterior axis (anterior or posterior compartments).
➤ Determine the autonomous (inside the clone) and non-autonomous (outside the clone) phenotypes induced by the mutant cells, especially the differentiation of bristles or outgrowth.
➤ Draw the outline of the interesting clones in the provided wing blade profile (Figure 14.3).

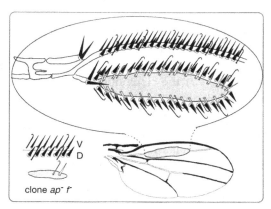

Figure 14.3. Phenotype of *ap* mutant clones. A scheme of an *ap⁻* clone in the D compartment. The clone, identified by the *f* marker in trichomes and bristles (empty bristles), induces the formation of an ectopic WM. In this case, a clone near the A WM, the clone induces an ectopic triple row of bristles. The outgrowth that these clones usually produce in the surrounding wild-type cells is not represented. Reprinted from *Cell*, Vol. 75, Diaz-Benjumea and Cohen, "Interaction between dorsal and ventral...", pp. 741–752, Copyright (1993), with permission from Elsevier.

EXPECTED RESULTS

➤ The D/V segregation occurs at 60 hours AEL. Clones induced in these experiments at 72 ± 12 hours of development (80 ± 12 hours *Minute* time) should be mostly either D or V with a similar frequency. Ventrally located *ap⁻* clones should not produce any phenotypic consequences and can be identified under the compound microscope by the *f* marker. Dorsally located *ap* clones can be easily scored because of their obvious phenotypic effects. They induce the formation of an ectopic wing margin (WM) in the periphery of the clone (Figure 14.3). This ectopic WM reproduces the pattern of the closest located normal WM, and thus will appear as a triple row of bristles in the anterior-most located clones and a double row in the posterior-most located clones. Note that in this ectopic WM, ventral bristles belong to the clone and are f^- while dorsal bristles are f^+.

➤ Dorsally located clones produce an outgrowth of the wild-type surrounding cells. The extent of this outgrowth depends on the location of the clone in the wing. Clones located close to the endogenous WM show little outgrowth, whereas clones located far from the WM produce more extensive outgrowth.

➤ A fraction of the dorsal f^- clones will not show any phenotype. These clones are the product of recombination events proximal to f (between f and the centromere) but distal to *ap*. Thus, they will be f^- ap^+ with no phenotypic consequences.

DISCUSSION

The expression of the *ap* gene in the wing is restricted to dorsal cells. In *ap* mutant clones in the dorsal compartment, cells differentiate ventral bristles. Hence we infer that *ap* determines dorsal identity while cells that do not express *ap* acquire a ventral identity by default. *ap⁻* clones have an unusual non-autonomous effect over the surrounding wild-type cells. Firstly, cells at the boundary of *ap* mutant clones differentiate into an ectopic WM. Wild-type cells adjacent to the clone boundary differentiate into the dorsal row of bristles of the ectopic WM. *ap* mutant cells at the clone border differentiate into the ventral row of bristles. Hence we conclude that short-range cell interactions between *ap*-expressing and *ap*-nonexpressing cells determine the WM fate and lead to the differentiation of the characteristic pattern of bristles of the WM. Secondly, wild-type cells in the proximity of the clone over-proliferate and cause an outgrowth of the wing tissue. This outgrowth duplicates the characteristic structures of the normal D wing. Hence we conclude that the boundary of *ap* expression acts in development as an organizing center that promotes the proliferation and patterning of the rest of the wing cells. Thus, the *ap* gene may have three roles in wing development:

➤ *ap* expression determines D identity.
➤ *ap* expression is maintained by lineage, thus the border of *ap* expression coincides with the D/V cell lineage restriction.
➤ Short-range cell interactions between cells at both sides of the *ap* expression boundary establish an organizing center that has a long-range effect over the rest of the wing cells.

This organizing center promotes growth and patterns the whole wing. These data explain why, although *ap* is expressed only in D cells, in *ap* mutants the whole wing is lost, since the D/V organizing center at the WM is not established.

TIME REQUIRED FOR EXPERIMENTS

To amplify the *Minute* stock and to get enough virgin females will take close to one month. To get a significant number of clones will take another month. To dissect the wings, to mount the slides, and to analyze the clones under the compound microscope will take one or two days. Practical instructors could simplify the experiments so that students would just perform wing mountings and clonal analysis. However, if an inquiry-style lab is adopted, complete experiments from Alternative Exercises may be carried out.

POTENTIAL SOURCES OF FAILURE

Overcrowding in fly culture delays larval development. A mistake in timing the larvae to be irradiated will affect the frequency of clones. The irradiation of larvae at earlier stages gives a much lower frequency of clones. As already discussed, an X-ray apparatus is required to carry out these experiments. The location of *ap* on the right arm of the second chromosome is proximal to any known FRT insertion, thus, in this experiment, it is not possible to follow the FLP/FRT protocol (see Chapters 15 and 16).

TEACHING CONCEPTS

Clonal analysis. This involves labelling a cell at a precise time of development and analyzing its progeny in the adult. Such analyses indicate the number of times a cell divides, which structures the progeny will develop into, and whether there are development restrictions to the fate of the cell progeny (cell lineage analysis). If the cuticular marker has a morphogenetic mutant allele in *cis* (in the same chromosome arm) gene function may be studied at a cellular level (morphogenetic mosaics).

Compartment. A group of cells related by lineage that segregate from other cells. Compartments have polyclonal origin.

Organizing center. This concept comes from pioneering experiments by Spemann and Mangold (1924) in amphibian embryos (see Hamburger 1988). They established that the "organizer" is a small group of cells that exerts an organizing effect on its environment in such a way that, if transplanted to an ectopic position, causes there the formation of a secondary embryonic axis by recruiting and changing the fate of surrounding cells.

Selector gene. A gene whose expression determines the activation of a specific developmental pathway.

ALTERNATIVE EXERCISES

PROPOSED EXPERIMENT

Similar results to those obtained in clones of *ap* mutant cells can be obtained by misexpressing *ap* in the V compartment of the wing (Rincón-Limas et al., 1999). Inducing clones of *ap*-expressing cells in the V compartment will transform V fate into D fate and generate in the periphery of the clone an ectopic WM and an outgrowth. In this case, the D bristles of the ectopic WM form from *ap⁻* cells, while the V bristles will differentiate from wild type cells at the periphery of the clone.

These clones can be induced by applying a combination of GAL4/UAS and FLIPOUT techniques (Pignoni and Zipursky, 1997). Virgin females from *y hsFLP122; Act5C> y⁺ > GAL4* should be crossed with *UAS-ap* males. Seventy-two-hour AEL larvae should be heat shocked in a water bath at 35°C for 6 minutes. Frequency of clones depends on the temperature and the length of the heat shock. The expression of the *FLP* recombinase under the control of the heat-shock promoter will induce recombination between the two *FRTs* (>) which will result in loss of the *y⁺* marker in individual cells, as well as placing *GAL4* under the control of the *Act5C* promoter that is ubiquitously expressed (Figure 14.4). The expression of the *GAL4* gene will activate *ap* expression (*Act5C> GAL4/UAS-ap*) in individual cells. Clones should be identified in the male progeny by the loss of the *yellow⁺* (*y*) marker that will produce bristles with a light yellow pigmentation.

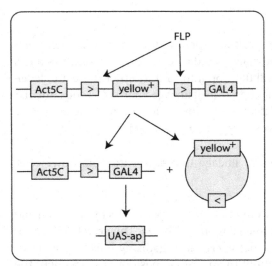

Figure 14.4. Scheme for producing clones of *ap* expressing cells. The heat shock activates the expression of the *FLP* recombinase that acts on the two *FRT* sites. As a result of the recombination both one *FRT* and the *yellow*+ gene are lost and the *GAL4* gene is emplaced under the control of the *Act5C* promoter. Individual cells in which this event happens become *yellow*⁻ and will express *Act5C>GAL4>UAS-ap*.

Fly stocks are also available to produce clones of *engrailed* loss-of-function alleles to analyze their function in the posterior compartment (García-Bellido and Santamaría, 1972) or to produce clones of *engrailed* expressing cells in the anterior compartment (Zecca, Basler, and Struhl, 1995).

QUESTIONS FOR FURTHER ANALYSIS
- ➤ Why does the amount of non-autonomous outgrowth depend on the position of the induced clone within the D compartment?
- ➤ If *Drosophila* selector genes act in a combinatorial way, what would be the outcome of eliminating *apterous* and *engrailed* gene functions in the same clone? How would you perform this clonal analysis? As *apterous* is very near the centromere, at what frequency would you expect to see double mutant clones?

REFERENCES

Cohen, B., McGuffin, E., Pfeifle, C., Segal, D., and Cohen, S. M. (1992). *apterous*, a gene required for imaginal disc development in *Drosophila* encodes a member of the LIM family of developmental regulatory proteins. *Genes Dev.*, 6, 715–29.

Díaz-Benjumea, F. J., and Cohen, S. M. (1993). Interaction between dorsal and ventral cells in the imaginal disc direct wing development in *Drosophila*. *Cell*, 75, 741–52.

Ferrús, A. (1975). Parameters of mitotic recombination in *Minute* mutants of *Drosophila melanogaster*. *Genetics*, 79, 589–99.

García-Bellido, A. (1971). Parameters of the wing imaginal disc development of *Drosophila melanogaster*. *Dev. Biol.*, 24, 61–87.

García-Bellido, A. (1975). Genetic control of wing disc development in *Drosophila*. In *Cell Patterning*, ed. P. Lawrence, Ciba Found Symp. 29, pp. 161–82. Amsterdam: Elsevier.

García-Bellido, A., and Santamaría, P. (1972). Developmental analysis of the wing disc in the mutant *engrailed* of *Drosophila melanogaster*. *Genetics*, 72, 87–104.

García-Bellido, A., Ripoll, P., and Morata, G. (1973). Developmental compartmentalisation of the wing disk of *Drosophila*. *Nat. New Biol.*, 245, 251–3.

García-Bellido, A., and Dapena, J. (1974). Induction, detection and characterization of cell differentiation mutant in *Drosophila*. *Molec. Gen. Genet.*, 128, 117–30.

Hamburger, V. (1988). *The Heritage of Experimental Embryology: Hans Spemann and the Organizer*. Oxford: Oxford University Press.

Morata, G., and Lawrence, P. A. (1975). Control of compartment development by the *engrailed* gene in *Drosophila*. *Nature*, 255, 614–7.

Morata, G., and Ripoll, P. (1975). *Minutes*: Mutants of *Drosophila* autonomously affecting cell division rate. *Dev. Biol.*, 42, 211–21.

Pignoni, F., and Zipursky, S. L. (1997). Induction of *Drosophila* eye development by *decapentaplegic*. *Development*, 124, 271–8.

Rincón-Limas, D. E., Lu, C. H., Canal, I., Calleja, M., Rodríguez-Esteban, C., Izpisúa-Belmonte, J. C., and Botas, J. (1999). Conservation of the expression and function of *apterous* orthologs in *Drosophila* and mammals. *Proc. Natl. Acad. Sci. USA*, 96, 2165–70.

Roberts, D. B. (ed.) (1986). *Drosophila: A Practical Approach*. Oxford: IRL Press.

Spemann, H., and Mangold, H. (1924). Über die Induktion von Embryonalanlagen durch Implantation artfremder Organisatoren. *Arch. Mikrosk. Anat. Entwicklungsmech.*, 100, 599–638.

Strickberger, M. W. (1962). *Experiments in Genetics with Drosophila*. New York: John Wiley and Sons.

Zecca, M., Basler, K., and Struhl, G. (1995). Sequential organizing activities of *engrailed, hedgehog* and *decapentaplegic* in the *Drosophila* wing. *Development* 121, 2265–78.

15 *Extramacrochaetae*, an example of a gene required for control of limb size and cell differentiation during wing morphogenesis in *Drosophila*

A. Baonza

OBJECTIVE OF THE EXPERIMENT The goal of these experiments is to study the effects on cell proliferation and vein patterning caused by mutations in the *emc* gene during development of the imaginal wing discs. To this end, two different experiments will be performed:

1. Twin analysis. Clones of *emc* mutant cells will be generated by mitotic recombination at the same time that "twin" clones which are *emc*$^+$ are generated (control). With this experiment we will be able to analyse the difference in size (number of cells), shape, and distribution of *emc* mutant clones in comparison to control twins.
2. *Minute* analysis. *emc*^{1} M^+/*emc*^{1} M^+ mutant cells will be induced in a M^-/M^+ mutant background. In a M^-/M^+ heterozygous background M^+/M^+ cells exhibit a growth advantage that allows clones initiated early in development to occupy large wing territories. This experiment, which supports the notion that size control is not based on cell proliferation rate control, will allow us to study the effects caused by the presence of a large region of *emc* mutant cells in the growing wing.

DEGREE OF DIFFICULTY Moderate.

INTRODUCTION

The development of the imaginal wing discs of *Drosophila melanogaster* is a classic model system for analysis of the cellular and genetic bases of cell proliferation and pattern formation. Each wing disc gives rise to one wing and one-half of the adult thorax. These discs arise from a group of 20–40 cells that are segregated from the embryonic cells during early embryogenesis. These founder cells proliferate extensively during the larval stages until the beginning of metamorphosis. During this period of cell proliferation the morphological features of the adult wing (size, vein, and sensory element patterns) are defined. The adult wing consists of a folded epithelium with the dorsal and ventral cell layers attached to each other by their basal membranes. Individual cells

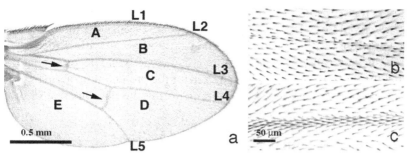

Figure 15.1. (a) Vein pattern of an adult wing of *Drosophila*. Five longitudinal veins (L1–L5) and two transverse veins (arrows) are visible. The longitudinal veins define five inter-vein regions (A–E). The cells that form the veins are more compacted and pigmented than the inter-vein cells (see figures b and c). (b and c) Each vein has a dorsal and a ventral component. It is possible to distinguish each component because only one component protrudes from the wing surface (producing a corrugation) and this consists of cells that are more compacted and pigmented (see c) than the other surface (b). The relative position of the corrugated and non-corrugated aspects of every vein constitutes an invariant pattern. Vein L2 and the proximal part of L4 have the corrugated aspect in their ventral surfaces, whereas veins L3, L5, and distal part of L4 are corrugated dorsally.

can be easily identified because they differentiate a unique hair or trichome. The veins appear as longitudinal stripes of cells which are more compacted and pigmented than the inter-vein cells (Figure 15.1a, b, and c). The wing has an invariant pattern consisting of five longitudinal and two transverse veins.

In the embryonic wing anlage, two populations of cells are already defined that constitute two clonal compartments, one anterior (A) and one posterior (P). During growth the discs become further subdivided by a new clonal restriction (see Chapter 14). Mutations in genes required for the definition of the compartments affect the final size of the wing and the vein pattern, indicating the importance of compartment boundaries as major references for the ordered growth and patterning of the wing discs (see Chapter 14). In addition to the restriction associated with these major boundaries, clonal analysis has revealed that veins also often coincide with lineage restriction borders, hence, inter-vein regions can be considered as autonomous units of proliferation.

The function of several genes and intercellular signalling pathways, which are not involved in the definition of the compartment boundaries, is required for the normal patterning and growth of the wing discs. The activity of these genes and signalling pathways depends on the correct definition of compartment. The function of these genes is necessary in order to translate the primary references established in the boundaries to the definition of the final pattern and size of the wing. Among these genes is *extramacrochaetae (emc)*. The *emc* gene encodes a non-basic helix-loop-helix (HLH) protein that antagonises the activity of other bHLH proteins (Ellis, Spann, and Posakony, 1990, Garrell and Modolell, 1990). It has been shown that EMC works as a repressor in different processes during the development of *Drosophila* (Cubas et al., 1991, Cubas and Modolell, 1992, Cubas, Modolell, and Ruiz-Gómez, 1994). One of the processes where the function of *emc* has been comprehensively analysed is in the development of the peripheral nervous system. In this process, *emc* is involved in the repression of sensory precursor cells (Cubas et al., 1991; Cubas and Modolell 1992). Mutations in *emc* cause the appearance of ectopic bristles or macrochaetae in the adult notum (the name *extramacrochaetae* is derived from this phenotype) (Botas et al., 1982). *emc* mutant alleles also prevent the growth of the inter-vein region and disrupt the vein pattern (de Celis, Baonza, and García-Bellido, 1995; Baonza and García-Bellido, 1999). The latter phenotypes indicate that the function of *emc* is also required for the proliferation of inter-vein cells, and consequently, to define the vein patterning during the development of the wing discs.

MITOTIC RECOMBINATION CLONES

Most of the genes involved in wing disc development are also required during embryogenesis; thus, mutations in these genes are frequently recessive lethal and the homozygous individuals die during embryogenesis. In order to analyse the function of these genes in wing disc development, it is necessary to study the phenotype of homozygous mutant cells in mosaic organisms. One of the classic techniques to make mosaics is to induce mitotic recombination clones by X-rays (see Chapter 14). These clones are generated by irradiating larvae heterozygous for a mutant allele in cis with a genetic marker on the same chromosome arm (Figure 15.2a). In this case, the genetic marker is a recessive mutation in a gene that affects the differentiation of the adult cuticle: homozygous mutant cells differentiate three hairs instead of one (Figure 15.3a and c). This marker will allow us to distinguish the mutant cells from the surrounding wild-type cells in the adult wing. X-ray irradiation produces fractures in the DNA, causing (at low frequency) the interchange of chromatids in homologous chromosomes during the G2 stage of the cell cycle. The cells in which this event occurs will, following segregation of the chromatids, divide to form two different daughter cells: one wild-type and another homozygous for the mutant allele studied and the genetic marker (Figure 15.2; see also Chapters 14 and 16). The progeny of each daughter cell may give rise to a clone of genetically identical cells that can be examined in the adult.

A more efficient method of inducing clones by mitotic recombination is to use the FLP/FRT system (Xu and Rubin, 1993). This method is based on the introduction of two yeast sequences into *Drosophila*. The yeast recombinase enzyme (Flipase (FLP)) recognises and catalyses recombination of a yeast sequence known as FRT. In order to induce mitotic clones, the FRT sequence has to be localised in cis with the mutant allele of the gene that is the subject of analysis. On the homologous chromosome, in the same position, there is another FRT sequence. Recombination between the two FRT sequences occurs only when the FLP enzyme is present (Figure 15.2b). The gene that encodes this enzyme can also be placed under the control of a heat shock promoter. A 37°C heat shock then induces the expression of this gene and therefore the mitotic recombination. In the experiments proposed we will use both approaches (X-rays and FRT system).

MATERIALS AND METHODS

EQUIPMENT AND MATERIALS

The equipment, materials, and reagents per student and practical group used in this practice are identical to those shown in Chapter 14.

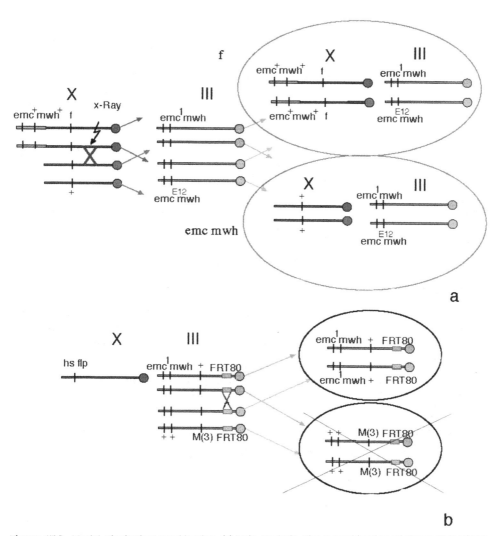

Figure 15.2. Model of mitotic recombination. (a) Twin analysis. The recombination of chromatids of two homologous X-chromosomes induced by X-rays may give rise to two different daughter cells. One cell may receive two copies of the emc^+ mwh^+ duplication. In this cell, the duplication rescues the emc and mwh mutations on the third chromosome. This cell is also homozygous for $forked(f)$ and it is possible to identify its offspring using this marker. If this occurs, appropriate recombination and segregation, the other cell loses the duplication of emc and mwh on the X chromosome and becomes mutant for these two genes. (b) *Minute* technique. The induction of mitotic recombination by *flipase* in hs-flp; $M(3)i^{55}$ FRT80B/ emc^1 mwh FRT80B larvae gives rise to one cell that loses the *Minute* mutation but is emc mwh mutant and one *Minute* homozygous cell that does not proliferate and dies. The emc mwh M^+ homozygous cell exhibits a proliferative advantage with respect to the M^+/M^- surrounding cells and will give rise to a large clone of cells that are genetically identical.

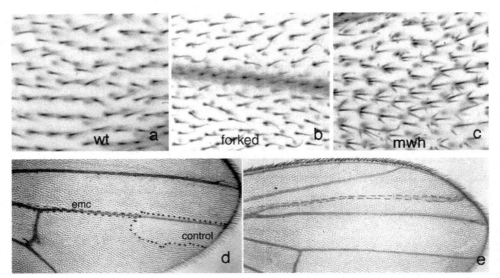

Figure 15.3. (a) Wild-type trichomes (hairs) are produced by the cells that form the adult wing of *Drosophila*. Each cell differentiates a unique trichome. (b) Phenotype caused by a mutant allele of *forked* (*f*). Trichomes appear curled (compare with a). (c) Phenotype of *mwh* mutant trichomes. Each cell differentiates 3 or 4 trichomes. (d) Clones of *emc¹ mwh/ emc^E12 mwh* mutant cells (dashed line) are very elongated and appear mainly over the vein. Compare the shape and size of the mutant clone with the control twin clone (dotted line). The mutant clone contributes mostly to veins and is smaller than the control clone. (e) Occasionally, *emc¹ mwh/emc^E12 mwh* mutant clones appear in the middle of inter-vein region, but in these cases they are also very elongated and differentiate as vein tissue. Parts d–e of this figure are reprinted from de Celis et al. (1995). Copyright (1995), with permission from Elsevier.

Biological material

FLY STRAINS. The strains to be used are the following:

> *DpM2 mwh⁺ f ³⁶ᵃ/FM6; emc¹ mwh/ TM2*
> *emc^E12 mwh / TM2*
> *hsflp; M(3)i ⁵⁵ FRT80B/ TM6*
> *emc¹ mwh FRT80B/ TM2*

DESCRIPTION OF ALLELES

➤ *emc¹* (3-0.0) : a hypomorphic loss-of-function allele.

➤ *emc^E12* (3-0.0): a deficiency covering the entire *emc* locus.

➤ *DpM2* is a duplication of *emc* and *mwh* present at the tip of X chromosome

➤ *mwh* (3-0.3): a mutant allele that causes each cell to produce multiple hairs (Figure 15.3c). This allele is used as a cuticular marker.

➤ *forked* (*f ³⁶ᵃ*, 1-56.7): another cuticular marker which produces the differentiation of curled hairs (Figure 15.3b).

BALANCER CHROMOSOMES. *TM2* has a dominant mutation in the *Ubx* gene causing a very weak transformation of the haltere to wing (see Chapter 11). This balancer is difficult to see. *FM6* has a dominant mutation in *Bar* gene, resulting in eyes that are kidney shaped. *TM6B* can be recognised because the flies are shorter (as are the pupae) and there is a group of long bristles (4–5) in the upper part of the thorax.

STOCK REQUEST. Fly stocks can be requested from the Bloomington Drosophila Stock Center (http://fly.bio.indiana.edu), and the author.

REAGENTS
Reagents are the same as in Chapter 14.

PREVIOUS TASKS FOR STAFF
Maintenance of fly strains and preparation of fly food and solutions as in Chapter 14.

PROCEDURES
Mounting of wings. Adult wings of the appropriate genotype must be dissected in water under the dissecting microscope. Wash the isolated wings twice with 100% ethanol and then remove the ethanol and add a solution of lactic acid/ethanol (1:1). Transfer the wings in that solution to a slide and mount them for microscopic examination (see also Chapter 14).

LARVAE IRRADIATION. Cut vials with fly food at about the edge of the food. The upper part of the vials should be fixed to the lower part with tape. We will use these bottles or vials to irradiate the larvae with X-rays. An X-ray machine that allows irradiation at the following doses is required: 1,000 R (approximately 1,000 cGy); 300 R (approximately 300 cGy)/min, 100 KV, 15 mA.

Heat shock induction of clones. To induce clones with the FRT system a water bath that maintains a 37°C temperature is needed. The vials will be partially immersed in the water during the heat-shock period.

OUTLINE OF THE EXPERIMENTS

EXPERIMENT 1. TWIN ANALYSIS
Mitotic recombination on the X chromosome of $DpM2\ mwh^+\ f^{36a}/+$; $emc^1\ mwh/emc^{E12}$ *mwh* flies will give rise to one cell that will lose the duplication and therefore will be mutant for the heteroallelic combination $emc^1\ mwh\ /emc^{E12}mwh$, the strongest cell viable combination of *emc* alleles, while the sister cell will be homozygous for $Dp\ M2$ $mwh^+\ f^{36a}$, and be marked with *forked* (Figure 15.2a and 15.3b). To do this experiment, set up a cross with the following stocks (symbols are explained in Chapter 11):

1. Females $DpM2\ mwh^+\ f^{36a}/FM6$; $emc^1\ mwh/TM2$ will be crossed to male emc^{E12} *mwh/TM2.*
2. After 4 days larvae of the cross will be irradiated with X-rays. Mitotic recombination in the X chromosome in larvae which are $DpM2\ mwh^+\ f^{36a}/+$; $emc^1\ mwh/\ emc^{E12}$ *mwh* gives rise to emc^1/emc^{E12} cells labelled with *mwh* and twin wt cells labelled with *forked* (Figure 15.2a and 3).

Table 15.1. X-ray schedule

Mon	Tue	Wed	Thur	Fri	Sat	Sun	Mon
B1	B1 to B2	B2 to B3	B3 to B4	B4 to B5			
			X-ray B1	X-ray B2	X-ray B3	X-ray B4	X-ray B5

3. After 10–11 days females of the *DpM2 mwh$^+$ f^{36a}/+*; *emc^1 mwh/emc^{E12} mwh* genotype will be selected and mounted for analysis under microscope.

Protocol for the twin analysis

1. 10–20 *emc^1 mwh/TM2* males will be crossed with 30–40 *DpM2 mwh$^+$ f^{36a}/ FM6*; *emc^{E12} mwh/TM2* females in cut vials with fly food.
2. The cross will be maintained at 25°C for 1 day, the next day the parents will be transferred to a new bottle. The old bottle is designated "cross day 1" and the new bottle "cross day 2" (B1 to B2 in Table 15.1).
3. Repeat the process for several days and use a similar nomenclature. Always leave the old bottle with the larvae at 25°C.
4. Three days later after the first cross, irradiate the larvae of the "cross day 1" bottle. For the irradiation remove the upper part of the cut vials and leave the lower part with the larvae under the X-ray source. The next day repeat the same process with the bottle cross day 2 (X-ray B2 in Table 15.1).
5. After 11–14 days select females of *DpM2 mwh$^+$ f^{36a}/+*; *emc^1 mwh/ emc^{E12} mwh* genotype and store them in the solution to preserve flies.
6. The next day, dissect the wings under the microscope and mount them in the mounting solution (lactic acid/ethanol 1:1). It is possible to use the following schedule.

EXPERIMENT 2. *MINUTE* ANALYSIS

In this experiment we will assay the effect upon wing growth of large *emc* mutant clones. In order to generate large territories of *emc* mutant cells we will use the *Minute* technique and the *emc* allele *emc^1*. This allele causes an insufficiency for *emc* weaker than the heteroallelic combination *emc^1/emc^{E12}*. The combination of the M^+ condition and a weaker loss-of-function allele of *emc* gives rise to large *emc* mutant clones. There are multiple *Minute* mutations throughout the four chromosomes of *Drosophila*. These are dominant mutations that cause a defect in the production of ribosomal protein. Heterozygous M^-/M^+ flies are of normal size but their development is delayed when compared to wild-type flies (Morata and Ripoll, 1975). This delay is a consequence of a lower proliferation rate of the M^-/M^+ cells. Clones of M^+/M^+ cells in an M^-/M^+ background exhibit a proliferative advantage that allows clones initiated early to occupy large territories.

In order to generate *emc mwh M^+* homozygous cells in M^-/M^+ background the *Minute* mutation must be recombined in *cis* with the FRT on the homologous chromosome (Figure 15.2b). The following stocks will be crossed: females *hs-flp; M(3)i^{55}*

FRT80B/ TM6 by males *emc¹ mwh FRT80B/ TM2*. As the *hs-flp* is on the X chromosome and we will use females with this chromosome, all the flies will carry a copy of the *flipase*. Three days after the cross the larvae will be heat-shocked to induce *flipase* expression. The Flipase enzyme will recombine the two FRT sequences in *hs-flp; M(3)i⁵⁵ FRT80B/ emc¹ mwh FRT80B* larvae so that after the chromosome segregation and cell division, one cell will be *emc¹ mwh M⁺ FRT80B* homozygous. This cell will have a proliferating advantage over the surrounding *M⁺/M⁻* cells and will give rise to a large clone. The other daughter cell resulting from the segregation of the mitotic recombination (*M⁻ FRT80B / M⁻ FRT80B*) will die, since the *Minute* mutation is recessive cell lethal (Figure 15.2b).

Protocol for the Minute analysis

1. Cross 10–20 males *emc¹ mwh/TM2* to 30–40 females *hs-flp; M(3)i⁵⁵ FRT80B/TM6* in vials with fly food.
2. Proceed as in the preceding protocol but instead of irradiating the larvae heat shock them at 37°C for 60 minutes.

Data recording. As described in Chapter 14.

EXPECTED RESULTS

TWIN ANALYSIS (EXPERIMENT 1)

Different characteristics of the *emc¹/emc^{E12}* mutant cells must be analysed:

1. Size of mutant clones compared with control twin clones. Mutant *emc¹ mwh/emc^{E12} mwh* clones are much smaller than their twin control clones (*f*). The average ratio of mutant clone/twin clone is 0.2–0.4 while in the control it is close to 1. Frequently *f emc⁺* control clones appear without the expected *emc¹ mwh /emc^{E12} mwh* clone. Because we have used X-rays that produce random fractures in the DNA, we would expect that the recombination events between the duplication *DpM2* and the marker *f* would give rise to *f* control clones but not *emc¹ mwh/emc^{E12} mwh* mutant twins. However, the chromosomal location of the *f* marker relative to the *DpM2* ensures that most of *f* control clones have an *mwh* twin (90% in control experiment). The relative absence of mutant clones, compared to their control twins, can be explained by the poor viability of these mutant cells (de Celis et al., 1995).
2. Distribution and shape of the mutant clones compared to the control twin. *emc¹ mwh/emc^{E12} mwh* mutant clones are very elongated and appear mainly along veins. By contrast the control twin clones appear all over the wing blade and present different shapes (Figure 15.3d) (de Celis et al., 1995).
3. Vein differentiation. *emc¹ mwh/emc^{E12} mwh* mutant cells differentiate as vein tissue. As we mentioned before these clones appear preferentially along the veins, and therefore these mutant cells differentiate into normal veins. Clones that are not localised over the normal veins differentiate as ectopic veins. These clones are usually localised in the middle of an inter-vein region (Figure 15.3e) (de Celis et al., 1995).

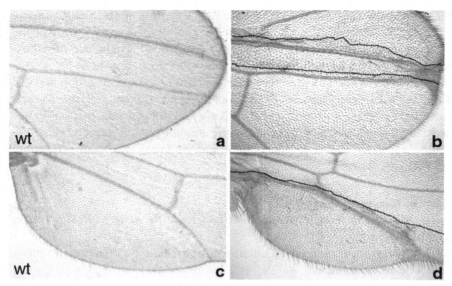

Figure 15.4. *emc¹ M⁺* clones. (a) High magnification of the region between veins 3 and 4 (region A) of a wild-type wing. (b) A dorsal and ventral *emc¹ mwh M⁺* mutant clone causes a reduction of the size of the region between veins 3 and 4. Compare this figure with A. (c) Region between vein 5 and posterior margin of a wild-type wing. (d) The same region shown in c occupied by a large *emc¹ mwh M⁺* mutant clone. Notice the difference in size.

MINUTE ANALYSIS (*emc¹ M⁺* CLONES) (EXPERIMENT 2)

The *emc¹ mwh M⁺* clones are much larger than the *emc¹ mwh/emc^{E12} mwh* clones and can occupy a complete inter-vein region. The size of the inter-vein region with *emc¹ mwh M⁺* clones is smaller than control regions. When a mutant clone completely fills an inter-vein region, this region can disappear and the adjacent veins become fused (Figure 15.4). These effects are stronger when the clones occupy the ventral and dorsal compartments. Large clones not only cause a reduction of the size in the mutant region, but can also affect the size of adjacent wild-type regions and occasionally the size of a neighbouring compartment. *emc¹ mwh M⁺* clones frequently differentiate into ectopic veins and cause the displacement of normal veins to the border of the clone (Figure 15.4) (de Celis et al., 1995; Baonza and García-Bellido, 1999).

DISCUSSION

We have observed different phenotypes associated with *emc* mutant cells, including the following:

➤ Cell lethality (lack of mutant clones in the twin analysis)
➤ Growth of mutant cells restricted to vein territories
➤ Reduction or failure of growth of the inter-vein region
➤ Ectopic vein differentiation

All these results suggest that *emc* function is required for the proliferation of the inter-vein region and consequently to define vein pattern.

We have seen that *emc* mutant clones also form ectopic veins, preferentially along middle inter-vein regions, but also occasionally in other regions of the wing. These results suggest a function of *emc* in the regulation of vein differentiation. However, not all *emc* mutant cells differentiate as veins, indicating that the *emc* insufficiency is not enough to induce vein differentiation. This result suggests that in order to induce vein differentiation the down regulation of *emc* has to be accompanied by the activation of another gene or genes. Only in the regions where these two events occur can the cells differentiate as veins. The fact that the *emc* clones grow along veins and differentiate ectopic veins suggests that the *emc* insufficiency generates a developmental condition similar to normal veins. In this context, it is possible to speculate that different levels of activity of *emc* are involved in defining the regions that proliferate versus the regions that will differentiate into veins. During wing disc development, the regions with high levels of *emc* proliferate actively whereas the regions with low levels of *emc* have less proliferation and are more likely to differentiate into veins when vein-promoting genes are activated.

TIME REQUIRED FOR THE EXPERIMENTS

The preparation and maintenance of stocks for the selection of females and males require more than two weeks before the stocks are ready. Selection of the virgins and males and setting up the crosses take about 20 minutes. The induction of the clones following the schedule described above will take one week. It is also necessary to maintain the crosses until the flies eclose; this takes 10–14 days.

POTENTIAL SOURCE OF FAILURE

As in Chapters 14 and 16.

TEACHING CONCEPTS

Limb size. During the proliferation of the wing discs every inter-vein region tends to proliferate independently. Therefore, in order to define the final size of the wing, it is necessary to control the size of each inter-vein region. Emc is required for this function.

Proliferation vs differentiation. The function of Emc-like proteins is necessary to regulate the balance between cell proliferation and differentiation.

Twin analysis. This technique permits analysis of the behaviour of a clone of mutant cells compared to a "sister" clone of normal cells. The advantage of this technique is that both mutant and control clones come from "sister" cells. Therefore, both clones should be identical. Any difference between clones has to be a consequence of the mutation analysed.

***Minute* technique.** This technique permits the generation of a large amount of mutant tissue in a mosaic fly. This technique is also very helpful for analysing mutations that cause poor cell viability. Frequently, the *Minute+* condition protects against the poor cell viability caused by some of these mutations.

FRT/*Flipase* system. This technique permits induction of mitotic recombination clones at high frequency.

Duplications can be used to induce a mutant clone. The use of duplications to induce clones of mitotic recombination permits analysis of clones of different heteroallelic combination, as we propose in this chapter. Using this technique also permits induction of clones double mutant for two genes that are localised in different chromosome arms.

ALTERNATIVE EXERCISES

PROPOSED EXPERIMENT

emc is one example of a gene that is involved in the regulation of the final size of an organ. Using the approach described in this chapter one could study the function of other genes similar to *emc* in wing development. Among these genes are members of the epidermal growth factor receptor signalling pathway, such as *vein, egfr*, and *rhomboid*. Alleles suggested for clonal analysis of these genes are $egfr^{top4A}$ and vn^{ddd3}.

QUESTIONS FOR FURTHER ANALYSIS

➤ What phenotype do you expect from *Achaetous* (*Ach*) mutant clones in the wing blade? *Achaetous* is an *emc* gain-of-function allele.

➤ Should one of the genes determining the vein fate and inter-vein size code for a bHLH protein? How would you design an experiment to test this question?

➤ How would you carry out a $Minute^+$ experiment to verify the existence of compartments in the wing disc?

ADDITIONAL INFORMATION

Homologues of *emc* have been identified in several species, such as human, mouse, and rat. These molecules constitute the Id family of helix-loop-helix (HLH) proteins. It is unknown what the function of these proteins is during the development of these species, however, the function of Id proteins in mammaliam cells has been studied. It has been proposed that these proteins regulate the balance between cell growth and differentiation by negatively regulating the function of basic helix-loop-helix (bHLH) transcription factors (such as MyoD). Recent studies have revealed that Id can bind to cell-cycle regulatory proteins other than bHLH proteins. Some of these proteins have a key role in the normal cell-cycle progression, such as pRB (retinoblastoma tumour suppressor protein) family proteins and Ets-family transcription factors.

Other interesting results obtained in tissue culture demonstrate that Id genes are expressed at high levels in proliferating cells, and are low or absent in nonproliferating or terminally differentiated cells (Zebedee and Hara, 2001).

REFERENCES

Baonza, A., and García-Bellido, A. (1999). Dual role of *extramacrochaetae* in cell proliferation and cell differentiation during wing morphogenesis in *Drosophila*. *Mech. Dev.*, 80, 133–46.

Botas, J., Moscoso del Prado, J., and García-Bellido, A. (1982). Gene-dose titration analysis in the search of trans-regulatory genes in *Drosophila*. *EMBO J.*, 1, 307–10.

Cubas, P., de Celis, J. F., Campuzano, S., and Modolell, J. (1991). Proneural cluster of *achaetae-scute* expression and the generation of sensory organs in the *Drosophila* imaginal discs. *Genes Dev.*, 5, 996–1008.

Cubas, P., and Modolell, J. (1992). The *extramacrochaetae* gene provides information for sensory organ patterning. *EMBO J.*, 9, 3385–93.

Cubas, P., Modolell, J., and Ruiz-Gómez, M. (1994). The helix-loop-helix *extramacrochaetae* protein is required for proper specification of many cell types in the *Drosophila* embryo. *Development*, 120, 2555–65.

de Celis, J. F., Baonza, A., and García-Bellido, A. (1995). Behavior of *extramacrochaetae* mutant cells in the morphogenesis of the *Drosophila* wing. *Mech. Dev.*, 53, 209–21.

Ellis, H. M., Spann, D. R., and Posakony, J. W. (1990). Extramacrochaetae, a negative regulator of sensory organ development in *Drosophila*, defines a new class of helix-loop-helix proteins. *Cell*, 61, 27–38.

Morata, G., and Ripoll, P. (1975). *Minutes*: Mutant of *Drosophila* autonomously affecting cell division rate. *Dev. Biol.*, 42, 211–21.

Garrell, J., and Modolell, J. (1990). The *Drosophila extramacrochaetae* locus, an antagonist of proneural genes that, like these genes, encodes a helix-loop-helix protein. *Cell*, 61, 39–48.

Xu, T., and Rubin, G. M. (1993). Analysis of genetic mosaics in developing and adult *Drosophila* tissues. *Development*, 117, 1223–37.

Zebedee, Z., and Hara, E. (2001). Id proteins in cell cycle control and cellular senescence. *Oncogene*, 20, 8317–25.

GENERAL REFERENCES

Greenspan, R. J. (ed.) (1997). *Fly Pushing*. New York: Cold Spring Harbor Press.

Roberts, D. B. (ed.) (1986). *Drosophila: A Practical Approach*. Oxford: IRL Press.

FlyBase: A database of the Drosophila genome http://fbserver.gen.cam.uk:7081/

16 Hedgehog transduction pathway is involved in pattern formation of *Drosophila melanogaster* tergites

M. Marí-Beffa

OBJECTIVE OF THE EXPERIMENT Lewis Wolpert originally proposed that a gradient of a diffusible molecule could control pattern formation depending on its concentration (Wolpert, 1969). Hedgehog (Hh) is one such widely accepted morphogenetic signal. In this exercise, we will study the function of the Hh transduction pathway in the control of pattern formation during the development of tergites (the dorsal cuticle of each abdominal segment) of *Drosophila melanogaster*.

DEGREE OF DIFFICULTY Moderate. The experiments described are relatively easy, inexpensive, and can be carried out quickly.

INTRODUCTION

Pattern formation is one of the fundamental topics in Developmental Biology. Lewis Wolpert proposed a theoretical explanation of this process in his positional information model (Wolpert, 1969). Although a previous related model was also published (von Ubisch, 1953), the positional information model has been widely applied to a variety of developing systems. On the basis of previous results from hydra (Chapter 1) and insect segments (Locke, 1959; Lawrence, 1966; Stumpf, 1966, 1968), Wolpert (1969) suggested the existence of a gradient of a diffusible substance. This diffusible substance would be differentially interpreted into positional values. Depending on its position and how each cell interpreted the concentration of the substance, a variety of cell types could then differentiate. In order to explain his model better, Wolpert (1969) proposed the French flag model (Figure 16.1). The different colours in a cellular flag would appear as the differential expression of genes induced by the concentration of a diffusible molecule, or morphogen, distributed in a gradient away from a source or organiser (Figure 16.1).

The tergites of *Drosophila melanogaster* are a good system for studying pattern formation. The tergites differentiate as a hard cuticle in the dorsal surface of each segment of the abdomen of the adult fly. Tergites show a variety of pattern elements that

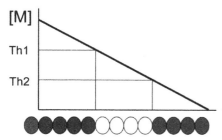

Figure 16.1. The "French flag" model. A signal, or morphogen (M), diffuses from an organiser (grey cell). Depending on the concentration of M and two interpretation thresholds (Th1 and Th2) three "types" of cells may be induced: blue, white or red. (See also color plate 4).

can be used to understand how positional information is read. That is, each cuticular structure can be understood as being one of the colours in Wolpert's French flag model. Bristles (or chaetae, elements of the peripheral nervous system; see Chapters 22 and 23), pigmented and unpigmented trichomes, and denuded cuticle are all organised into a stereotyped pattern (Figure 16.2a). These pattern elements and the flexible cuticle-joining segments act as positional landmarks. The sclerotised cuticle (undashed region

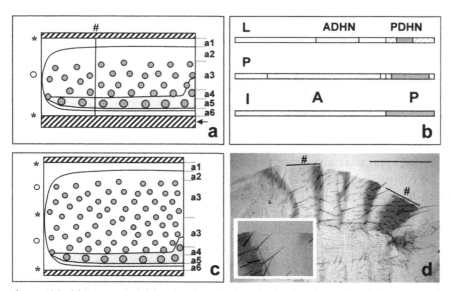

Figure 16.2. (a) Pattern of a left hemitergite. a1–a6: territories showing each cuticular type. Dashed region: posterior compartments. The line designated by # is the anterior–posterior transect shown in Fig. 16.2b. Circles represent the sockets of bristles. The grey region shows the pigmented posterior band. The compartment boundary is indicated by an arrow. (b) Growth pattern of histoblasts (grey) during pupariation. A: anterior compartment. P: posterior compartment. ADHN; anterior dorsal histoblast nest. PDHN: posterior dorsal histoblast nest. L: larva; P: pupa; I: imago. Dashed region: posterior LEC. Anterior is left. (c) Ipsilateral fusion followed by cauterisation of PDHN during larval development. No *engrailed* expressing cells (dashed region) are observed. a1 and a2 from the posterior segment and a6, a5, and a4 from the anterior segment are absent. (d) Right margin of an open mounted abdomen in which the tergites have been previously cut through the dorsal midline. Medio-ventral region is down. Anterior is left. # corresponds to the transect in Fig. 16.2a. The insert shows an unfolded intersegmental membrane. Bar = 200 μm.

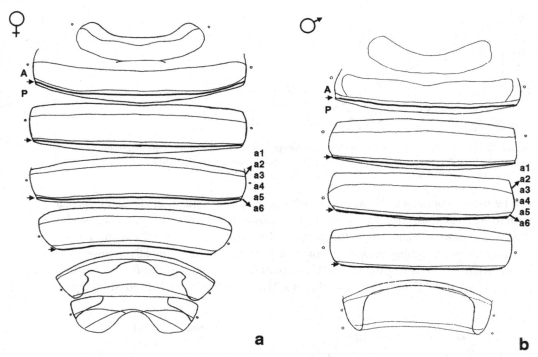

Figure 16.3. Schemes of the tergites of an abdomen from a female (left) and a male (right). The broader lines (small arrows) show the A/P boundaries. The thinner lines include bristles (a3, a4 and a5) and trichome-differentiating regions. The distribution of cuticular types is shown in tergite 4. Arrows indicate territories showing a2 and a6 cuticular types. a1 is devoid of trichomes. a6 posterior limit cannot be seen without *en* staining (*en-lacZ*) or clonal analysis.

in Figure 16.2a) approximately adjusts to the anterior (A) compartment, isolated from posterior (P) compartment by lineage restriction, as studied by mitotic recombination clones; see Chapters 14 and 15; (Kornberg, 1981; Figure 16.3). In the normal arrangement, the anterior compartment can be divided into a number of different regions characterised by pattern elements. Each region or cuticular type has been named with a code (a1 to a6, not to be confused with the A1 to A8 nomenclature of abdominal segments), as shown in Figure 16.2a (Struhl, Barbesh, and Lawrence, 1997a). The posterior limit of a6 and the anterior limit of a1 can be detected only by the localization of *engrailed* gene expression (see Alternative Exercises) or by using mitotic recombination techniques (a method developed by García-Bellido and Merriam (1971) (see also Bryant and Schneiderman, 1969) (see Chapter 15). In each segment, there are two independent, right and left medio-laterally symmetrical hemitergites. Each hemitergite develops from two separated nests of histoblasts (tergite precursor cells) (Robertson, 1936), one anterior (giving rise to the anterior compartment) and one posterior (forming the *engrailed* gene-expressing posterior compartment). The histoblasts are imaginal epidermal cells that form next to the larval epidermal cells (LECs) and remain quiescent during larval

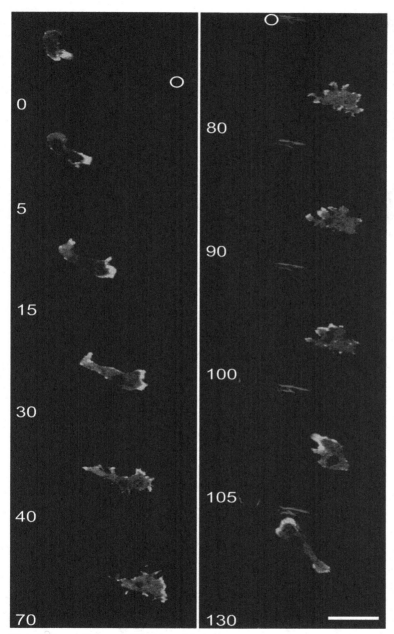

Plate 1. For caption see p. 61, Figure 4.6.

Plates 1-7 are available for download in colour from www.cambridge.org/9780521179768

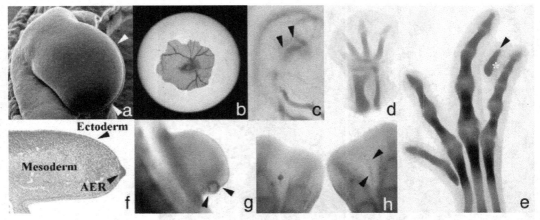

Plate 2. For caption see p. 86, Figure 7.1.

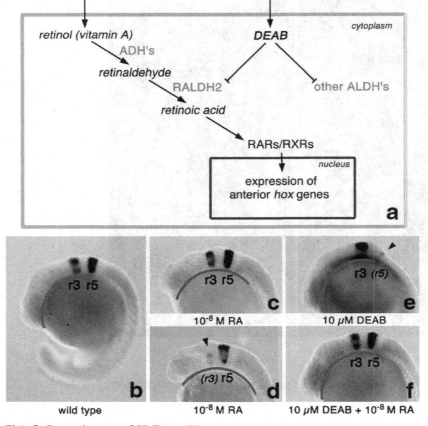

Plate 3. For caption see p. 207, Figure 17.1.

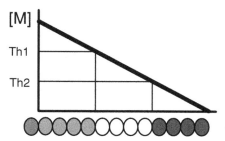

Plate 4. For caption see p. 191, Figure 16.1.

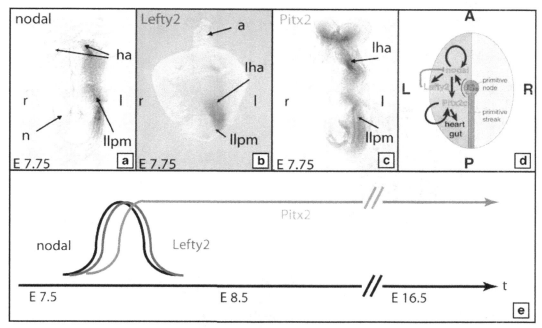

Plate 5. For caption see p. 218, Figure 18.1.

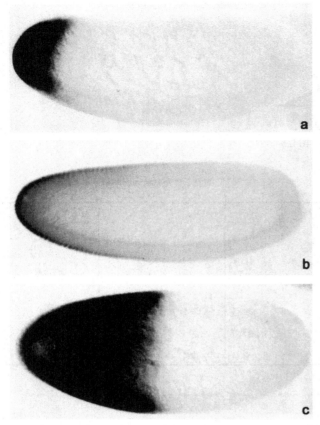

Plate 6. For caption see p. 240, Figure 19.3.

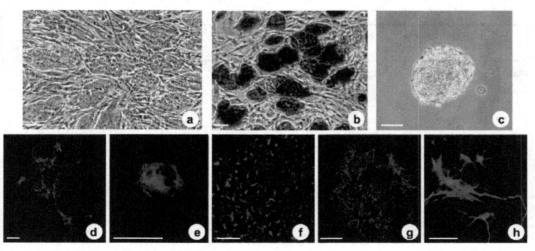

Plate 7. For caption see p. 326, Figure 25.3.

development. After pupariation they proliferate and substitute the LECs that die during metamorphosis. The posterior dorsal histoblast nest (PDHN) expresses *engrailed* (*en*) and is located next to *engrailed*-expressing LECs (Struhl et al., 1997*a*; Kopp et al., 1997) (Figure 16.2b). Cells in *engrailed* and *invected* (*inv*) mutant clones in the posterior compartment are transformed into anterior compartment cells in the tergites (Lawrence et al., 1999*a*), which suggests a role of these two genes in the establishment of P compartment identity (see Chapter 14).

When regions including, or very near, the posterior histoblast nests in the larva cuticle (Roseland and Schneiderman, 1979) are cauterised (heated by the incandescent tip of a tungsten needle), many perturbations result, including ipsilateral fusions of contiguous hemitergites (Figure 16.2c) (Santamaría and García-Bellido, 1972). Ipsilateral fusions, affecting the same side, right or left, are characterised by the continuum of a3 bristles from one segment to the next. This is accompanied by the absence of pigmentation, the absence of large- and medium-sized bristles in one segment, and the most anterior regions missing bristles in the next segment (regions a6, a5, a4, a2, and a1 as suggested by Struhl et al., 1997*a*). Some authors have suggested that a morphogen may act from neighbouring LECs and posterior histoblasts to control these patterns (Madhavan and Madhavan, 1982). Ipsilateral fusions suggest that the histoblasts in the posterior compartment induce the formation of these regions in the neighbouring anterior compartments. Indeed, ipsilateral fusions are correlated with the absence of *engrailed* expression (Marí-Beffa, unpublished results) (Figure 16.2c). Previous theoretical works proposed that compartment boundaries might act as a morphogen source (Meinhardt, 1983).

In order to uncover the genes involved in these inductions, several authors started analyzing the function of segment polarity genes, involved in the parasegmental patterning in the embryo and whose function depends on cell communication. One of these genes, *patched* (*ptc*), encodes a protein with 12 predicted transmembrane domains, known to act as a member of the Hh receptor complex (for a review see Ingham, 1998). The viable conditions of *patched* loss-of-function mutants suggested that this gene was involved in the patterning induced by the posterior compartment (Marí-Beffa, unpublished results). The study of the null alleles of *ptc*, *ptc^{IIw}* (Lawrence et al., 1999*a*) and *ptc^{6P}* (Marí-Beffa, unpublished results), as well as the hypomorphic allele *ptc^{S2}* (Lawrence, Casal, and Struhl, 2002) in mitotic recombination clones, all proved to be relevant to understanding the function of this gene in the adult fly. Because Ptc is part of the Hh receptor complex, modulating the amount of Ptc affects how cells perceive the amount of signalling molecule they receive, thus enabling them to change their differentiation. This results in the formation of specific cuticular structures in areas where they are not normally found.

In this chapter, the series of experiments described in this section, and originally carried out by the author in Antonio García-Bellido's lab in 1991, are recreated as a laboratory exercise. These experiments help to decipher the mode of action of Hedgehog and its transduction pathway in the patterning of the anterior compartment of *Drosophila melanogaster* tergites.

MATERIALS AND METHODS

EQUIPMENT AND MATERIALS

Per student and practical group

See Chapters 11, 14, and 23.

Glass wells can be obtained from Labor (staining blocks).

Biological material

FLY STRAINS

FRT 42D pwn ptc^{6P}/CyO

y w hsflp122; FRT42D cn sha/ CyO

ptc $^{tuf-1}$ ltd

Oregon-R

DESCRIPTION OF ALLELES

➤ The ptc^{6P} (2–59) amorphic (loss of function) allele encodes a nonfunctional, truncated protein (Strutt et al., 2001).

➤ ptc^{tuf-1} is a viable hypomorphic (reduction of function) allele of the *patched* gene. This allele shows an insertion in the promoter region producing a partial loss of function. For a proper definition of the allele class (amorph, hypomorph,...), see Müller (1932).

➤ *pawn (pwn)* (2–55.4) is a cell marker with phenotypes of size reduction of bristles and trichomes.

➤ *shavenoid (sha)* (2–62) is another cell marker with missing or very short trichomes.

➤ *CyO (Curly of Oster)* is a balancer chromosome carrying a dominant mutation, *Curly*, which causes the wing to bend upwards (see Chapters 14, 16, 19, 21, and 26).

➤ *Oregon-R* is a wild type strain.

STOCK REQUEST. Mutant fly strains can be requested directly from the author or from the Bloomington Drosophila Stock Center (FlyBase) (http//fly.bio.indiana.edu).

REAGENTS

See Chapters 11, 14, and 23.

PREVIOUS TASKS FOR STAFF

Maintenance of fly strains and preparation of fly food are described in Chapter 14. Special care must be taken when handling flies. FLY ROOMS MUST BE ISOLATED SO THAT RELEASE OF MUTANT OR TRANSGENIC FLIES TO THE ENVIRONMENT IS PREVENTED. If necessary fly traps must also be used. Moreover, discarded flies must be autoclaved, or heated in an oven, to destroy the genetic material. Solutions are also described in Chapters 11, 14, and 23.

PROCEDURES

Distinguishing males from females and isolation of virgin females are described in Chapter 11. The procedure for setting up fly crosses is shown in Chapter 14.

Induction of clones by heat shock. In order to induce clones with the Flipase/FRT system, a 37°C incubator is required. The fly vials must be incubated at this temperature for 60 minutes. A detailed description of the rationale behind the induction of mitotic recombination clones is presented in Chapters 14 and 15.

Mounting of abdomens

1. Transfer the flies, previously fixed in SH, to a glass well with water.
2. Separate the abdomen from the notum using fine-tipped forceps.
3. Transfer abdomens to 1% KOH.
4. Heat abdomens at 100°C for 10 minutes on a hot plate.
5. Under the microscope, extract the abdomen contents by gently pressing from the sides with flat-tipped forceps, thus folding the abdomen along the dorsal and ventral midline. Students should first practice this technique with wild type flies until it can be done without ruining the bristles on the abdomen.
6. Transfer the abdomens to water and wash twice for 5 minutes each. At this point, abdomens may be cut through the dorsal midline with microsurgery scissors, holding the abdomen open by the posterior edge. The rest of the abdomen content can be eliminated with the fine-tipped forceps.
7. Transfer the abdomens twice to 96% alcohol for 5 minutes each.
8. Open the abdomens as shown in Figure 16.2d (so that intersegmental regions can be better observed; see insert in the figure) over a slide with a drop of lactic-ethanol. Add another drop and gently cover the abdomens with a coverslip so that bubbles are not produced.
9. Cover the preparation with a paper towel and gently press until the edges are completely dry.
10. Seal the coverslip with synthetic resin.
11. Let the preparation dry with a weight on top to completely flatten the abdomen.

OUTLINE OF THE EXPERIMENT

EXPERIMENT 1. ANALYSIS OF A VIABLE HETEROALLELIC COMBINATION OF *Patched*

1. Cross 20 virgin females from stock *FRT42D pwn ptc^{6P}/CyO* with 15 males from the strain *ptc^{tuf-1}*.
2. Transfer the flies to a new vial every second day.
3. Screen the progeny flies that do not carry the *CyO* chromosome (showing straight wings). The frequency of adult *ptc^{tuf-1}/ptc^{6P}* is reduced (according to the expected Mendelian segregation) due to a late pupal semilethality. Most of the flies die as pharate adults inside the pupal case. However, these flies show all morphological characters, except pigmentation (although, in many cases, they can even develop normal pigmentation), and can still be used in this study.
4. Extract the pharate adults, opening the pupal case at its frontal margin and gently pulling the fly out with the fine-tipped forceps. The genotype must be verified by looking for the mutant eye phenotype (size reduction) of *ptc^{tuf-1}/ptc^{6P}* flies.

Table 16.1. Heat Pulse Regime

	1	2	3	4	5
Monday	Trans. 5 to 1			Heat pulse 4	
Tuesday		Trans. 1 to 2			Heat pulse 5
Wednesday			Trans. 2 to 3		
Thursday	Heat pulse 1			Trans. 3 to 4	
Friday		Heat pulse 2	Heat pulse 3		Trans. 4 to 5

"Trans. X to Y" means transfer the flies from vial X to vial Y.

5. Fix both the experimental and Oregon-R control flies (directly from the vial) in SH and store them in a labeled Eppendorf tube.
6. Mount the abdomens and label the slides. The label should include the following data: genotype, name of researcher, and date.

Data recording. Draw the observed phenotype of ptc^{tuf-1}/ptc^{6P}, ptc^{tuf-1}/CyO siblings and Oregon-R wild type dorsal abdomen in the provided schematic outline of tergites (Figure 16.3). Five flies of each genotype are enough to obtain statistical significance. A morphometric study might be performed. a5, a4, a3, and a2 width must be estimated by measuring the length between its limits in the A/P axis at a fixed position in the D/V axis of the abdomens (Figure 16.2a). a6 size can be estimated by the size of the trichome region posterior to the pigmented cuticle (see Alternative Exercises). Constant criteria must be taken when measuring the anterior limit of the pigmented region due to its gradual decline. A proper method for measuring a6 size is provided in Alternative Exercises. Observer students blind to the experiments should be used to score the results.

EXPERIMENT 2. CLONAL ANALYSIS
Although the original experiment was carried out by X-ray induction of mitotic recombination clones, we will proceed with the Flipase/FRT system (Xu and Rubin, 1993) (See also Chapter 15).

1. Cross 20 virgin females from stock y w $hsflp122$; $FRT42D$ cn sha/CyO with 15 males from the stock FRT $42D$ pwn ptc^{6P}/CyO. The appropriate number of females depends on the size of the vial. Incubate the cross at 25°C.
2. Transfer the flies to a new vial daily.
3. Incubate the vials at 37°C for 60 minutes following the regime in Table 16.1.
4. Screen for males (flp) and females ($flp/+$) of the genotypes y w $hsflp122/(+)$; $FRT42D$ pwn $ptc^{6P}/FRT42D$ cn sha. These flies do not carry the balancer chromosome CyO showing wild-type wings. Mitotic recombination clones will be obtained according to the events described in Chapter 15. Briefly, $FRT42D$, pwn, and ptc are cis ordered distal to the centromere on the right arm of the second chromosome; thus, cells homozygous mutant for pwn ptc^{6P} and sha, as the corresponding

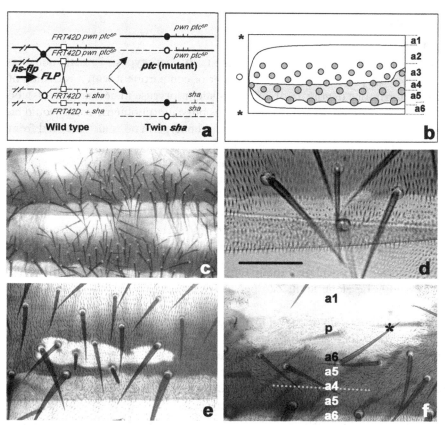

Figure 16.4. (a) Scheme of mitotic recombination induced by Flipase. After heat shock, *flipase* transcription is induced, causing recombination over the FRT sequences. The resulting cells may be homozygous for either *pwn ptc*^{6P} (*ptc* mutant clone) or *sha* (twin control clone). These mutations are all distal to the FRT42 sequence which is proximal in the right arm of the second chromosome. (b) Schematic phenotype of *ptc*$^{tuf-1}$/*ptc*^{6P} left hemitergites. a4, a5, and a6 territories are expanded. Observe the wiggly profile of the posterior pigment boundary (grey). (c) Detail of an abdomen of a *ptc*$^{tuf-1}$/*ptc*^{6P} female fly. (d-f) Clones in territories a6, which shows a mutant ectopic bristle (d), a5-a4 (e) and a3 (f). The phenotypes are described in the text. P, a4 to a6 in (f) are induced cuticular types. The asterisk * denotes a large bristle with rotated polarity. The dotted line shows the place of polarity and pattern change in the induced mirror image duplication. Bar = 50 μm.

twin clone, may be obtained after mitotic recombination induced by Flipase (Figure 16.4a).

5. Fix the flies in SH.
6. Mount the abdomen following the procedure described earlier. Label the preparations.

Data recording. The data recording is as described in Chapter 14. Find and draw the profiles and describe the phenotypes both inside and outside the clone *pwn ptc*^{6P} using the schemes in Figure 16.3. For a control, find *sha* clones (missing or small trichomes) and describe their phenotypes.

EXPECTED RESULTS

EXPERIMENT 1

The heteroallelic combination ptc^{tuf-1}/ptc^{6P} results in phenotypes of bristle and pigments shifting in the posterior region of the anterior compartment with a widening of a6, a5, and a4 territories (Figure 16.4b and c). A slight increase in a6 size may also be observed as a dominant phenotype in $ptc^{6P}/+$ flies but a morphometric study would be required for measuring this size difference. The phenotype in a6 is also inferred from the relative position of large bristles and pigment posterior boundary, which shows a wiggly profile (Figure 16.4b). Further experiments to better describe this phenotype are proposed in Alternative Exercises.

EXPERIMENT 2

We will limit our study to the posterior-most regions in the anterior compartment (a6, a5, and a4, according to Struhl et al., 1997a). The phenotypes expected are as follows:

1. Clones in a6 territory will eventually differentiate ectopic medium-sized bristles (Figure 16.4d). The clone limits can be easily distinguished by the phenotype of pwn trichomes.
2. Clones in a5 and/or a4 do not develop pigment, and may differentiate some mutant bristles (Figure 16.4e).
3. Clones in a4 and/or posterior a3 are similar to clones 2, except for cell polarity modifications in the posterior regions of the clone.
4. Clones in a3 (which may include posterior a2) show both an autonomous (i.e. inside the clone) transformation to posterior compartment and a non-autonomous (i.e. outside the clone) phenotype of polarity change and pattern transformation. The anterior–posterior transect of the mutant transformation may be a1, a2, a1, P, a6, a5, a4, a5, and a6 (Figure 16.4f). ptc^{6P} clones differentiate into P and a6 cuticular types. The non-autonomous phenotype suggests that the ectopic A/P boundary behaves as a new organiser controlling both pattern and polarity.
5. Clones in a2 might be inferred from denuded regions in abdomens with induced mitotic recombination (Lawrence et al., 1999a).

sha does not show a bristle or pigment phenotype (Lawrence et al., 1999b). Twin analysis (see Chapter 15) may provide internal control about size and shape of the clones, but not about bristle differentiation. The twin control clones do not show the ptc^- clonal phenotype mentioned above.

DISCUSSION

Hedgehog is a secreted morphogen, produced by the posterior compartment cells. In the absence of Hh, the Ptc-Smo complex is inactive. The basic function of *patched* is to repress the Hedgehog transduction pathway in the A compartment. This repression is liberated when Ptc receives Hh signal from the posterior compartment. In the most accepted model, Patched protein interacts with the Smoothened protein as a Hedgehog

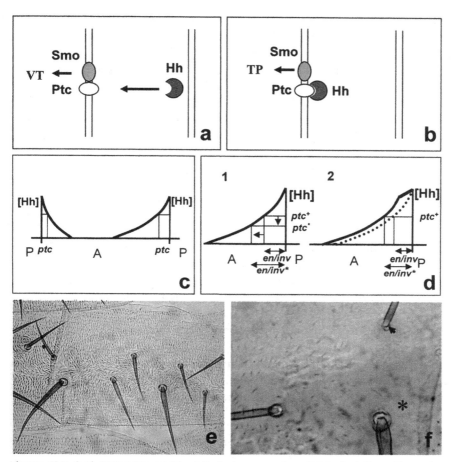

Figure 16.5. (a-b) Molecular model of Hedgehog reception. Patched at the cell membrane binds Smoothened, inducing its modification/degradation by vesicular trafficking (VT) (a). When Hedgehog binds Patched, Smoothened is released from modification/degradation and transduces the signal (TP: transduction pathway) (b). (c) Two Hedgehog gradients occur in the developing tergites: one anterior and one posterior. As a result of high levels of Hh, and possibly dependent on activation thresholds, *patched* gene (*ptc*) is also transcribed at high levels. (d) Two hypotheses may explain *ptc^{tuf-1}/ptc^{6P}* phenotype of a6 (probably *engrailed* dependent) expansion: (1) the responding cells become more sensitive to Hedgehog induction. This might mean that the activation threshold for responding genes (probably *engrailed*) is decreased. (2) Patched sequesters Hedgehog protein by binding it at the cell membrane. In *ptc* viable mutant flies the absence of sequestration results in an increase of the Hedgehog gradient. (e-f) Details of a third (e) and a fifth (f) tergite of an *en^{ES}/ptc^{IN}* male fly. Large bristles and trichomes without pigmentation may differentiate (e). Moreover, bristles (arrow) and pigmentation without trichomes may also differentiate (asterisk) in male A5 tergites (f).

reception complex in the cell membrane. The complex regulates intracellular trafficking, targeting Smo for a modification/degradation event (Denef et al., 2000; Martin et al., 2001) (Figure 16.5a). When Hedgehog binds Patched, Smoothened is released from this modifying/degradation pathway and transduces the signal inside the nucleus along a pathway to finally regulate gene transcription (Figure 16.5b). *patched* and *engrailed* (also expressed in the A compartment of imaginal discs) are genes that respond to Hh in a concentration-dependent manner in the A compartment. Studies on tergite development suggest similar conclusions (Struhl et al., 1997b; Lawrence et al., 2002). Several lines of evidence suggest that Hedgehog functions as a morphogen produced in the P cells for regulating *ptc* (Figure 16.5c) and probably also *engrailed/invected* (Figure 16.5d) gene expression. Hedgehog may also regulate a6 size in the anterior compartment. Moreover, Hh transduction pathway quantitatively regulates the size of a5, a4, and, probably, the differentiation of a2 type (Lawrence et al., 1999a). The viable *ptc* mutant phenotype leads to a reduction in a3 territory size, which could be the ground state of the anterior compartment. The following conclusions can be obtained from this work:

1. Clones deficient for *engrailed* and *invected* do not differentiate the a6 cuticular type (Lawrence et al., 1999a). The ectopic differentiation of a6 cuticular type in *patched* partial loss-of-function heteroallelic combinations (ptc^{tuf-1}/ptc^{6P}), suggested by the relative position of large bristles and posterior pigment boundary, and the differentiation of an a6/p-like cuticular type in *patched* null clones ($pwn\ ptc^{6P}/pwn\ ptc^{6P}$), both may be due to a widening and over-expression of *engrailed/invected* genes. Indeed, *ptc⁻ en⁻ inv⁻* clones differentiate normal pigmentation (Lawrence et al., 1999a).
2. Similar phenotypes to *ptc⁻* clones are observed when over-expression of *engrailed* is induced (Kopp et al., 1997).
3. The a6, a5, a4, a2, and a1 types depend on an induction from the P compartment, as suggested by the *ptc⁻* clones in the a3 region, which produces ectopic pattern differentiation and polarity reversal. This result further suggests that the A/P boundary acts as an organiser center.

These results can be explained by Wolpert's positional information model in two different ways:

1. a6 extension may be due to a reduction in interpretation thresholds. *ptc* transcription is regulated downstream of the Hedgehog pathway. *ptc* is expressed in the a6 and a1 regions in a double gradient (Struhl et al., 1997b; Kopp and Duncan, 2002), suggesting that Hh is also distributed in a double gradient as shown in Figure 16.5c. However, *ptc⁻* clones show a mutant phenotype all along the tergite. This suggests that the Hh reception complex can be induced in all histoblast cells. The partial reduction in *patched* function can also be conceived as an increase in sensitivity of the responding cells in the anterior compartment to Hedgehog concentration. In other words, more Smoothened protein is released from the modification/degradation pathway for a given Hedgehog concentration. This could be understood, under Wolpert's model, as a decrease in the interpretation threshold: with the same Hh concentration outside the cell, the cell interprets a higher Hh concentration. If the gradient of Hh is maintained and a reduction in the interpretation threshold occurs, a widening of *en/inv*-induced a6 territory would be obtained (Figure 16.5d.1). *ptc⁻* clones show a phenotype similar to *en* over-expression. If the *ptc⁻* cells constitutively activate the Hh pathway (Smo is released to transduce the

signal), *en/inv* would be over-expressed/activated. Under these experimental conditions, the cells differentiate into the a6/p-like cuticular type. The interpretation thresholds of Wolpert's model would depend on the activity of the transduction pathway of the morphogen receptor.

2. Ptc sequesters the Hh protein, preventing further diffusion near the anterior–posterior compartment boundary (Chen and Struhl, 1996). Normally, *ptc* is highly expressed near the A/P compartment boundary. Thus, in *ptc* hypomorphic conditions, the Hh gradient would show the profile depicted in Figure 16.5d.2 due to the loss of Hh sequestering. We cannot rule out that this function is also occurring during tergite development. However, in *ptc*[6P] clones that are null for signalling and sequestering functions (Guerrero, pers. comm.), there is no obvious pattern perturbation of a6, a5, or a4 cuticular types outside the clone, except for an infrequently seen slight increase in the size of one or two bristles. This suggests that sequestering of the Hh protein may be less important in tergite formation.

These genetic conclusions support the notion that Hh acts as a morphogen modulating a6 differentiation during tergite development. Moreover, the Hh transduction pathway is involved in tergite pattern formation by controlling a5 and a4 size in a quantitative manner. In imaginal discs, secondary signals occur for transmitting the A to P induction in a "signal relay" process (see Chapter 4). The loss of non-autonomous phenotypes in a4 or a5, when *ptc*[6P] clones are located in a6, suggests that this "signal relay" process may be even less important in developing tergites, so that a5 and/or a4 could also depend on in situ Hh concentration. In any case, the molecular basis of a possible "signal relay" process from a6 to a5/a4 is unknown.

TIME REQUIRED FOR THE EXPERIMENTS

Fly stock amplification requires 20 days. Crosses and phenotypic analysis require one month.

POTENTIAL SOURCES OF FAILURE

Viable *patched* mutant and wild-type phenotypes may overlap so that either a statistic analysis of the phenotypic frequency (penetrance) or a morphometrical study of the intensity of the mutant phenotype (expressivity) must be carried out. Ten flies for each genotype could be enough. The clones in a6 may be hidden in the folds of the intersegmental membrane so that different drawings must be done to reconstruct the original shape of each clone or measure the size of posterior cuticular types.

TEACHING CONCEPTS

Morphogen: diffusible signal produced by an organiser, in our case the anterior–posterior compartment boundaries, which controls cell differentiation depending on its concentration.

Signal transduction pathway: see Chapter 5.

Genetic repression: negative action over the activity of a gene exerted by another gene.

Genetic function below detection level: some genes may function below the in situ hybridization level of detection.

Partial loss-of-function alleles of a gene are useful for understanding its function: viable alleles of *ptc* provide evidence of its function in pattern formation of *Drosophila* tergites.

ALTERNATIVE EXERCISES

PROPOSED EXPERIMENTS

Experiment 1. It has been suggested that, in viable *ptc* mutant flies, a6 might have increased in anterior–posterior size as seen by the relative position of large bristles and posterior pigment boundary. This phenotype might depend on the differential repression of pigments and bristles by a6 controlling genes. *engrailed/invected* may either directly or indirectly repress pigment- and bristle-forming genes (for the latter, see Chapters 22 and 23) to control a6 fate. This would also explain the phenotype of *ptc⁻* clones, which probably differentiate bristles due to partial overexpression of *en/inv*. The over-expression of *engrailed* in the *en^{ES}* hypermorphic (gain of function) allele also suggests that *engrailed* independently represses pigment, bristle, and trichome differentiation in each segment. Pigmentation is more sensitive to the presence of *engrailed* and disappears in most tergites (Kopp et al., 1997) (Figure 16.5e). However, trichomes are more sensitive to repression by *engrailed* in some regions of the A5 segment in males, and disappear in pigmented cuticles with bristles (Figure 16.5f).

A possible explanation could be that patterning mechanisms in the dorsal anterior histoblasts do not regulate the proposed cuticular types (Struhl et al., 1997a) but rather regulate those genes that control each pattern element independently (for a molecular explanation of bristles patterning in tergites see Marí-Beffa, de Celis, and García-Bellido, 1991; for pigment formation Wittkopp et al., 2002). Bristle precursor cells are singled out in a posterior-to-anterior wave (Marí-Beffa et al., 1991, and unpublished results) correlated with bristle size. This process, as well as the organization of the different arrows in a5, a4, and a3, may be organised in a posterior-to-anterior wave of sensory organ precursors singling out as dependent on *en*-expressing/*en*-nonexpressing interface at the p-a6/a5 boundary. A theoretical model is proposed in Chapter 27. Indeed, after cauterisation of PDHN, the posterior bristles may be highly reduced in size to very small, suggesting that the wave may have been perturbed (Marí-Beffa, unpublished results). This wave might depend on the *achaete-scute* complex gene regulation by *prepattern* genes (see Chapter 23).

Experiment 2. In order to measure a6 size (the length between the A/P and the posterior pigment boundaries), clones homozygous for cell markers might be induced in both wild type and *ptc^{tuf}/ptc^{6P}* individuals. This can be obtained by heat shock induction of mitotic recombination clones in both *y w hsflp122/(+); FRT42D sha/FRT42D* and *y w hsflp122/(+); ptc^{tuf}/ptc^{6P}; mwh FRT80B/FRT80B* flies. These flies can be obtained from crosses between males *FRT42D sha/CyO* and females *y w hsflp122; FRT42D*, on the one hand, and males *ptc^{6P}/CyO; mwh FRT80B/TM2* and females *y w hsflp122; ptc^{tuf}/CyO; FRT80B*, on the other. Fly stocks can be requested from the author.

DATA RECORDING. Draw clones in a6 and measure the distance between the posterior-most limit of anterior compartment clones (those which do not reach the posterior limit of trichomes and are located near the pigments) and the a5/a6 boundary. Compare the size of a6 in wild type (*sha*) and *ptc^{tuf}/ptc^{6P}* (*mwh*).

QUESTIONS FOR FURTHER ANALYSIS

➤ Do the cuticular types show a profound developmental basis or are they mere anatomical descriptive concepts?

➤ How is it possible that the *patched* gene functions in regions where it is not detected by histochemical techniques?

➤ Are a5 and a4 territories size regulated by Hh concentration?

➤ What phenotype do you expect in flies heteroallelic for hypomorphic alleles instead of the amorphic allele *ptc^{6P}*?

➤ Is the French flag model valid to explain pattern formation in developing tergites? Obtain your own database to answer this question.

➤ What phenotype would you expect from gain of function alleles of *patched*? *Hs-ptc* or GAL4/UAS system can be used to test your hypothesis.

ACKNOWLEDGEMENTS

The author is endebted to Professor Antonio García-Bellido for his direction, support, and patience during the experimental work shown in this chapter, to I. Guerrero and G. Struhl for fly stocks, and to I. Guerrero and J. Castelli-Gair for critical reading of the manuscript. The author is also endebted to B. Fernández and I. Durán for technical support. This work was supported by fellowships from Fundaciones Juan March and BBV.

REFERENCES

Bryant, P. J., and Schneiderman, H. A. (1969). Cell lineage, growth, and determination in the imaginal leg discs of *Drosophila melanogaster*. *Dev. Biol.* 20, 263–90.

Chen, Y., and Struhl, G. (1996). Dual roles for Patched in sequestering and transducing Hedgehog. *Cell*, 87, 553–63.

Denef, N., Neubuser, D., Pérez, L., and Cohen, S. M. (2000). Hedgehog induces opposite changes in turnover and subcellular localization of Patched and Smoothened. *Cell*, 18, 521–31.

García-Bellido, A., and Merriam, J. R. (1971). Parameters of the wing imaginal disc development of *Drosophila melanogaster*. *Dev. Biol.* 24, 61–87.

Ingham, P. W. (1998). Transducing Hedgehog: The story so far. *EMBO J.*, 17, 3505–11.

Kopp, A., Muskavitch, M. A. T., and Duncan, I. (1997). The roles of *hedgehog* and *engrailed* in patterning adult abdominal segments of *Drosophila. Development*, 124, 3703–14.

Kopp, A., and Duncan, I. (2002). Anteroposterior patterning in adult abdominal segments of *Drosophila. Dev. Biol.*, 242, 15–30.

Kornberg, T. (1981). Compartments in the abdomen of *Drosophila* and the role of the *engrailed* locus. *Dev. Biol.*, 86, 363–72.

Lawrence, P. (1966). Development and determination of hairs and bristles in the milkweed bug *Oncopeltus fasciatus* (*Lygaediae, Hemiptere*). *J. Cell Sci.*, 1, 475–98.

Lawrence, P. A., Casal, J., and Struhl, G. (1999*a*). *hedgehog* and *engrailed*: Pattern formation and polarity in the *Drosophila* abdomen. *Development*, 126, 2431–9.

Lawrence, P. A., Casal, J., and Struhl, G. (1999*b*). The Hedgehog morphogen and gradients of cell affinity in the abdomen of *Drosophila*. *Development*, 126, 2441–49.

Lawrence, P. A., Casal, J., and Struhl, G. (2002). Towards a model of the organisation of planar polarity and pattern in the *Drosophila* abdomen. *Development*, 129, 2749–60.

Locke, M. (1959). The cuticular pattern in an insect, *Rhodnius prolixus* Stål. *J. Exp. Biol.*, 36, 459–77.

Madhavan, M. M., and Madhavan, K. (1982). Pattern regulation in tergite of *Drosophila*: A model. *J. Theor. Biol.*, 95, 731–48.

Marí-Beffa, M., de Celis, J. F., and García-Bellido, A. (1991). Genetic and developmental analyses of chaetae pattern formation in *Drosophila* tergites. *Roux's Arch. Dev. Biol.*, 200, 132–42.

Martin, V., Carrillo, G., Torroja, C., and Guerrero, I. (2001). The sterol-sensing domain of Patched protein seems to control smoothened activity through Patched vesicular trafficking. *Curr. Biol.*, 11, 601–7.

Meinhardt, H. (1983). Cell determination boundaries as organizing regions for secondary embryonic fields. *Dev. Biol.*, 96, 375–85.

Müller, H. J. (1932). Further studies on the nature and causes of gene mutations. *Proc. 6th Int. Cong. Genet.*, 1, 213–55.

Robertson, C. W. (1936). The metamorphosis of *Drosophila melanogaster*, including an accurately timed account of the principal morphological changes. *J. Morphol.*, 59, 351–99.

Roseland, C. R., and Schneiderman, H. A. (1979). Regulation and metamorphosis of the abdominal histoblasts of *Drosophila melanogaster*. *Roux's Arch. Dev. Biol.*, 186, 235–65.

Santamaría, P., and García-Bellido, A. (1972). Localization and growth pattern of the tergite anlage of *Drosophila*. *J. Embryol. Exp. Morphol.*, 28, 397–417.

Struhl, G., Barbash, D. A., and Lawrence, P. A. (1997*a*). Hedgehog acts by distinct gradient and signal relay mechanisms to organise cell type and polarity in the *Drosophila* abdomen. *Development*, 124, 2155–65.

Struhl, G., Barbash, D. A., and Lawrence, P. A. (1997*b*). Hedgehog organises the pattern and polarity of epidermal cells in the *Drosophila* abdomen. *Development*, 124, 2143–54.

Strutt, H., Thomas, C., Nakano, Y., Stark, D., Neave, B., Taylor, A. M., and Ingham, P. W. (2001). Mutations in the sterol-sensing domain of Patched suggest a role for vesicular trafficking in Smoothened regulation. *Curr. Biol.*, 11, 608–13.

Stumpf, H. F. (1966). Mechanisms by which cells measure their position within the body. *Nature*, 212, 430–31.

Stumpf, H. F. (1968). Further studies on gradient-dependent diversification in the pupal cuticle of *Galleria mellonella*. *J. Exp. Biol.*, 49, 49–60.

Xu, T., and Rubin, G. M. (1993). Analysis of genetic mosaics in developing and adult *Drosophila* tissues. *Development*, 117, 1223–37.

Von Ubisch, L. (1953). *Entwicklungsprobleme*. Jena: Gustav Fischer.

Wittkopp, P. J., True, J. R., and Carroll, S. B. (2002). Reciprocal functions of the *Drosophila* Yellow and Ebony proteins in the development and evolution of pigment patterns. *Development*, 129, 1849–58.

Wolpert, L. (1969). Positional information and the spatial pattern of cellular differentiation. *J. Theor. Biol.*, 25, 430–1.

SECTION VII. IN SITU HYBRIDIZATION

17 Retinoic acid signalling controls anteroposterior patterning of the zebrafish hindbrain

G. Begemann

OBJECTIVE OF THE EXPERIMENT The zebrafish, *Danio rerio*, is a successful addition to the collection of vertebrate model systems that offers several attractive features: mating pairs are readily available and spawn large numbers of eggs on a regular basis. Development is rapid and occurs entirely outside the mother, and embryos and larvae are completely transparent, allowing the observation of the circulatory system and internal organs in the living organism. Up to a few thousand mutants have now been isolated through mutagenesis screens that allow a systematic dissection of vertebrate developmental genetics, mirroring the success of screens in *Drosophila* and *C. elegans*.

Segmentation of the body along the anteroposterior axis is a feature found in arthropods, annelids, and chordates. In the zebrafish, as in other vertebrates, segmentation is apparent at embryonic stages in the subdivision of the trunk paraxial mesoderm into somites. Although less obvious, the anterior neural tube is also segmentally arranged, as can be seen in the partition of the prospective hindbrain, the rhombencephalon, into seven rhombomeres.

Here the mechanisms that establish anteroposterior identity among rhombomeres will be analysed at the gene expression level. In particular, the effects of exogenous manipulation of retinoic acid (RA) signalling on the development of the central nervous system will be studied in the zebrafish embryo.

DEGREE OF DIFFICULTY Experiments require breeding pairs of zebrafish and some competence in handling fish and setting up pair matings. Both experiments require basic molecular techniques and care in minimising RNAse contamination.

INTRODUCTION

In vertebrates, anteroposterior fates in the central nervous system (CNS) are determined during embryogenesis. Currently there is good evidence that the developing CNS is regionalised by a two-step mechanism: first, an activating signal induces differentiation towards a general anterior neural fate, while a subsequent signal modifies this pattern to form posterior structures such as hindbrain and spinal cord (reviewed by Lumsden and Krumlauf, 1996).

One factor that exerts a posteriorising activity on the CNS is *all-trans*-retinoic acid (RA), an oxidative product of vitamin A. RA binds to and activates nuclear RA receptors, which regulate the transcriptional activity of target genes (Figure 17.1a) (reviewed by Maden, 2002). One of the enzymes that catalyses the last step of RA synthesis, the aldehyde-dehydrogenase RALDH2, is expressed during gastrulation in the prospective paraxial mesoderm, and subsequently in its derivative, the somites. It is thus present at the right place and time to generate the necessary signalling activity required for posteriorisation of the hindbrain. As predicted by the two-step model, inactivation of this gene in the mouse and zebrafish embryo leads to anteriorisation of the hindbrain to different degrees, such that cells in the posterior hindbrain have taken on the fate of more anterior cells. These animal models have confirmed the role of RALDH2 as a source for RA during development of the neural tube.

The anteroposterior pattern of the CNS is set up prior to the formation of rhombomeres and is revealed by segment-specific expression of genes in the hindbrain. Among these genes, the transcription factor *krx-20* serves as a marker for the territory that gives rise to rhombomeres 3 and 5. In the following experiments, alterations in hindbrain patterning will be uncovered by analysing *krx-20* gene expression upon manipulation of RA signalling.

MATERIALS AND METHODS

EQUIPMENT AND MATERIALS
Per practical group
EQUIPMENT FOR OBTAINING AND HANDLING EGGS

Aquarium for maintaining the stock, in a room heated to approx. 24–28°C

Incubator set to 28.5°C

Fish net

Plastic containers (3–10 liters) for pair matings

Glass marbles

Petri dishes (82 mm)

Plastic Pasteur pipettes (opening diameter 2 mm)

Plastic vials (50 ml) with screw caps

Gilson pipettes

Yellow and blue tips

Pairs of fine-tipped forceps (e.g. Dumont number 5)

EQUIPMENT FOR WHOLE MOUNT IN SITU HYBRIDISATION

Microtubes (1.5 ml)

Staining dishes or multi-well plates

Table-top microcentrifuge

Water bath or oven at 68°C.

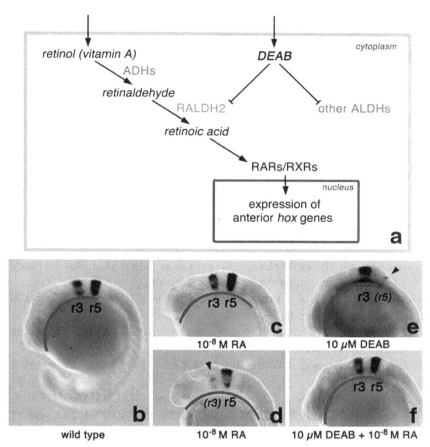

Figure 17.1. Retinoic acid (RA) signalling and anteroposterior patterning of the zebrafish neural plate. (a) Schematic overview of RA signalling. Vitamin A (retinol) enters the cell and is oxidised by alcohol dehydrogenases (ADHs) to retinaldehyde, which then is oxidised to RA by retinaldehyde dehydrogenases (RALDHs), of which RALDH2 mediates anteroposterior CNS patterning. RA binds to and modulates the activity of the retinoic acid receptors, which ultimately regulate the expression of target genes, including *hox* genes. Application of DEAB inhibits the second step of RA biosynthesis, but may also block the activity of other aldehyde dehydrogenases. (b) Expression of *krx-20* at 20 hpf marks rhombomeres (r) r3 and r5. (c,d) Treatment with RA between 5 and 24 hpf posteriorises cell fates in the prospective hindbrain, here evident through a variable reduction of r3 (c; arrowhead in d), while r5 appears unaffected. (e) Inhibition of RA synthesis results in a posterior expansion of r3 (and r4, visible as an unstained gap between r3 and r5) and dramatic reduction of r5 (arrowhead). (f) Combined treatment with RA rescues the patterning defects induced by DEAB (note similarity to wild-type expression), demonstrating that DEAB reduces RA signalling. The white arch marks the length of the territory anterior to the r5/r6 boundary in wild type embryos (b) and overlays it on top of the other phenotypes (c–f); this territory is posteriorised in (c,d) and slightly anteriorised in (e). (See also color plate 3).

Biological material

Wild-type *Danio rerio* (from pet shop or research facility), 6 to 18 months old, ideally 10–15 of each sex.

PREVIOUS TASKS FOR STAFF

Reagents and solutions. For all solutions, use distilled autoclaved water (dH$_2$O). Wear gloves and work with a fume hood when handling retinoic acid, diethyl pyrocarbonate (DEPC), formamide, paraformaldehyde and methanol. (E.G. DEPC IS A SUSPECTED CARCINOGEN AND SHOULD BE HANDLED WITH CARE. IT IS IRRITATING TO EYES, RESPIRATORY SYSTEM AND SKIN. WEAR SUITABLE PROTECTIVE CLOTHING, SUIT-ABLE GLOVES AND EYE/FACE PROTECTION. IN CASE OF CONTACT WITH EYES, RINSE IMMEDIATELY WITH PLENTY OF WATER AND SEEK MEDICAL ADVICE.) To minimise RNase activity in the water and PBS-Tween, stir DEPC into solution at 0.1% for several hours, autoclave for 15 minutes/liter to inactivate DEPC.

> 1 × embryo medium (EM): 5 mM NaCl, 0.33 mM CaCl$_2$, 0.33 mM MgSO$_4$, 0.17 mM KCl, 5% Methylene Blue (prepare a 30 × stock solution).
>
> *all-trans*-retinoic acid (Sigma): 10 mM stock solution in DMSO, store aliquots at −20°C (keep protected from light).
>
> 4-diethylaminobenzaldehyde (Fluka): 10 mM stock solution in DMSO, store at room temperature for up to 2 weeks.
>
> 0.4% Tricaine® stock (Sigma): 400 mg/100 ml, adjust to pH 7 with 1 M tris-HCl, pH 9.
>
> 4% Paraformaldehyde in PBS, pH 7.3 (store aliquots at −20°C). HANDLE PARA-FORMALDEHYDE AS A POSSIBLE CARCINOGEN. HARMFUL IN CONTACT WITH SKIN OR IF SWALLOWED. TOXIC BY INHALATION. WEAR SAFETY GLASSES AND GLOVES; SUPPLY EFFECTIVE VENTILATION.
>
> PBS/0.1% Tween-20/0.1% DEPC (use a stock solution of 20% Tween-20).

PREPARATION OF LABELLED ANTISENSE RNA. Antisense probes can be prepared ahead of the class (see WISH-protocol).

> T3 and T7 RNA-polymerase (Roche)
>
> DIG-labelling kit (Roche)
>
> Plasmids containing zebrafish *krx-20* and *raldh2* genes (available from the author).

WHOLE MOUNT IN SITU HYBRIDISATION (WISH)

> Proteinase K stock, 15 mg/ml (Roche)
>
> Formamide buffer: add the following and make to 50 ml with DEPC-water, store at −20°C:
>
> > 50% Formamide (deionised)
> >
> > 5 × SSC 12.5 ml (20 × SSC stock)
> >
> > 0.1% Tween-20
> >
> > 9 mM citric acid (1 M stock)
>
> FORMAMIDE MAY CAUSE HARM TO THE UNBORN CHILD. IN CASE OF ACCIDENT OR IF YOU FEEL UNWELL, SEEK MEDICAL ADVICE IMMEDIATELY.

Table 17.1. Plasmids for in vitro transcription

Plasmid	Restriction enzyme	RNA-Polymerase	RNA-volume for WISH (per tube)
krx-20	Xba I	T3	1 μl
raldh2	Kpn I	T7	3 μl

Hybridisation buffer: identical to formamide buffer, but also add:
  1 mg/ml tRNA (50 mg/ml stock)
  50 μg/ml Heparin (100 mg/ml stock)
Blocking solution: 2% sheep serum/2 mg/ml BSA in PBS-Tween.
Anti-digoxigenin-AP: anti-digoxigenin antibody fab fragments, conjugated with alkaline phosphatase (Roche).
Alkaline phosphatase (AP) solution:
  100 mM Tris HCl pH 9.5
  50 mM $MgCl_2$
  100 mM NaCl
  0.1% Tween-20
  water (add to make 50 ml)
Staining solutions (Roche):
  135 μg/ml Nitro Blue Tetrazolium (NBT)
  105 μg/ml 5-Bromo-4-chloro-3-indolyl phosphate (BCIP) in AP solution

PROCEDURES

Egg collection. For obtaining eggs, see Nüsslein-Volhard and Dahm (2002), Chapter 1, or *The Zebrafish Book* (Westerfield, 1995), Chapter 2. The latter book is conveniently available via the zebrafish web portal ZFIN (zfin.org/zf_info/zfbook/zfbk.html) and also contains basic protocols for zebrafish husbandry. The setting up of pairs of male and female zebrafish will take 10–30 minutes. Males are more reddish in colour, while females exhibit bright white and blue stripes and have thicker bellies. Collecting embryos takes about 30 minutes. Incubations of zebrafish embryos are carried out in EM at 28.5°C.

Fixation of embryos. Embryos are fixed in microtubes in 4% paraformaldehyde for 4 hours at room temperature. Wash 2 × 5 minutes in DEPC-PBS-Tween and remove the chorion membranes manually under the dissecting microscope with fine-tipped forceps. Replace liquid with methanol. Store overnight at −20°C. Embryos are fine for WISH even after several months storage.

Whole mount in situ hybridisation (WISH) (Jowett, 1997)

1. Digest 1 μg of plasmid DNA with a suitable restriction enzyme (Table 17.1) and proceed with RNA-labelling reaction according to manufacturer's protocol (Roche). Precipitate RNA with 0.5 volumes 7.5 M NH_4OAc (ammonium acetate) and 3 volumes 100% EtOH, for 30 minutes at room temperature. Centrifuge 15 minutes,

wash 1 × 70% EtOH (made with DEPC-H$_2$O), dry 5 minutes. Resuspend in 100 μl DEPC-H$_2$O and store aliquots at −80°C.

2. Rehydrate fixed embryos: In a microtube, wash 1 × 5 minutes in 50% methanol/PBS-Tween, 2 × 5 minutes in PBS-Tween. Refix with 4% paraformaldehyde for 20 minutes, rinse and wash 2 × 5 minutes, 1 × 10 minutes with PBS-Tween.

3. Dilute Proteinase K stock 1:10 in DEPC-PBS-Tween. Incubate embryos in 4 μl/1 ml PBS-Tween at room temperature for 1 minute. This time should be sufficient for 1-day old embryos, but is batch-dependent. Rinse 2 × with PBS-Tween and refix for 20 minutes. Rinse, wash 2 × 5 minutes, 1 × 15 minutes, 1 × 25 minutes in PBS-Tween.

4. Pre-hybridisation: Wash 1 minute in 500 μl PBS-Tween/formamide buffer (1:1), then wash 1 minute in 250 μl formamide buffer, and replace with 250 μl hybridisation buffer. Prehybridise at least 4 hours or overnight at 68°C.

5. Hybridisation: Mix 1–3 μl of RNA-probe (Table 17.1) with 30 μl hybridisation solution and heat to 68°C. Remove as much pre-hybridisation as possible, leaving the embryos submerged. Add probe/hybridisation solution mixture onto the embryos and leave at 68°C overnight.

6. Rinse, wash 2 × 15 minutes in 500 μl formamide solution at 68°C. Transfer to room temperature, add 500 μl PBS-Tween and mix, rinse, and then wash 2 × 15 minutes in PBS-Tween.

7. Wash 1 × 30 minutes in PBS-Tween + 0.5% blocking solution.

8. Incubation with anti-DIG-AP antibody: Replace blocking solution with 1:1,000 dilution of anti-DIG-AP antibody in 0.5% blocking solution. Gently agitate for 4 hours at room temperature.

9. Remove antibody solution and wash embryos at least 4 × 20 minutes in PBS-Tween at room temperature, or overnight at 4°C with as many washes as convenient.

10. Colour reaction: wash 3 × 5 minutes in fresh AP solution. Transfer embryos into staining dish. Replace last wash with 4.5 μl NBT and 3.5 μl BCIP to each 1 ml of staining solution (WEAR GLOVES). Allow colour to develop in the dark and monitor regularly. Expect strong staining within 10–60 minutes.

11. Stop colour reaction by washing 2 × 2 minutes with excess PBS-Tween. Refix for 1 hour. Transfer embryos stepwise into glycerol in PBS-Tween (30%, 50%, 70%) for microscopic observation and permanent storage at 4°C in the dark.

OUTLINE OF THE EXPERIMENTS

EXPERIMENT 1A. THE EFFECTS OF ALTERED RA SIGNALLING ON ZEBRAFISH HINDBRAIN PATTERNING

In zebrafish the consequences of a loss of RA signalling have been studied using mutants in the *raldh2* gene, or by inhibiting the activation of RA receptors with potent antagonists (Begemann et al., 2001; Grandel et al., 2002). However, the mutants cover only part of the spectrum of complete loss of RA phenotypes and obtaining antagonists is costly. Alternatively, RA signalling can be inhibited by exposure to 4-diethylaminobenzaldehyde (DEAB), which acts as an inhibitor of cytosolic aldehyde dehydrogenases, including RALDH2, and thus ultimately suppresses the activation of RA target genes (Begemann et al., 2004). Because of this, the developmental disruptions caused by DEAB could, in principle, also be a result of the inhibition of aldehyde dehydrogenases other than RALDH2. This experiment examines this possibility and tests the effects of both DEAB and excess RA on zebrafish neural plate patterning.

1. Sort embryos by developmental stage and divide them into four batches of 10–20 embryos in 25 ml EM, and keep at 28.5°C. Incubate each batch from 4.5 hours post-fertilisation (hpf) (equals dome stage/30% epiboly) onwards in
   (a) 10^{-8} M RA
   (b) 10 μM DEAB
   (c) 10 μM DEAB/10^{-8} M RA
   (d) 0.1% DMSO (control)
2. Fix the embryos at the 20-somite stage. For convenience, development can be slowed by transferring embryos to 18°C for the night. This cannot be done until after the end of gastrulation (100% epiboly), as low temperatures during gastrulation cause overall cell death. Perform WISH against *krx-20*, a gene specifically expressed in rhombomeres (r) r3 and r5.

EXPERIMENT 1B. DETERMINING THE DEVELOPMENTAL STAGES DURING WHICH RA SIGNALLING PATTERNS THE HINDBRAIN

This experiment aims to identify at which developmental stages of zebrafish embryogenesis production of endogenous RA is required to pattern the hindbrain.

Sort embryos by developmental stage, divide them into eleven batches of 10–20 embryos in 25 ml EM and keep at 28.5°C. Treat each batch with 10 μM DEAB (in EM) for two hours, starting from different points of development, according to the scheme in Table 17.2. If not enough embryos are available, assay only every other time point. As a control treat one batch with 0.1% DMSO from 1.5–14 hpf. Remove the test solutions by washing twice in 40 ml EM, then return embryos in 25 ml EM to the incubator until they have reached the 20-somite stage. Fix the embryos and proceed as in Experiment 1a.

EXPERIMENT 2. THE EXPRESSION PATTERN OF RETINALDEHYDE DEHYDROGENASE 2 (*raldh2*) IN THE ZEBRAFISH EMBRYO

In this experiment, the pattern of *raldh2* expression during embryonic development will be correlated with the timing of RA signalling that is required for r3/r5 patterning (determined in Experiment 1b). 10–20 embryos each are fixed at typical stages of development (4, 6, 8, 10, 15, and 20 hpf). This developmental series is used to follow the expression of *raldh2* over time by WISH.

Data recording. For all experiments, stained embryos in 70% glycerol will be observed under the dissecting microscope and may be mounted on coverslips and photographed. The results of Experiment 1b will be annotated by completing the columns in Table 17.2.

Table 17.2. Experimental scheme for 2-hour treatments of embryos with DEAB

Age (hpf) at 28.5°C	Corresponding developmental stage at 28.5°C	Results after 2-hour DEAB treatment (Numbers of embryos with effects on krx-20 expression in r5)	
		Number of embryos with reduced r5	Total number of embryos
1.5	16-cell stage		
2	64-cell stage		
3	1000-cell stage		
4	sphere stage		
5	40% epiboly		
6	shield stage		
7	60% epiboly		
8.3	80% epiboly		
10	100% epiboly		
12	6-somite stage		

EXPECTED RESULTS

THE EFFECTS OF ALTERED RA SIGNALLING ON ZEBRAFISH HINDBRAIN PATTERNING (EXPERIMENTS 1A, B)

In this experiment, control embryos express *krx-20* in r3 and r5 (Figure 17.1b), which are of similar size. Continuous high levels of RA signalling lead to a reduction in *krx-20* expression in r3. Moreover, the anteroposterior extent of the CNS anterior to r5 is variably reduced (compare white arches in Figures 17.1b–d), suggesting that ectopic RA specifies cell fates in the midbrain and hindbrain towards more posterior fates.

In contrast, inhibiting RALDH2 activity for 2-hour intervals during early developmental stages has the following result: all embryos treated up to 7 hpf (i.e., commencing from 1.5 to 5 hpf) and at any stage past 8.3 hpf will develop r3 and r5 indistinguishable from wild-type. However, under reduced RA synthesis between 6 and 9 hpf (i.e., commencing at 6 and 7 hpf), many embryos will develop with an enlarged r3, while r5 is variably reduced in a posterior-to-anterior fashion, and often small bilateral patches of *krx-20* remain at the presumptive r4/r5 boundary (Figure 17.1e). These effects are reversible when RA is applied simultaneously.

Continuous exposure to DEAB from 4 hpf onwards covers the critical phase and eliminates r5-specific expression posteriorly (Figure 17.1e), while simultaneous exposure to RA alleviates this effect (Figure 17.1f).

THE EXPRESSION OF RETINALDEHYDE DEHYDROGENASE 2 (*raldh2*) IN THE ZEBRAFISH EMBRYO (EXPERIMENT 2)

raldh2 mRNA is first detectable at 30% epiboly in an open ring along the blastoderm margin. During gastrulation *raldh2* is expressed by involuting cells at the gastrula margin; these cells will form mesendoderm. *raldh2* expression is excluded from the dorsal

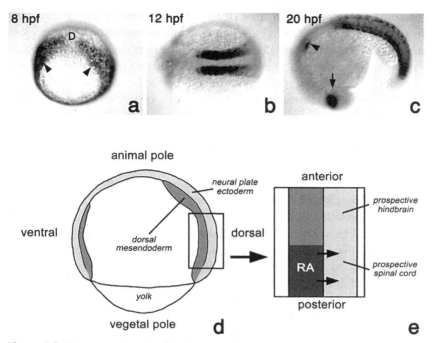

Figure 17.2. RA patterns the zebrafish rhombencephalon through RALDH2 activity during gastrula stages. (a) Gastrulating embryos at 8 hpf express *raldh2* in the germ ring (arrowheads), lateral to the dorsal (D) side; these cells will develop as paraxial mesodermal cells. (b) At 12 hpf strong expression is found in the somitic mesoderm; dorsal view. (c) *raldh2* expression remains strong at 20 hpf in the somites, but is also found in the eye (arrow) and paraxial head mesenchyme (arrowhead). Anterior is to the top (a) and to the left (b,c). (d) Lateral schematic view of an embryo at 75% completion of epiboly. Invaginating mesendodermal cells (dark grey) converge dorsally towards the animal pole and are in contact with the overlying ectoderm (light grey). Box depicts schematic in (e). (e) *raldh2* (dark area) is expressed in cells at the margin of involuting endomesoderm, in a lateral position to the axis. It is thought that RA produced in these cells acts locally on the overlying neuroectoderm to posteriorize the neural plate, thus generating hindbrain and spinal cord fates.

axial cells in the embryonic shield, and at 60% epiboly is down-regulated on the ventral side of the embryo (Figure 17.2a). Throughout the gastrula period, expression persists in posterior and lateral mesoderm and remains excluded from notochord precursors. Throughout somitogenesis stages expression remains strong in forming somites (Figures 17.2b and 17.2c), and later is also detected in the eye and the branchial arch primordium (Figure 17.2c) (Begemann et al., 2001; Grandel et al., 2002).

DISCUSSION

Application of DEAB to developing zebrafish embryos results in alterations to pre-rhombomeric gene expression domains. Changes in hindbrain fates of r3 and r5 have been tested using *krx-20* as a molecular marker. Two observations can be made: DEAB affects the anteroposterior pattern in the hindbrain by reducing the number of cells converted towards an r5 cell fate, and these phenotypes are reversible when RA is applied at the same time. It is thus likely that DEAB acts by inhibiting mainly retinalde-hyde dehydrogenases during these stages. When embryos are treated with excess RA, specific teratogenic defects occur: anterior structures of the neural plate are reduced at the expense of posterior structures. This results in an expansion of the hindbrain and spinal cord anteriorly through a respecification of anterior cells, particularly apparent as a reduction in anterior structures, which decreases the overall length of the brain (Figures 17.1c and 17.1d).

These experiments show that anteroposterior identity exhibits an early plasticity that is governed by the activity of RA signalling along the prospective hindbrain region. In its absence the hindbrain fails to be specified towards cell fates posterior to r4 and the resulting phenotype represents the state of the hindbrain prior to posteriorisation.

The experiments show that RA synthesis is required roughly during a 3-hour period of gastrulation. Posteriorisation of the neural plate towards r3–r5 fates is therefore an early event that should be correlated with the activity of RALDHs both temporally and spatially. Of the known vertebrate enzymes, only RALDH2 is expressed during gastrulation. As determined in Experiment 2, *raldh2* is first expressed in the marginal zone of the blastula and during gastrulation is restricted to the paraxial mesoderm of the prospective trunk region. A model can now be proposed where RA is produced by RALDH2 in the involuting mesendoderm and acts locally on the overlying neural plate (Figure 17.2e). These results further suggest that the hindbrains of the two *raldh2* loss-of-function mutants, in which mainly cell fates posterior to r6 are affected, must experience RA signalling from sources other than zygotic *raldh2* expression, either from an as-yet unknown enzyme or from maternally provided RALDH2 protein.

Anteroposterior patterning of the hindbrain and spinal cord is brought about by a combination of *hox* gene expression, such that each rhombomere is specified by its own *hox* code. Several anteriorly expressed *hox* genes contain RA responsive cis elements, which are regulated by RA-activated receptors. Accordingly, the changes in anteroposterior fates in the neural plate observed after exposure to DEAB or RA seem to be a consequence of the failure of coordinate expression of anterior *hox* genes. Indeed, in *raldh2* mutants, several *hox* genes fail to be expressed correctly.

TIME REQUIRED FOR THE EXPERIMENTS

Setting up pair matings takes 30 minutes (day 1); collecting eggs, cleaning, and sorting into age-synchronized batches takes 1 hour; DEAB and RA-treatments last the day (day 2). Fixation of embryos, dechorionation, and transfer into methanol require 6 hours (day 3). Subsequently embryos are stored at −20°C for at least one day.

Preparation of RNA-probes for WISH may be done in advance by the teachers and takes 1–1.5 days. The WISH procedure requires 3 days, with overnight steps for hy-bridisation and for washing after antibody incubation.

POTENTIAL SOURCES OF FAILURE

Exposure of RA to light and freezing of DEAB inactivates these substances. Thawed aliquots of fixative should be kept in the fridge and used within 5 days. The WISH procedure can be problematic. Weak signals or high background staining is indicative of problems with RNA degeneration. The amount and quality of the RNA probe should be monitored on a clean agarose gel and a strong band with little or no smear below should be present. Make sure that students work with RNAse-free solutions up to the antibody incubation step.

TEACHING CONCEPTS

Disruption of normal development by teratogens. The availability of mutants that lead to abnormalities in development has allowed an analysis of the molecular processes governing development. However, environmental factors also have an important influence on development. Exogenous agents, whose presence during critical stages disrupts development, are called teratogens. Examination of the defects caused by teratogens can therefore successfully contribute to understanding the molecular mechanisms of both teratogenesis and normal development.

Anteroposterior patterning of the neural plate. During gastrulation the neural plate is regionalised along the anteroposterior axis. In a two-step process, the neural plate initially develops towards an entirely anterior fate. Further signals then modify this pattern to determine posterior identities, such that a central nervous system develops with clear regionalisation into a tripartite brain anteriorly and a spinal cord posteriorly. RA is one such factor that controls genes that pattern the neural plate along the anteroposterior axis.

ALTERNATIVE EXERCISES

QUESTIONS FOR FURTHER ANALYSIS

➤ In these experiments the contribution of RA signalling to the patterning of the hindbrain was analysed. Identify and use molecular markers of more posterior spinal cord fates in the CNS to find out if they are determined during the same developmental stages or later, and determine the posterior limit of RA-dependency.

➤ Use the same approach to ask questions about the requirement for RA signalling for the development of the pectoral fins. Having determined the expression pattern of RALDH2, investigate which tissues could be involved in specifying pectoral fin development.

ACKNOWLEDGEMENTS

I thank Trevor Jowett for the *krx-20* plasmid, Rita Hellmann and Katharina Mebus for excellent technical assistance, and Axel Meyer for support. Gerrit Begemann is funded by the Deutsche Forschungsgemeinschaft (BE 1902/3-1).

REFERENCES

Begemann, G., Schilling, T. F., Rauch, G. J., Geisler, R., and Ingham, P. W. (2001). The zebrafish *neckless* mutation reveals a requirement for *raldh2* in mesodermal signals that pattern the hindbrain. *Development*, 128, 3081–94.

Begemann, G., Marx, M., Mebus, K., Meyer, A., and Bastmeyer, M. (2004). Beyond the neckless phenotype: Influence of reduced retinoic acid signaling on motorneuron development in the zebrafish hindbrain. *Dev. Biol.*, 271, 119–29.

Grandel, H., Lun, K., Rauch, G. J., Rhinn, M., Piotrowski, T., Houart, C., Sordino, P., Kuchler, A. M., Schulte-Merker, S., Geisler, R., Holder, N., Wilson, S. W., and Brand, M. (2002). Retinoic acid signalling in the zebrafish embryo is necessary during pre-segmentation stages to pattern the anterior–posterior axis of the CNS and to induce a pectoral fin bud. *Development*, 129, 2851–65.

Jowett, T. (1997). *Tissue "in Situ" Hybridization: Methods in Animal Development.* New York: John Wiley and Sons.

Lumsden, A., and Krumlauf, R. (1996). Patterning the vertebrate neuraxis. *Science*, 274, 1109–15.

Maden, M. (2002). Retinoid signalling in the development of the central nervous system. *Nature Reviews Neuroscience*, 3, 843–53.

Nüsslein-Volhard, C., and Dahm, R. (2002). *Zebrafish, A Practical Approach.* Oxford: Oxford University Press.

Westerfield, M. (1995). *The Zebrafish Book: A Guide for the Laboratory Use of Zebrafish (Danio rerio).* Oregon: University of Oregon Press.

18 Left–right asymmetry in the mouse

M. Blum, A. Schweickert, and C. Karcher

OBJECTIVE OF THE EXPERIMENT From the outside vertebrates appear bilaterally symmetrical, with paired sensory organs and a symmetrical axial skeleton. The abdominal and thoracic organs, however, assume asymmetric positions with respect to the midline (*situs solitus*). The apex of the heart points to the left, the right and left lung differ with respect to lobation, the stomach and spleen are located on the left, the liver is located on the right, and the small and large intestines coil asymmetrically. Laterality is created during early embryogenesis. Embryological and genetic experiments in chick, frog, zebrafish, and mouse have revealed a conserved asymmetric signaling cascade, which relays asymmetric cues from the embryonic midline to the lateral plate mesoderm and the forming organs.

The objective of the experiments in this chapter is to visualize both molecular and morphological asymmetries in mouse embryos during development. Eight- and nine-day-old embryos will be isolated and asymmetric left-sided transcription of the homeobox transcription factor *Pitx2* will be analyzed by whole-mount in situ hybridization (ISH). Morphological asymmetries will be observed both in embryos and adult mice.

DEGREE OF DIFFICULTY ISH requires basic molecular biological skills, in particular the handling of RNA. A prerequisite for isolation of mouse embryos is a facility to keep laboratory mice and a licence to kill experimental animals. Experience in recognizing plugs is required to set up timed matings. In addition, it is preferable that students have some previous work experience in dissection techniques using forceps, scissors, and the stereomicroscope.

INTRODUCTION

Left–right asymmetry is generated during embryogenesis, distinguished by three phases. The initial breakage of the bilateral symmetry of the gastrula/neurula embryo (phase 1) results in transient asymmetric gene expression in the left lateral plate

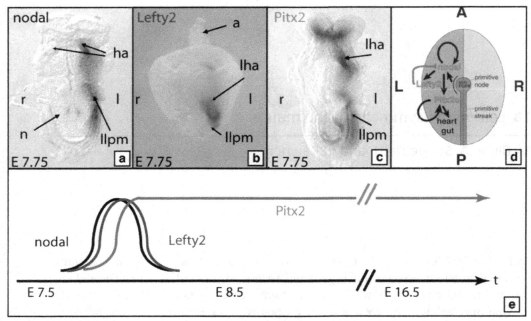

Figure 18.1. Molecular asymmetries: the Nodal cascade. Whole-mount in situ hybridization of E7.75 mouse embryos reveals asymmetric mRNA transcription of the genes *nodal* (a), *Lefty2* (b) and *Pitx2* (c) in the left LPM (llpm) and left heart anlage (lha). Genetic interactions are schematically depicted in (d), the dynamic time course is outlined in (e). l, left; r, right. (See also color plate 5).

mesoderm (LPM) in phase 2 (Figure 18.1), from where the asymmetric cue(s) translocate to the forming organs which undergo asymmetric morphogenesis in phase 3 (Figure 18.2).

Asymmetric gene expression (phase 2) is the best-characterized phase (Capdevila et al., 2000; Mercola and Levin, 2001; Hamada et al., 2002). Three asymmetrically expressed genes have been identified in all vertebrates, *nodal, Lefty2*: and *Pitx2*.* Asymmetric transcription of these genes in the left LPM in E7.75 mouse embryos is shown in Figures 18.1a–c; interactions between these factors (the so-called "nodal cascade") are schematically depicted in Figure 18.1d. The gene *nodal* encodes a secreted growth factor, which belongs to the superfamily of transforming growth factors (TGF) β. Its asymmetric transcription is first visible in embryos that possess 2–3 pairs of somites (E7.75). *nodal* spreads rapidly throughout the left LPM because of a positive feedback loop of Nodal signaling on *nodal* transcription. At the same time, the target genes *Lefty* and *Pitx2* become activated. Lefty2, a secreted protein related to Nodal, acts as a feedback inhibitor of Nodal signal transduction, which interrupts the signaling cascade by blocking the Nodal receptor. This negative interaction terminates *nodal* transcription after a very short time period, around the 6–8 somite stage (E8.0–E8.25), only 6–8 hours after its first induction. As *Lefty2* depends on Nodal signaling as well, its transcription in the left LPM vanishes around the same time. The second target gene, *Pitx2*, however,

---

* A fourth gene, *Lefty1*, in addition is asymmetrically expressed on the left side of the floor plate.

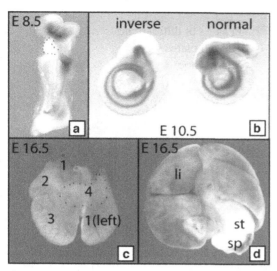

Figure 18.2. Morphological asymmetries. (a) At E8.5 the linear heart tube (outlined with arrowheads) loops to the right side. Ventral view of an embryo stained for *Pitx2* mRNA transcription. (b) Turning of the embryo along the anterior–posterior axis positions the tail on the right side of the head in normal embryos. Occasionally, inverse turning is observed. Lateral views of embryos stained with a *shh* probe to visualize the position of the tail more clearly. (c) Ventral view of a wild-type lung at E16.5 with 4 lung lobes on the right (1–4) and one lobe on the left side. (d) Ventral view of wild-type liver, stomach and spleen at E16.5. Stomach and spleen are positioned on the left side of the embryo.

stays on in the absence of Nodal signaling, because of a positive feedback loop of Pitx2 on its own transcription (Schweickert et al., 2000). Asymmetric transcription of *Pitx2* can be observed during morphogenesis of the heart, lung, and gastro-intestinal tract; i.e., Pitx2 relays the asymmetric Nodal signal to the forming organs. In phase 3, asymmetric organ development is accomplished; however, the molecular pathway(s) of this phase have not been elucidated in detail as yet.

MATERIALS AND METHODS

EQUIPMENT AND MATERIALS
Per student
EMBRYO PREPARATION

Stereomicroscopes
10-cm glass petri dishes
Scissors
20-μl pipette
Yellow tips (sterile)

Forceps (Fine Science Tools (FST) #3
  and #5)
Crushed ice
Glass vials

PREPARATION OF PROBE

Crushed ice
Pipettes (20 μl, 200 μl, and 1,000 μl)
Yellow and blue tips

Reaction tubes 1.5 ml (autoclaved)
4°C microfuge

IN SITU HYBRIDIZATION

   5-ml glass vials (+ lids) with   Beakers of different sizes
      appropriate racks   12-well plates
   Sterile pipettes (3–5 ml, disposable)   Microscope

Per practical group

PREPARATION OF PROBES

   Electrophoresis equipment
   Gel documentation system

IN SITU HYBRIDIZATION

   Shaker
   Camera
   Water bath (70°C)

Biological material. A common laboratory mouse strain is C57BL/6J. They breed well, are long-lived, and can be obtained from the Jackson Laboratory (JAX® Mice strain C57BL/6J, Stock Number 000664).

REAGENTS AND KITS

   MAXIscript™ In vitro Transcription Kit (Ambion)
   DIG labeling mix (Roche)
   Formamide
   Glycin
   Glutaraldehyde
   Substrate: BM-Purple (Roche)
   100% ethanol
   Anti-Digoxigenin Antibody, AP conjugated (Roche)
   PBS Dulbecco's w/o sodium bicarbonate; PBS (Invitrogen)
   PBS Dulbecco's w/o calcium and magnesium, w/o sodium bicarbonate, PBS⁻
      (Invitrogen)

PREVIOUS TASKS FOR STAFF

Prior to the experiments, timed matings have to be set up. To be on the safe side it is advised to stock embryos of the respective stages in the freezer. The plasmid used for probe preparation needs to be prepared and linearized by restriction digest. Also, the probe should be synthesized beforehand and stored at −80°C as a backup in case the *in vitro* transcription fails. These preparations require little time, but need to be carefully planned ahead. Except for the ones that have to be freshly made, it is best to prepare all the solutions in advance, especially the hybridization mix and the blocking solution, because their preparation is time consuming.

Setting up timed matings. To get embryos of a defined stage, female mice have to be mated to males in the afternoon (two females and one male per cage). Pregnant females

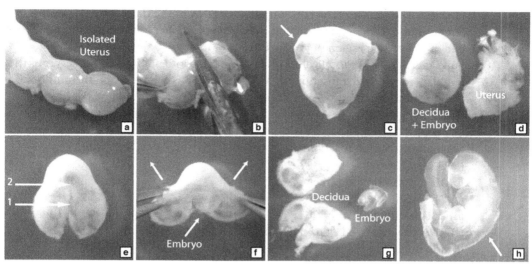

Figure 18.3. Isolation of E8.5 mouse embryos. For details see text.

are recognized the next morning by the presence of a copulation plug in the vagina (vaginal plug). Noon on the day of the plug is defined as stage 0.5 days post coitum (dpc) or embryonic day E0.5. Separate pregnant females and clearly mark the date of plug on the cage.

Isolation of mouse embryos (Figure 18.3). Pregnant mice are killed by cervical dislocation. Open the body cavity and remove the uterus. Place into a 10 cm petri dish with ice-cold PBSw. Remove the mesometrium and fat tissue as much as possible and then separate the individual swelling (Figure 18.3a, b). Transfer to a 5-cm PBS-agarose dish (see below). Remove the uterus using two forceps (FST#3). Hold the uterine tissue with one forceps at one of the openings (arrow in Figure 18.3c), and by carefully tearing with the other one extend the cut to the other side. Now the uterus can be easily removed from the decidua without damaging the embryo (Figure 18.3c, d). Change to finer forceps (FST#5) and longitudinally cut into the proximal (mesometrial) third of the decidua (Figure 18.3e, cut #1). The embryo is located in a central cavity in the distal half of the decidua. Using #5 forceps extend the cut in the decidual tissue on both sides between the points #1 and #2 as marked in Figure 18.3e, taking care not to harm the embryo proper. Pull the two halves of the deciduum apart and shell out the embryo (Figure 18.3f, g). Remove residual extra-embryonic tissues (Figure 18.3h). Transfer the embryos into 4% paraformaldehyde using a pipette with wide opening (i.e. disposable pipettes). Fix the embryos for 1 h at RT or overnight at 4°C. After fixation, wash 3 times in PBSw (see below) and dehydrate by transferring through a methanol series into 100% methanol. The embryos can be stored at −20°C for several months.

Preparation of labeled antisense RNA probe. Probes are generally prepared with MAXIscript™ In vitro Transcription Kit (Ambion) as Digoxigenin-labeled RNA. All solutions should be RNAse-free.

1. Linearize the plasmid. The concentration of DNA in the digest should be 1 $\mu g/\mu l$. For antisense RNA use a restriction site in the polylinker at the amino-terminal (5′) end of the coding region of your sequence of interest.
2. Full-length (1.9 Kb) *Pitx2* (isoform c) cDNA, cloned into pBluescript KS, serves as the template for *in vitro* transcription. A plasmid can be obtained from the authors. To avoid read-through into vector sequences, the plasmid should be linearized by restriction digest with XbaI (for antisense RNA), and dissolved in TE at a concentration of 1 $\mu g/\mu l$.
3. Set up transcription reaction:

   (a) DNA   1 μg
   (b) 10 × transcription buffer   2 μl
   (c) DIG labeling mix   2 μl
   (d) RNAse inhibitor   1 μl
   (e) RNA polymerase (T7 enzyme mix for *Pitx2* antisense RNA) 2 μl
   (f) Add ddH$_2$O to   20 μl

4. Incubate for 2 hours at 37°C.
5. Check RNA on a 1% agarose gel (1 μl of the reaction).
6. Add 1 μl of DNAse I to digest the template DNA.
7. Add 30 μl ddH$_2$O to the DNAse I–treated transcription reaction to bring the volume to 50 μl.
8. Add 5 μl 5 M ammonium acetate and mix by vortexing.
9. Add 3 volumes of 100% ethanol.
10. Incubate at −20°C for at least 30 min.
11. Spin down the RNA at maximum speed in a 4°C microfuge for 30 min.
12. Dissolve the RNA in 30–40 μl of 50% formamide/50% nuclease-free water.
13. Check 1 μl on a 1% agarose gel.
14. Store the RNA at −80°C.

Data recording. Photograph the gel. A typical reaction is shown in Figure 18.3a. Take a picture at an early time point (when the dye has migrated just about 1–2 cm), because RNAses in the gel or running buffer may degrade the probe (compare at later time points).

Solutions
EMBRYO PREPARATION

PBS-agarose dishes: Pour a bottom layer of 1% agarose in PBS (microwave) into 5 cm plastic petri dishes.

PBSw: 0.1% Tween-20 in PBS (add 500 μl Tween-20 to 500 ml PBS). 4% paraformaldehyde

(a) Dissolve 4 g of PFA in 80 ml PBS at 65°C
(b) Add 2 drops of 10 N NaOH
(c) Add PBS to 100 ml

Methanol series in PBS (25%, 50%, 75%)

IN SITU HYBRIDIZATION
Methanol series
PBSw
Proteinase K stock solution: 10 mg/ml in H_2O
Hybridization mix (50 ml):
(a) 0.5 g Boehringer blocking reagent
(b) 25 ml formamide
(c) 12.5 ml 20 × SSC, pH7
Heat to 65°C for about 1 h. Once dissolved add:
(d) H_2O 6 ml
(e) 10 mg/ml yeast RNA (heat 2 min at 65°C to dissolve)   5 ml
(f) 50 mg/ml heparin   100 μl
(g) 20% Tween-20   250 μl
(h) 10% CHAPS   500 μl
(i) 0.5 M EDTA   500 μl
Filter the solution. The hybridization solution can be aliquoted and stored at −20°C.
SSC: 10 × stock solution: 1.5 M NaCl, 0.15 M Na-citrate.
Maleic acid buffer (MAB): 100 mM maleic acid, 150 mM NaCl; pH 7.5 (necessary to
dissolve completely).
Blocking solution (2.5 ml are needed for each sample): 10% goat serum (heat in-
activated 30 min at 65°C), 1% Boehringer blocking reagent in PBSw. Dissolve by
heating the mixture to 65°C. Filter the solution. The solution can be aliquoted
and stored at −20°C.
Bovine Serum Albumin (fraction V)
AP-Buffer: 100 mM Tris pH 9.5, 100 mM NaCl, 50 mM $MgCl_2$.

OUTLINE OF THE EXPERIMENTS

EXPERIMENT 1: ANALYSIS OF ASYMMETRIC GENE TRANSCRIPTION BY ISH

This experiment aims to visualize molecular asymmetries in embryos that otherwise
appear perfectly symmetrical. In principle, any of the asymmetric genes would serve this
purpose. As *nodal* and *Lefty2* are expressed only very transiently, however, and exact
timing of embryos is sometimes difficult to obtain due to variabilities even within one
litter, the experiment is performed with *Pitx2*.

The steps are outlined schematically in Figure 18.4. Early somite stage embryos are
isolated from timed matings (see Materials and Methods). *Pitx2* mRNA transcripts are
detected in a hybridization reaction. Antisense RNA (probe) is synthesized *in vitro* by
transcribing a plasmid containing the cloned cDNA (see Materials and Methods). This
probe is non-radioactively labelled with digoxigenin (DIG) at uracil residues during *in
vitro* transcription (Figure 18.4a). Fixed and pretreated embryos are incubated with the
labelled probe, and non-hybridized RNA is removed by repeated rinses and washes
(Figure 18.4b). To detect the hybridized and labelled antisense RNA probe embryos

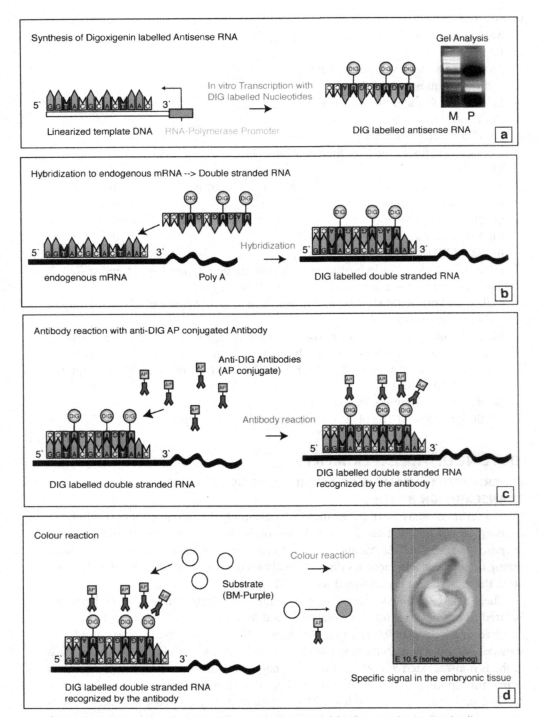

Figure 18.4. Schematic outline of a whole-mount in situ hybridization experiment. For details, see text, Experiment 1. M, size marker; P, RNA probe.

are incubated with an antibody that detects the digoxigenin moiety linked to uracil residues (Figure 18.4c). The antibody is conjugated to the enzyme alkaline phosphatase (AP), which is used to reveal the localization of *Pitx2* transcripts in the embryo in a color reaction after removing excess antibody by repeated washes (Figure 18.4d, step 1b).

General warnings

➤ Do not let the embryos dry at any stage, as the amount of background will increase. It is preferable to leave the embryos in a small volume of a given solution and to add the next solution.

➤ Filter all solutions to remove particles that will stick to the embryos.

➤ Use gloves and aerosol barrier tips for changing the solutions from the fixation step to the end of hybridization

➤ Until pre-hybridization, perform all steps (fixations, rinses, washes) on ice, except the Proteinase K treatment.

Day 1: hybridization

1. Prepare fresh 4% paraformaldehyde–0.2% glutaraldehyde in PBS – (about 5 ml will be needed for each sample after Proteinase K treatment).
2. Prepare fresh 4.5 μg/ml Proteinase K in PBSw.
3. Prepare fresh 2 mg/ml glycine in PBSw.
4. Rehydrate the embryos through 75%, 50%, and 25% methanol series in PBSw. Incubate each step for 5 min on ice.
5. Wash 3 times for 5 min with PBSw on ice.
6. Change into 1 ml 4.5 μg/ml Proteinase K in PBSw.
7. Incubate for 10 min (E5.5–E9.5) or 15 min (>E9.5) at RT. Make sure to thaw the Proteinase K stock completely and vortex to dissolve precipitates at the bottom of the tube.
8. Stop Proteinase K digestion by washing in freshly prepared 2 mg/ml glycine in PBSw.
9. Rinse in PBSw.
10. Wash 2 times in PBSw for 5 min each.
11. Refix in 5 ml of 4% paraformaldehyde/0.2% glutaraldehyde in PBS – for 15 min.
12. Rinse in PBSw.
13. Wash 3 times in PBSw for 5 min each.
14. Wash in 1 ml of 50% PBSw/50% hybridization mix, followed by 100% hybridization mix for about 3 min each without rocking.
15. Replace 900 μl of fresh hybridization mix in each glass vial.
16. Prehybridize samples for 3 h at 65°C.
17. Heat 200 ng (approximately 1 μl from a successful transcription reaction; compare Figure 18.4a) of the RNA probe in 100 μl of hybridization mix at 95°C for 5 min.
18. Add the probe/hybridization mix to the embryos.
19. Hybridize overnight at 70°C in a water bath, rocking.

Day 2: antibody reaction

1. Remove hybridization solution and add 800 μl of prehybridization solution.
2. Wash for 5 min at 70°C.
3. Add 400 μl of 2 × SSC, pH 4.5 (without removing hybridization solution).
4. Repeat step (3) two times more.
5. Remove the mix and wash twice for 30 min each in 2 × SSC pH 7.0 at 70°C.
6. Wash twice for 10 min each in Maleic Acid Buffer at room temperature (RT).
7. Wash twice for 30 min each in Maleic Acid Buffer at 70°C.
8. Wash twice for 10 min each in PBSw.
9. Wash 5 min in PBSw at RT.
10. Gently rock the embryos in 1 ml blocking solution for at least 2 h at 4°C.
11. Dilute the anti-DIG antibody 1/10,000 from a stock of 150 units/200 μl (Boehringer) in blocking solution and gently rock for at least 2 h at 4°C.
12. Replace blocking solution with the diluted antibody and gently rock overnight at 4°C.

Day 3: detection

1. Rinse embryos with 0.1% BSA in PBSw.
2. Wash 5 times with 5 ml 0.1% BSA in PBSw (fill to the top to minimize air bubble) for 45 min each with rocking.
3. Wash twice for 30 min each in PBSw.
4. Warm BM-Purple (staining solution) to RT.
5. Wash the embryos twice in AP1 buffer for 10 min each with rocking.
6. Prepare 12 well plates (1 ml BM-Purple per sample/well) and place the embryos into individual wells.
7. Incubate embryos overnight at RT. This procedure works well with most probes, including *Pitx2*. With some probes it is recommended to incubate at 4°C overnight and visualize the staining the following day at RT.
8. Stop the staining reaction by washing at least three times in PBSw.
9. Dehydrate through a methanol series (25, 50, 75, and 2 × 100%) and store in methanol at −20°C (staining intensifies in methanol).
10. To photograph embryos stepwise rehydrate to PBSw.

Data recording. Observe the stained embryos under the microscope. Count and record the number of somites. Photograph embryos from a ventral and dorsal perspective. In addition to overviews, show detailed views of the LPM and the heart region, and sites of symmetrical *Pitx2* transcription in the head.

EXPERIMENT 2: ASYMMETRIC MORPHOGENESIS

Asymmetric organ development is achieved by a series of morphogenetic processes in response to the Nodal signaling cascade (Hamada et al., 2002). The first signs of asymmetric morphogenesis can be observed in E8.5 mouse embryos, when the linear heart tube loops to the right and adopts the shape of an S (Figure 18.3a). Between E8.5 and E9.0 the embryo rotates clockwise along the anterior–posterior axis, such that the tail now is found on the right side of the head (Figure 18.2b). These events are easily observed in 9.5-day-old mouse embryos (preparation, see Material and Methods).

At around E16.5 organ laterality has fully developed. The apex of the heart points to the left side, asymmetric lobe formation has resulted in a lung with four lobes on the right and a single lobe on the left side (Figure 18.2c), and asymmetric rotation and coiling of the digestive tract have placed the stomach and spleen on the left side of the abdominal cavity (Figure 18.2d). The end result may be observed and studied in the females sacrificed for the collection of embryos as well. Students interested in analyzing the organ situs in adult mice should proceed as follows:

1. After removal of the uterus locate the stomach on the left side of the abdominal cavity. In close proximity the spleen is found.
2. To investigate the thoracic organs open the rib cage by cutting the sternum (scissors).
3. The apex of the heart, which itself occupies a medial position, clearly points to the left. Remove the heart (scissors) and locate left and right atria and ventricles. Thereby the lung becomes exposed. Note the single lobe on the left and the four lobes on the right side.

Data recording. Photograph the organs *in situ* and after removal from the body cavities.

EXPECTED RESULTS

ASYMMETRIC GENE EXPRESSION
The expected result is a staining pattern comparable to the one depicted in Figure 18.1c. Asymmetric staining in the left LPM should be clearly visible. In addition, staining on the right side of the linear heart tube may be observed, depending on the quality of the probe and the duration of the staining reaction.

If *nodal* or *Lefty* were used as probes, asymmetric gene expression may be visible only if the embryos used in the experiment were at the 3–7 somite stage. Embryos stained for *nodal* transcription should show expression at the primitive node as well, which in most cases should be considerably stronger on the left side. *Lefty* probes which do not distinguish between *Lefty1* and *Lefty2* (i.e. probes which contain coding sequences in addition to 5'-UTR) should stain the left floor plate (*Lefty-1*) in addition to the left LPM (*Leftx-2*). Left-asymmetric staining in the floor plate, however, becomes obvious only in histological sections.

ASYMMETRIC ORGAN MORPHOGENESIS
Asymmetric organ placement and morphogenesis should be readily detectable. In very rare cases a spontaneous inversion of this arrangement (situs inversus) may be found.

DISCUSSION

The experiments allow the following conclusions:

1. Asymmetric organ morphogenesis is preceded by asymmetric gene expression on the left side of the embryo (LPM) at about 8 days of development.
2. The first signs of morphological asymmetries are observed at E8.5 and become obvious at E9.5 (asymmetric turning of the embryo and looping of the linear heart tube to the right side).
3. Organ placement follows a consistent pattern (situs solitus).
4. Asymmetric *Pitx2* transcription is observed before and during asymmetric organ morphogenesis, suggesting that this homeobox transcription factor is involved in mediating the transfer of the original asymmetric signal to the forming organs.

TIME REQUIRED FOR THE EXPERIMENTS

The experiments require one week of work in the laboratory. On Day 1 the probe is prepared and embryos are isolated. On Days 2–4 the ISH experiment is performed. On Day 5 the results are evaluated and the embryos are photographed.

POTENTIAL SOURCES OF FAILURE

Mice may not be pregnant even though a vaginal plug was clearly observed. Probe preparation may fail due to inexperience of students (RNAses!). If both embryos and probes can be provided, the ISH experiment itself works in most cases, even if students have very little laboratory experience.

TEACHING CONCEPTS

Gene expression patterns as indicators of specification prior to overt differentiation. Left and right LPM are indistinguishable by anatomical, morphological, and cell biological criteria. Asymmetric transcription precedes asymmetric organ morphogenesis.

Signaling cascades. The "Nodal cascade" is an example of a signaling pathway involving both positive (Nodal signal transduction on *nodal* transcription; Pitx2 on *Pitx2* transcription) and negative (Lefty on Nodal signal transduction) feedback regulation.

Homeobox genes. Homeobox genes play important roles in axis formation in all animals studied. The anterior–posterior axis is set up by clustered homeobox genes (Hox genes). *Pitx2* is a unique gene in left–right axis formation, as it is present during early stages (specification), and stays on during organ development.

ALTERNATIVE EXERCISES

QUESTIONS FOR FURTHER ANALYSIS

- ➤ The asymmetric expression pattern of *Pitx2* is very dynamic between E7.75 and E9.5. Describe expression in individual embryos and correlate the pattern with developmental age (somite number).
- ➤ What other expression sites of *Pitx2* can you identify besides the left LPM?
- ➤ Observe and describe the asymmetric morphogenesis of the heart and its associated vessels in E8.5 and E9.5 embryos and adult mice.

ADDITIONAL INFORMATION

SYMMETRY BREAKAGE

The "Nodal flow hypothesis" postulates that an extracellular morphogen becomes asymmetrically localized to the left side of the primitive node by rotary movement of primary (9 + 0) cilia localized on node cells (Hamada et al., 2002). Motility of node cilia and asymmetric fluid flow have been demonstrated, and mutants affecting cilia invariably alter organ situs. A variation of this model proposes that directional fluid flow might be sensed by mechanosensory cilia, resulting in asymmetric calcium ion influx at the node ("two cilia model;" Tabin and Vogan, 2003). A radically different hypothesis claims asymmetric cytoplasmic ion transport as a general very early step in the orientation of the embryonic left–right axis (Levin et al., 2002). This hypothesis is based mainly on experiments in frog and chick embryos which demonstrate a role of ion channels and gap junctional communications in the generation of laterality, and on asymmetric mRNA localization of the H^+/K^+-ATPase ion exchanger in the first few cell divisions in the frog *Xenopus* (Mercola and Levin, 2001; Levin et al., 2002).

EVOLUTIONARY CONSERVATION

The "Nodal cascade" is conserved in all vertebrates studied to date (fish, amphibia, birds, mammals). Differences might exist in the mechanism of symmetry breakage (see above), and have been demonstrated for the transfer of the original, unidentified, asymmetric signal from the midline (node) to the left LPM. Analyses of the secreted growth factor FGF8 in chick, mouse, and rabbit embryos have shown that in embryos which develop from a flat blastodisc (chick, rabbit), FGF8 acts as a right determinant at the node, which prevents activation of the "Nodal cascade" on the right side. On the other hand, in the mouse embryo, which develops from a cup-shaped egg cylinder, FGF8 is required for the induction of *nodal, Lefty*, and *Pitx2* on the left side (Boettger, Wittler, and Kessler, 1999; Meyers and Martin, 1999; Fischer, Viebahn, and Blum, 2002).

ACKNOWLEDGEMENTS

Work in our laboratory is supported by grants from the Deutsche Forschungsgemeinschaft (BL 285/3-7 and SFB 495).

REFERENCES

Boettger, T., Wittler, L., and Kessel, M. (1999). FGF8 functions in the specification of the right body side of the chick. *Curr. Biol.*, 9, 277–80.

Capdevila, J., Vogan, K. J., Tabin, C. J., and Izpisúa-Belmonte, J. C. (2000). Mechanisms of left-right determination in vertebrates. *Cell*, 101, 9–21.

Fischer, A., Viebahn, C., and Blum, M. (2002). FGF8 acts as a right determinant during establishment of the left-right axis in the rabbit. *Curr. Biol.*, 12, 1807–16.

Hamada, H., Meno, C., Watanabe, D., and Saijoh, Y. (2002). Establishment of vertebrate left-right asymmetry. *Nat. Rev. Genet.*, 3, 103–13.

Levin, M., Thorlin, T., Robinson, K., Nogi, T., and Mercola, M. (2002). Asymmetries in H(+)/K(+)-ATPase and cell membrane potentials comprise a very early step in left-right patterning. *Cell*, 111, 77–89.

Mercola, M., and Levin, M. (2001). Left-right asymmetry determination in vertebrates. *Ann. Rev. Cell. Dev. Biol.*, 17, 779–805.

Meyers, E. N., and Martin, G. R. (1999). Differences in left-right axis pathways in mouse and chick: Functions of FGF8 and SHH. *Science*, 285, 403–6.

Schweickert, A., Campione, M., Steinbeisser, H., and Blum, M. (2000). Pitx2 isoforms: Involvement of Pitx2c but not Pitx2a or Pitx2b in vertebrate left-right asymmetry. *Mech. Dev.*, 90, 41–51.

Tabin, C. J., and Vogan, K. J. (2003). A two-cilia model for vertebrate left-right axis specification. *Genes Dev.*, 17, 1–6.

SECTION VIII. **TRANSGENIC ORGANISMS**

19 Bicoid and Dorsal: Two transcription factor gradients which specify cell fates in the early *Drosophila* embryo

S. Roth

OBJECTIVE OF THE EXPERIMENT The early *Drosophila* embryo is patterned by two types of morphogen gradients which are organized by maternally expressed genes. Intracellular morphogens spread within the egg cytoplasm from mRNA sources that are tightly localized to the anterior or posterior egg cortex. Extracellular morphogens spread within the extraembryonic space (perivitelline space) surrounding the embryo and depend on localized cues within the eggshell or extracellular matrix. *bicoid*, an intracellular morphogen, is required to specify the head and thorax of the embryo. Spätzle, an extracellular morphogen, activates the Toll receptor at the surface of the embryo, ultimately leading to a concentration gradient of the Dorsal protein, which is required to establish the dorsoventral axis of the embryo.

The following experiments explore the morphogen concept by showing how changes in concentrations of the transcription factors Bicoid and Dorsal affect the cell fates of the early embryo. Alterations of cell fates in mutant as compared to wild-type embryos will be monitored by looking at gastrulation and cuticle patterns. The observed shifts in cell fates will be correlated to the expression of zygotic genes which are targets of concentration-dependent activation and/or repression by Bicoid or Dorsal. Together, these experiments should provide an understanding of how each gradient specifies a polar sequence of stripe-like expression domains, which in turn determine the pattern of cell fates along the two major body axes.

DEGREE OF DIFFICULTY The experiments require that instructors and staff have some basic knowledge about the handling of flies (keeping fly stocks, distinguishing males and females, recognizing the dominant marker mutations of the balancer chromosomes, setting up crosses). Furthermore, each student needs access to both a dissecting microscope (ideally with transmitting light) and to a compound microscope. At least one of the compound microscopes should be equipped with either a camera lucida, a photocamera or an electronic camera (video or digital) so that pictures from stained embryos can be produced for subsequent analysis.

INTRODUCTION

Since the beginning of the 20th century gradients of morphogenetic substances have been used to explain the generation of cellular diversity in development (Wolpert, 1996). However, the morphogen concept was not widely accepted in the first half of the century because both clear molecular demonstrations of the existence of morphogens and theoretical concepts explaining gradient formation were missing. Renewed interest emerged through theoretical work. Turing showed in his 1953 paper that simple chemical reactions can cause stable asymmetric distributions of substances which, as he postulated, might influence development (reprinted as Turing, 1990). He was the first to coin the term "morphogen," by which he meant "form-generating substances." The term gained a more specific and definite meaning after Wolpert introduced the concept of positional information in the late 1960s. Wolpert stressed the fact that in most developmental systems, pattern formation precedes cell differentiation (Wolpert, 1969, 1971). Naïve cells first are instructed about their position within a developmental field and then acquire different cell fates according to this positional information. The simplest way of providing positional information relative to a fixed boundary is by a morphogen mechanism. In the framework of these ideas a morphogen is defined by two crucial properties. First, its distribution is graded. Starting from a localized source (at the boundary) the morphogen concentration decreases monotonically so that the value of its concentration at any point of the gradient provides a measure of the distance from the boundary. Second, distinct ranges of morphogen concentrations elicit different developmental fates in the cells or nuclei that are exposed to the gradient. The combination of these two properties (graded distribution and concentration-dependent action) within a single molecule provides an elegant and powerful mechanism for specifying a pattern of cell fates as a function of physical distance.

Further theoretical work in the 1970s provided models for gradient formation and interpretation (Crick, 1970; Gierer and Meinhardt, 1972; Lewis, Slack, and Wolpert, 1977; Meinhardt, 1982). However, although experimental evidence for the existence of morphogens accumulated (for example Sander, 1976; Tickle, Summerbell, and Wolpert, 1975), their molecular nature remained largely elusive until the role of maternal genes in *Drosophila* axis formation was elucidated (for a recent review of history and new developments, see Ephrussi and St. Johnston, 2004; Nüsslein-Volhard, 2004; Driever, 2004).

Axis formation in the *Drosophila* embryo is mediated by four systems defined genetically by groups of maternal genes with similar or identical phenotypes. Three of the systems provide positional information for regions along the anteriorposterior axis, while pattern along the dorsoventral axis is determined by a single system (St Johnston and Nüsslein-Volhard, 1992). The anterior and posterior systems work via mRNAs localized at the termini of the egg cytoplasm. The maternal key component for anterior patterning is *bicoid* (*bcd*, Frohnhöfer and Nüsslein-Volhard, 1986). Mutations in *bcd* lead to embryos which lack all head and thorax structures. *bcd* mRNA is localized at the anterior pole of the egg through polar transport and anchoring processes which occur during oogenesis (Berleth et al., 1988; St. Johnston et al., 1989). After egg deposition the anteriorly anchored *bcd* mRNA is translated and Bcd protein diffuses within the egg cytoplasm (Figure 19.3). As a result, Bcd protein becomes distributed in an exponential concentration gradient with a maximum at the anterior tip, reaching lowest levels in the posterior third of the embryo (Driever and Nüsslein-Volhard, 1988*a*; Driever and Nüsslein-Volhard, 1988*b*). Bcd is a homeobox transcription factor that binds to enhancer elements of various segmentation genes, most importantly to that of gap genes which are required for the formation of head and thorax structures. Distinct concentration thresholds of Bcd lead to the activation of different gap genes. Thus, the concentration gradient of Bcd is translated into a unique pattern of gap gene expression domains, which, in turn, determine the sequence and identity of head and thoracic segments. The analysis of Bcd binding sites in the promoter of zygotic *hunchback* provided the first model of how gene activation might occur in a concentration-dependent manner (Driever and Nüsslein-Volhard, 1989; Driever, Thuman, and Nüsslien-Volhard, 1989; Struhl, Struhl and MacDonald, 1989).

The key components for anterior and posterior patterning are asymmetrically localized within the egg cytoplasm even before fertilization and egg deposition. In contrast, none of the components of the terminal and dorsoventral systems, which are produced by the oocyte or early (preblastoderm) embryo, is asymmetrically distributed (St Johnston and Nüsslein-Volhard, 1992). Spatial asymmetry relies entirely on extraembryonic cues present in the eggshell or extracellular matrix. The molecular nature of these cues is still not entirely known. In both the terminal and dorsoventral system, the actual morphogen is a diffusible cytokine-like protein with structural similarity to nerve growth factor (NGF), secreted into the fluid-filled perivitelline space (Casanova et al., 1995). Upon uniform secretion it is proteolytically processed and this generates the active ligand of a cell surface receptor that is present in the plasma membrane of the early embryo. Although the receptors are uniformly distributed, cytokine processing is spatially restricted such that the active ligands presumably form concentration gradients in the perivitelline space, in turn leading to the graded activation of the receptors.

For the dorsoventral system (for a review see Morisato and Anderson, 1995) the diffusible ligand in the perivitelline space is a proteolytically activated form of Spätzle (Morisato and Anderson, 1994; Schneider et al., 1994). Proteolytic activation occurs only at the ventral side of the embryo and depends on a cascade of four proteases. Activated Spz presumably forms a gradient with highest concentrations in a stripe along the ventral midline and decreasing concentrations towards the lateral sides of the embryo. Spz is believed to bind to the transmembrane receptor Toll at the surface of the embryo (Hashimoto, Gerttula, and Anderson, 1991; Hashimoto, Hudson, and Anderson, 1988). Toll belongs to an evolutionary conserved group of receptors that have an ancestral function in the innate immune response to pathogens (Hoffmann et al., 1999). These receptors control the nuclear import of transcription factors of the NF-κB/rel family. During dorsoventral axis formation in *Drosophila*, Toll activation leads to the nuclear import of the NF-κB/rel transcription factor Dorsal (Steward, 1987; Steward et al., 1988). As a result, the extracellular gradient of activated Spz is transformed into the nuclear concentration gradient of Dorsal (Roth, Stein, and Nüsslein-Volhard, 1989; Rushlow et al., 1989; Steward, 1989). Dorsal acts both as transcriptional activator and repressor of zygotic genes involved in the establishment of different cell fates along the dorsoventral axis. High nuclear concentrations of Dorsal induce the formation of the mesoderm, the ventralmost structure, and intermediate concentrations induce the neuroectoderm, whereas the absence of Dorsal is required for the formation of dorsal ectoderm and amnioserosa. It is likely that we know most of the target genes of Dorsal including their cis-regulatory regions, which are responsible for their concentration-dependent activation or repression by Dorsal (Stathopoulos et al., 2002). The dissection of these sequences has provided a picture of how the graded distribution of a transcription factor leads to a distinct set of trancriptional responses (Stathopoulos and Levine, 2002).

In the experiments described below, Bicoid and Dorsal are introduced as two examples of transcription factor localized in gradients. The phenotypes which result from the genetic manipulation of their distribution or their activity reveal how the embryonic axes of the early *Drosophila* embryo are patterned through morphogen mechanisms.

MATERIALS AND METHODS

EQUIPMENT AND MATERIALS

The equipment necessary for handling flies is described in Chapter 14.

Materials required for egg collection and cuticle preparation are described in Chapter 11. (See also Wieschaus and Nüsslein-Volhard, 1998)

Microscopes. Living embryos are observed with transmitted light on a dissecting microscope (25- to 50-fold magnification). For cuticle preparations a compound microscope should be used which is equipped with objectives for both dark-field and phase-contrast optics. As a substitute for true dark-field optics, low-power objectives ($\times$ 10 and $\times$ 16) can be used with the condenser in phase-contrast position (Ph 2 and Ph 3). Embryos from in situ hybridizations or antibody stainings should be analyzed with bright-field or Nomarski optics.

Biological materials (fly stocks). The following *Drosophila melanogaster* stocks are required. (The slash separates homologous and the semicolon non-homologous chromosomes. For nomenclature and further explanations regarding *Drosophila* genetics, see Greenspan, 1997):

The maternal mutations are kept over balancer chromosomes (*FM7* for first, *CyO* for the second and *TM3* for the third chromosome). These balancers are homozygous lethal and carry inversions which suppress recombination. In addition they carry recessive and dominant marker mutations and can therefore be easily recognized.

1. STOCKS FOR MANIPULATING THE BICOID (BCD) GRADIENT (FROHNHÖFER AND NÜSSLEIN-VOLHARD, 1986)

Oregon R (wild-type)

ru th st ri bcd^{E1} roe p^p e^s ca/TM3, Sb: null-allele of *bcd*

ru st bcd^{E2} e/TM3, Sb: null-allele of *bcd*

P{bcd$^+$ (5)} P{bcd$^+$(8)}/FM7: two wild-type copies of *bcd* on the first chromosome (Busturia and Lawrence, 1994)

2. STOCKS FOR MANIPULATING THE DORSAL (DL) GRADIENT

 b dl¹ pr cn vg^D/CyO: null-allele of *dl* (Nüsslein-Volhard, 1979*a*)

 Df(2L)TW119, cn bw/CyO: deficiency covering *dl* (Steward and Nüsslein-Volhard, 1986)

 b dl²/CyO: weak loss-of-function allele of *dl* (Nüsslein-Volhard, 1979*b*)

 ru th st ri roe p^p Tl^rm9/TM3: recessive lateralizing *Tl* allele (Anderson et al., 1985; Schneider et al., 1991)

 ru kls e Tl^rm10/TM3, Sb: recessive lateralizing *Tl* allele (Anderson, Jürgen, and Nusslein-Volhard, 1985; Schneider et al., 1991)

 Tl^10b, mwh e/TM3, Sb/T(1;3) OR60: dominant ventralizing *Tl* allele (Roth et al., 1989; Schneider et al., 1991)

3. STOCKS CARRYING REPORTER CONSTRUCTS. The following stocks contain transgenes in which cis-regulatory elements of zygotic segmentation and dorsoventral patterning genes control the expression of *β-galactosidase* (*LacZ*). Thus, the detection of *β-galactosidase* mRNA or protein can be used to monitor the expression of the corresponding genes.

 P{hb-LacZ}: *β-gal* expression in a broad anterior domain (Driever and Nüsslein-Volhard, 1989)

 P{ftz-LacZ}: *β-gal* expression in seven stripes along the AP axis (Hiromi, Kuroiwa, and Gehring, 1985)

 P{twi-LacZ}: *β-gal* expression in a ventral stripe (Jiang et al., 1991)

 P{sog-LacZ}: *β-gal* expression in lateral stripes (Markstein et al., 2002)

 P{dpp-LacZ}: *β-gal* expression in a dorsal stripe (Jackson and Hoffmann, 1994)

PROCEDURES

Setting up fly crosses. See Chapter 11 and 14.

Observing live embryos (see also Wieschaus and Nüsslein-Volhard, 1998). To observe gastrulating embryos, eggs should be collected in a three-hour interval using apple-juice–agar plates with a small dot of live yeast. Flies produce the largest number of eggs in the late afternoon or evening. An egg collection from 1 p.m. to 4 p.m. should yield enough material. The living embryos can be examined directly on agar plates with a dissecting microscope using transmitted light. They also can be transferred to a glass slide and observed with a compound microscope (bright-field setting). The embryos are surrounded by two eggshells: the inner vitelline membrane and the outer chorion. The chorion is air-filled and therefore opaque. It is made transparent by covering the embryos with halocarbon oil (e.g., Voltalef 3S or Halocarbon Oil 27, Sigma). Using chorion specializations as landmarks, the eggs can be oriented unambiguously. At the anterior pole the eggs carry a micropyle and at the dorsal side two respiratory (also called dorsal) appendages (Figure 19.1). The wild-type embryo develops in a fixed orientation with regard to these egg-shell markers. Its head always points towards the micropyle and its dorsal side towards the respiratory appendages. By convention, egg length is measured from the posterior pole (0% egg length posterior tip, 100% egg length anterior tip).

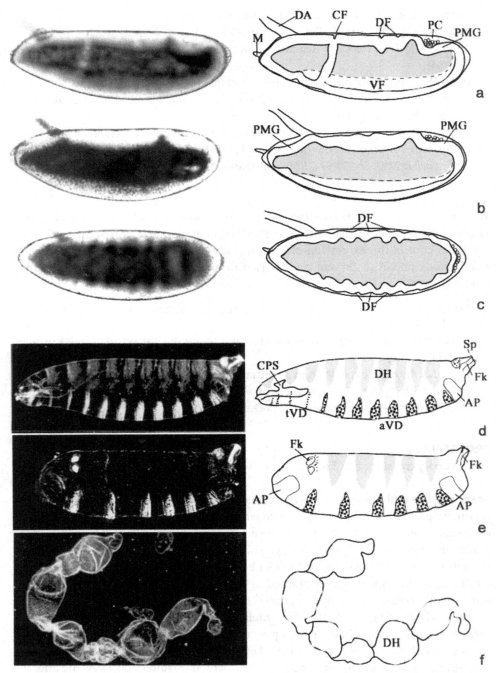

Figure 19.1. Gastrulation and larval cuticles of wild-type, *bicoid* and *dorsal* mutants (a–c). Gastrulating embryos. Left: Bright-field micrographs of living embryos. The chorion is transparent because the embryos are covered by halocarbon oil. Right: Schematic drawings of the same embryos highlighting characteristic chorion structures and morphogenetic movements. (a) Wild-type. (b) Embryo derived from a *bicoid* mutant female (*bcd^{E2}/bcd^{E1}*). (c) Embryo derived from *dorsal* mutant female (*cont'd on next page*)

Cuticle preparation. See Chapter 11.

Preparation of a dig-labelled lacZ-antisense probe. The pBs-LacZ plasmid contains the *E. coli LacZ* inserted in the polylinker of pBlueskript KS+ (the plasmid is available from the author's laboratory). Cut the plasmid with BamHI and use T3 polymerase to make an antisense probe. For probe production use the DIG RNA Labeling Kit from Roche or Maxiscript from Ambion (see also Chapter 20).

In situ **hybridisation.** See Chapter 20.

OUTLINE OF THE EXPERIMENTS

EXPERIMENT 1. CHANGING THE DOSE OF *bicoid* CAUSES SHIFTS IN THE ANLAGEN OF THE AP AXIS

Experiment 1a. Generating flies carrying different numbers of *bcd*$^+$ copies. The aim of the experiment is to generate flies that carry 0, 1, 2, 4, or 6 copies of *bcd*$^+$. Flies with 0 or 1 copies of *bicoid* are produced in the following cross (Frohnhöfer and Nüsslein-Volhard, 1986):

ru th st ri bcd^{E1} roe p^p e^s cal TM3, Sb ♀ (♂) × *ru st bcd^{E2} el TM3, Sb* ♂ (♀)

Put 10 virgin females (♀) and 10 males (♂) into each of 5 big food bottles. Keep the flies at 25°C and transfer to new vials every second day for one week. After 10 days, the progeny from the crosses will begin to hatch. Since *TM3, Sb/TM3, Sb* is lethal, the following genotypes will be observed:

ru th st ri bcd^{E1} roe p^p e^s cal TM3, Sb
ru st bcd^{E2} el TM3, Sb
ru th st ri bcd^{E1} roe p^p e^s cal ru + st + bcd^{E2} + + e +

Collect approximately 100 females of the genotype *bcd^{E1}/TM3, Sb* or *bcd^{E2}/TM3, Sb* (1 copy of *bcd*$^+$) and 100 females of the genotype *bcd^{E1}/bcd^{E2}* (0 copies of *bcd*$^+$) as virgins. In addition, collect 100 virgins from each of the following genotypes: Oregon R (2 copies of *bcd*$^+$), *P{bcd$^+$ (5)} P{bcd$^+$(8)}/FM7* (4 copies of *bcd*$^+$) and *P{bcd$^+$ (5)} P{bcd$^+$(8)}/P{bcd$^+$ (5)} P{bcd$^+$(8)}* (6 copies of *bcd*$^+$: Tübingen stock collection, Driever and Nüsslein-Volhard, unpublished; Busturia and Lawrence, 1994). The 100 females of each genotype will be used for two crosses. Cross 50 of these females to 25 males carrying *P{hb-LacZ}* (Driever and Nüsslein-Volhard, 1989) and the remaining 50 to 25 males carrying *P{ftz-LacZ}* (Hiromi et al., 1985). Transfer the females and males into

Figure 19.1 (cont.) (*dl^1/Df(2L)TW119*). Wild-type embryos are 400–500 μm in length. (d–f) Larval cuticles. Left: Dark-field micrographs taken from cuticle preparations of dechorionated and devitellinized larvae. Right: Schematic drawings of the same larvae highlighting characteristic cuticle landmarks. (a) Wild-type. (b) Larva derived from *bicoid* mutant female (*bcd^{E2}/bcd^{E1}*). (c) Larva derived from *dorsal* mutant female (*dl^1/Df(2L)TW119*). **AP**: anal pads. **CF**: cephalic fold. **CSP**: cephalopharyngeal skeleton. **DA**: dorsal (respiratory) appendage. **DF**: dorsal fold. **DH**: dorsal hairs. **Fk**: filzkörper. **M**: micropyle. **PMG**: posterior midgut. **Sp**: spiracles. **aVD**: abdominal ventral denticles. **tVD**: thoracic ventral denticles.

large food bottles and leave overnight at room temperature. The next day, prepare 10 egg-collection chambers and Petri dishes containing an apple-juice–agar medium and a dot of live yeast suspension. Transfer the flies to the egg-collection chambers resting on the apple-juice plates. The chambers should be placed in a dark box and kept at room temperature (approximately 22°C). The females will not start depositing eggs efficiently until after one day in the egg collection chamber. The plates have to be changed at least once per day to feed the flies sufficiently with fresh yeast and to keep the chambers free of hatching larvae (embryonic development is completed after 22 hours at 22°C). When the plates are regularly changed the flies may remain in the same egg-collection chamber for more than a week.

Experiment 1b. Observing the gastrulation of embryos from females with different numbers of *bcd*⁺ copies. *bcd* is an example of a purely maternal gene. The phenotype of the embryos depends only on the genotype of the mothers irrespective of the males used in the cross. To observe the live embryos, cover the three-hour egg collection plate with halocarbon oil, and view under the dissecting microscope using transmitted light (Wieschaus and Nüsslein-Volhard, 1998). Arrange the syncytial and cellular blastoderm embryos using a needle into rows such that the micropyle marking the anterior end points to the left and the respiratory appendages to the top. Syncytial blastoderm embryos are characterized by a broad uniform layer of cortical cytoplasm surrounding the opaque yolk. The cortical cytoplasm contains the nuclei which are visible only in the compound microscope. During the cellular blastoderm stage the cell membranes move down between adjacent nuclei. The advancing front of the in-growing membranes is visible as a line running parallel to the surface of the egg. Look at the plate every ten minutes. Embryos will progress through these two blastoderm stages in about two hours at room temperature. Gastrulation begins as soon as the cells on the ventral side of the embryo have completed cellularization. The gastrulation movements are asymmetric with regard to both the anteriorposterior and the dorsoventral axis. For this experiment we concentrate on anteriorposterior differences. In wild-type eggs the cephalic furrow forms as an oblique lateral infolding at approximately 65% egg length (Figure 19.1a). At the same time, the posterior plate forms carrying the pole cells towards the dorsal side of the embryo. Record the approximate position (in % egg length) of the cephalic furrow in embryos with 1, 2, 4, and 6 copies of *bcd*⁺. Describe the gastrulation movements of embryos lacking *bcd* (Figure 19.1b). Do you see a cephalic furrow? What happens at the anterior pole?

Experiment 1c. Cuticles of wild-type and *bcd* mutant embryos. The cuticle secreted at the end of *Drosophila* embryogenesis has many landmarks that can be used to examine aspects of anteriorposterior and dorsoventral patterning (Figures 19.1d and 19.2; Wieschaus and Nüsslein-Volhard, 1998). For wt cuticles, use a 12 h egg collection from Oregon R flies. With the help of a brush, transfer several hundred eggs into a Petri dish containing tap water. Leave overnight at RT. The larvae hatch into the water, preventing food uptake and growth. Using a plastic pipette, sieve the water containing the larvae through a wire basket. The larvae will remain in the basket and can then be transferred

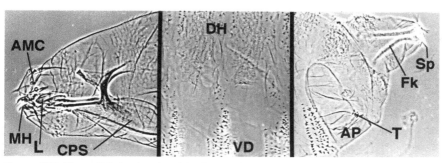

Figure 19.2. The cuticle landmarks of the *Drosophila* larva. Phase-contrast micrograph of larval cuticle. Left panel: head and thorax. Middle panel: trunk. Right panel: Telson. AMC: antenno-maxillary complex. AP: anal pads. CPS: cephalopharyngeal skeleton. DH: dorsal hairs. Fk: filzkörper. L: Labrum. MH: Mouth hook. Sp: spiracles. T : tuft. VD: ventral denticles.

using a clean brush to a drop of Hoyer's mountant-lactic acid (1:1) on a glass slide (see Material and Methods, cuticle preparations Chapter 11). The larvae from flies lacking *bicoid* (*bcd^{E1}/bcd^{E2}*) do not hatch. Therefore the eggshells have to be removed for inspection of the cuticle. Use an overnight egglay for cuticle preparation as described in Materials and Methods. Also check overnight egg lays from flies carrying multiple copies of *bicoid*. What do you observe?

Analyse the cuticle using dark-field and phase contrast settings on the compound microscope. First, make a schematic drawing of the wt cuticle (Figures 19.1d and 19.2). Try to identify the following structures: 8 abdominal denticle belts, the narrow denticle belts of the thoracic segments, the head skeleton with mouth hooks, antennal and maxillary organ, and the structure of the telson at the posterior end: filzkörper, spiracles, and anal pads. Then analyze the cuticle from embryos derived from *bcd* mutant females (Figure 19.1e). Which region of the anteriorposterior axis is affected? Which segments are missing? What do you see at the anterior end of the larvae? Compare cuticle pattern and gastrulation phenotype.

Experiment 1d. The expression of early segmentation genes in embryos from females with different numbers of *bcd^+* copies. Bicoid directly regulates the expression of *hunchback* (*hb*), the gap gene required for head and thorax formation in *Drosophila* (Driever and Nüsslein-Volhard, 1989; Struhl et al., 1989; Figure 19.3). *hb* cooperates with other gap genes to delimit the 7 stripes of expression of primary pair-rule genes, like *even-skipped* and *hairy*, which in turn regulate the expression of secondary pair-rule genes, like *fushi-tarazu* (*ftz*; for review see Rivera-Pomar and Jäckle, 1996). Here, we use reporter gene constructs to visualize the expression of *hb* (Figure 19.3) and *ftz* (Figure 19.4) in blastoderm embryos. The transgenes have been introduced into the embryo through the males. This does not affect the *bicoid*-related phenotypes which depend entirely on the genotype of the female.

Use three-hour egg collections of all crosses and age them for one hour. Dechorionate and fix the embryos and perform in situ hybridizations as described in Chapter 20 using a *LacZ* antisense probe. Keep the stained embryos in 50% glycerol and transfer

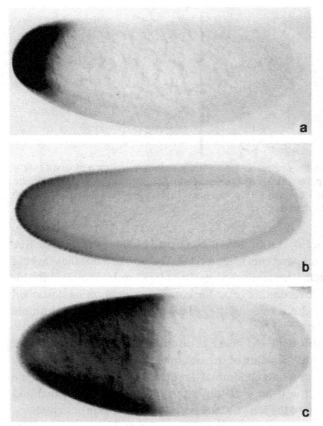

Figure 19.3. Maternal *bicoid* determines the expression domain of the anterior gap gene *hunchback*. (a) Optical midsection of a preblastoderm embryo. *bicoid* mRNA is concentrated at the anterior pole of the embryo. The mRNA distribution was detected by in situ hybridization. (b) Optical midsection of syncytial blastoderm embryo. Bicoid protein forms a concentration gradient with peak levels at the anterior pole. Bicoid protein was detected by antibody stainings. (c) Optical midsection of syncytial blastoderm embryo. The anterior gap gene *hunchback* is transcribed in a broad anterior domain. *hunchback* mRNA was detected by in situ hybridization. Modified from St. Johnston et al., 1989. Copyright (1989), the Company of Biologists Ltd. (See also color plate 6).

them into a drop of 87% glycerol on a glass slide. Place small coverslips to the left and right of the drop to function as spacers. Then cover the drop with a large coverslip. The embryos can now be analyzed using bright-field or Nomarski optics. To determine the expression of *hb* and *ftz*, select embryos at syncytial and cellular blastoderm stages, respectively. Make schematic drawings of the expression domains (Figure 19.4; Frohnhöfer and Nüsslein-Volhard, 1987). For more precise determination of the position of the expression borders use either camera lucida drawings or take photographs of optical sections (Nomarski optics). For each genotype analyze four embryos (Figure 19.4). Measure the length of the embryos and then calculate the position in % egg length of the posterior *hb* border and the first *ftz* stripe. Draw a graph which shows % egg length on the ordinate and number of bcd^+ copies on the abscissa. In this graph depict the position of the posterior *hb* border, the first *ftz* stripe, and the headfold as a function of

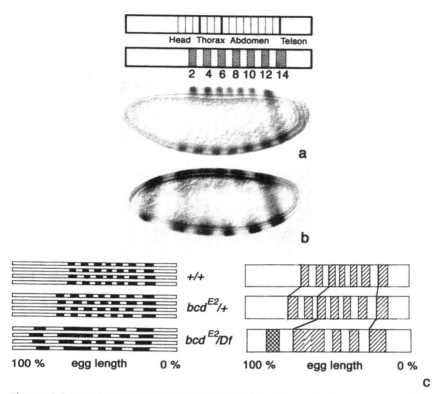

Figure 19.4. Mapping segmental anlagen with the help of pair rule gene expression. (a) The expression of *fushi tarazu* (*ftz*) in a wild-type embryo at cellular blastoderm stage. The upper two panels indicate the positions of the parasegmental anlagen 2–14 and the segments of head, thorax and abdomen in relation to the 7 stripes of *ftz* expression. (b) *ftz* expression in a cellular blastoderm embryo derived from *bicoid* mutant females. (c) Schematic drawings of *ftz* expression in wild-type embryos and in embryos from *bicoid* mutant females. Left: Each line represents one embryo, and for every genotype four embryos were analyzed. The maternal genotypes are indicated at the right. Right: The average of the data from the four embryos shown at the left. The anterior margins of the stripes corresponding to parasegment 2, 6 and 14 are connected by heavy lines to indicate the relative shift of head, thorax and abdomen. Modified from Frohnhöfer and Nüsslein-Volhard, 1987. Copyright (1987), with permission from Cold Spring Harbor Laboratory Press.

the amount of *bcd*. Do all three values change to the same degree dependent on the amount of *bcd*$^+$?

EXPERIMENT 2. A GRADIENT OF DORSAL DETERMINES THE CELL FATES ALONG THE DV AXIS

Experiment 2a. Generating flies which affect the shape of the maternal Dorsal gradient.

The presence of a morphogen mechanism can be tested by manipulations which change the shape of the postulated morphogen gradient. Generating an even distribution of the morphogen should result in loss of polarity and the differentiation of only one or few cell types. Different levels of the evenly distributed morphogen should promote the formation of specific cell types at the expense of all others. In the following crosses female flies are generated which produce embryos with altered distributions of nuclear Dorsal protein. Presumably, nuclear Dorsal is the readout of the extraembryonic Spätzle morphogen gradient. In the first two crosses, the amount of Dorsal protein itself is manipulated by using mutations in the *dorsal* gene. The subsequent two crosses use mutations in the transmembrane protein Toll, the postulated receptor for active Spätzle. Together the four crosses are used to generate females with 5 different (maternal) genotypes whose embryonic progeny will be analysed in Experiments 2b, 2c, and 2d.

CROSS 1. *dl¹* (Nüsslein-Volhard, 1979*a*) is a null-allele. We generate flies carrying *dl¹* in trans to a deficiency of *dorsal* (0 copies of *dl*). Flies carrying only one copy of *dl* (*dl¹/CyO*) show that like *bcd dl* is also dosage-sensitive.

$$b\ dl^1\ pr\ cn\ vg^D/CyO\ ♀\ (♂)\quad\times\quad Df(2L)TW119,\ cn\ bw/CyO\ ♂\ (♀)$$

Place 10 virgin females (♀) and 10 males (♂) into each of 7 big food bottles. Keep the flies at 25°C, transferring them to new vials every second day for one week. After 10 days the progeny from the crosses will begin to hatch. Because *CyO/CyO* is lethal, the following genotypes will be observed:

b dl¹ pr cn vg^D/CyO
Df(2L)TW119 cn bw/CyO
b dl¹ pr cn vg^D +/+ Df(2L)TW119 + cn + bw

Collect approximately 150 virgin females of the genotype *b dl¹ pr cn vg^D/CyO* (1 copy of *dl*) and 150 virgin females of the genotype *b dl¹ pr cn vg^D +/+ Df(2L)TW119 + cn + bw* (0 copies of *dl*).

CROSS 2. *dl²* is a partial loss-of-function allele of *dl* (Nüsslein-Volhard, 1979*b*)

$$b\ dl^2/CyO\ ♀\ (♂)\quad\times\quad Df(2L)TW119\ cn\ bw/CyO\ ♂\ (♀)$$

Cross 2 is set up in the same way as Cross 1. The following genotypes will be produced:

b dl²/CyO
Df(2L)TW119, cn bw/CyO
b dl² + +/+ Df(2L)TW119 cn bw

Collect approximately 150 virgin females of the genotype *b dl² + +/+ Df(2L)TW119 cn bw.*

CROSS 3. *Tl^{rm9}* and *Tl^{rm10}* are special recessive alleles of *Toll* (Anderson et al., 1985; Schneider et al., 1991). They have the unusual capacity to cause an intermediate activation of the Toll receptor around the entire dorsoventral circumference. This leads to intermediate levels of Dorsal in all nuclei around the circumference.

$$ru\ th\ st\ ri\ roe\ p^p\ Tl^{rm9}/TM3\ ♀\ (♂)\quad\times\quad ru\ kls\ e\ Tl^{rm10}/TM3,\ Sb\ ♂\ (♀)$$

Cross 3 is set up in the same way as Cross 1. The following genotypes will be produced:

ru th st ri roe p^p Tl^{rm9}/TM3, Sb

ru kls e Tl^{rm10}/TM3

ru th st + ri roe p^p + Tl^{rm9}/ru + + kls +++ e Tl^{rm10}

Collect approximately 150 virgin females of the genotype *ru th st + ri roe p^p + Tl^{rm9}/ ru + + kls +++ e Tl^{rm10}*.

CROSS 4. *Tl^{10b}* is a dominant mutation (Roth et al., 1989; Schneider et al., 1991). Females carrying one copy of *Tl^{10b}* produce strongly ventralized embryos. All nuclei around the embryonic circumference contain high levels of Dorsal.

Tl^{10b}, mwh e/TM3, Sb/T(1;3) OR60

Since all females which carry one copy of *Tl^{10b}* (*Tl^{10b}, mwh e/TM3, Sb* and *Tl^{10b}, mwh e/b/T(1;3) OR60*) are sterile, the stock is maintained by introducing a dominant male-lethal chromosome (*T(1;3) OR60*). *T(1;3) OR60/TM3, Sb* females are fertile. Collect approximately 150 virgin females of the genotypes *Tl^{10b}, mwh e/TM3, Sb* and *Tl^{10b}, mwh e/b/T(1;3) OR60*.

Use the 150 females of each genotype and 150 wt females (Oregon R) to set up three different crosses. Cross 50 females to 25 males carrying *P{dpp-LacZ}* (Jackson and Hoffmann, 1994), 50 to 25 males carrying *P{sog-LacZ}* (Markstein et al., 2002), and the remaining 50 to 25 males carrying *P{twi-LacZ}* (Jiang et al., 1991). Place the females and males in egg collection chambers as described in Experiment 1a. The chambers are kept at room temperature except for the crosses with *b dl^1 pr cn vg^D /CyO* (1 copy of *dl*) which are kept at 29°C. The dosage sensitivity of *dl* is enhanced at higher temperatures (Nüsslein-Volhard, 1979*a*).

Experiment 2b. The gastrulation phenotypes of dorsoventral mutants. Analyze the progeny of the crosses set up in Experiment 2a by following the procedure described in Experiment 1b. Three morphogenetic movements of gastrulation can be used to analyze dorsoventral patterning (Figure 19.1a). The most prominent one is the formation of the ventral furrow, a longitudinal cleft along the ventral midline of the embryo at 20–80% egg length. Ventral furrow formation leads to the invagination of the presumptive mesoderm, the ventralmost cell fate of the *Drosophila* embryo. The cephalic fold can also serve as a dorsoventral landmark, since in wt it is most prominent in lateral positions. Finally, when the posterior plate carrying the pole cells moves to the dorsal side, the cell layer in front of the posterior plate forms two small folds (the anterior and posterior dorsal fold).

When you are familiar with wt gastrulation look at the mutant embryos. Any of the three plates from one set of females can be used because the males do not influence the phenotype. For *dl^1 /CyO* (29°C) and *dl^2/Df(2L)TW119*, concentrate on the presence or absence of the ventral furrow. What happens to the embryo if *dl* is lacking entirely (*dl^1/Df(2L)TW119*, Figure 19.1c)? Compare gastrulation with reduced amounts of *dl* to gastrulation in which *dl* is absent. What is the main difference? Compare the absence of *dl* to *Tl^{rm9}/Tl^{rm10}* (*Tl^{rm9/rm10}*). What is the common feature of both gastrulations? Make schematic drawings of the early gastrulation stages, like those shown in Figure 19.1.

Experiment 2c. Cuticles of dorsoventral mutants (see Experiment 1c). Completely dorsalized embryos require more time compared to wt before they secrete their cuticle. Leave the plates from *dl^1/Df(2L)TW119* females for two days at RT before cuticle preparation. Use the wt cuticles of Experiment 1c. to familiarize yourself with dorsoventral cuticle landmarks (Figures 19.1 and 19.2). Try to distinguish the thin dorsal hairs from the thick ventral denticles. The former are derived from dorsal and the latter from lateral postions of the blastoderm fate map. Filzkörper and antennal/maxillary organs are derived from dorsolateral positions while the posterior spiracles and one structure of the head skeleton (the median tooth or labrum) are derived from dorsal positions. Make schematic drawings of the mutant cuticles like those shown in Figure 19.1. Generate a table in which you indicate the genotype and the cuticle elements which are present or absent. Try to correlate these data with your observations of the gastrulation phenotype. Be aware of the fact that the ventralmost structure of the dorsoventral axis, the mesoderm, does not contribute to the cuticle. Thus, its absence may only indirectly influence the appearance of the cuticle while its massive expansion might eliminate most of the cuticle.

Experiment 2d. The expression of zygotic dorsoventral patterning genes in mutant backgrounds affecting the shape of the Dorsal gradient. Like Bicoid, the transcription factor Dorsal directly regulates the expression of zygotic genes. The Dorsal target genes subdivide the embryonic DV axis into large domains similar to the way gap genes subdivide the AP axis (for review see Stathopoulos and Levine, 2002; Figures 19.5 and 19.6). At the ventral side, the expression of *twist (twi)* and *snail (sna)* is required to specify the mesoderm (Figures 19.5 and 19.6a). In lateral positions, genes like *short gastrulation (sog)* are required for the development of the neuroectoderm, which gives rise to the central nervous system, and to the ventral ectoderm, which produces the thick ventral denticles (Figure 19.6c). Finally, at the dorsal side, *decapentaplegic (dpp;* Figure 19.6e) and *zerknüllt (zen;* Figure 19.5) expression specify the non-neurogenic dorsal ectoderm and the amnioserosa, an extraembryonic cell layer which derives from a narrow stripe of cells along the dorsal midline. The dorsal ectoderm produces cuticle with dorsal hairs. The cells of the amnioserosa cover the embryo at the dorsal side and undergo apoptosis when dorsal closure occurs.

In the crosses described earlier, reporter constructs for all three zygotic genes have been used. Their expression pattern can now be analyzed in different maternal backgrounds using a *LacZ* antisense probe. Proceed as described in Experiment 1d. Analyze syncytial and cellular blastoderm stages for the expression patterns in wt and in the different maternal backgrounds. Correlate these data with those obtained from gastrulating embryos and from cuticle preparations. What types of regulatory interactions between Dorsal and its target genes would you postulate on the basis of your observations? (The shape of the Dorsal gradient in wt and some mutant embryos are shown in Figure 19.5.)

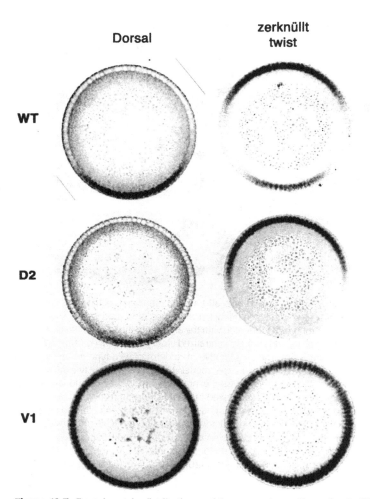

Figure 19.5. Dorsal protein distribution and the expression pattern of *zerknüllt* and *twist* in wild-type and mutant embryos. Wild-type and mutant embryos of syncytial or cellular blastoderm stage were stained for Dorsal (left) or simultaneously for Zerknüllt and Twist proteins (right). The embryos were embedded and sectioned. The transverse sections shown are derived from the trunk region. The dorsal side is facing up. **WT**: Dorsal protein forms a nuclear concentration gradient with peak levels at the ventral side. The nuclei of the ventralmost 15–20% of the embryonic circumference have highest Dorsal concentrations. In more lateral nuclei the Dorsal concentrations decrease rapidly such that the gradient is confined to the ventral 40% of the embryonic circumference. At the dorsal side the protein is excluded from the nuclei. *zerknüllt* is, like *decapentaplegic*, repressed by nuclear Dorsal and therefore confined to the dorsal side. *twist* is activated by high levels of nuclear Dorsal and therefore restricted to the ventral 20% of the embryonic circumference. **D2**: A partially dorsalized embryo produced from *spätzle^{67}/spätzle^{m7}* females. This phenotype is similar to that produced by *dl^1/+* at 29°C. The lateral extension and the ventral peak concentrations of the nuclear Dorsal gradient are reduced. Therefore, *zerknüllt* is expanded and *twist* is not expressed. **V1**: A strongly ventralized embryo produced from *Toll^{10b}/+* females. High nuclear Dorsal levels are found all around the circumference. *twist* is expressed in all cells. Cross sections are 150-200 μm in diameter.

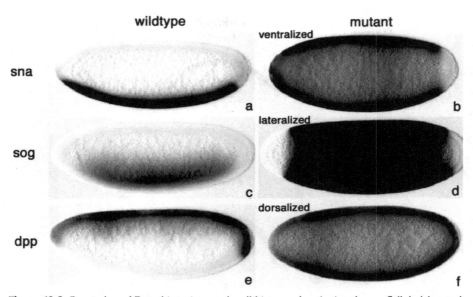

Figure 19.6. Expression of Dorsal target genes in wild-type and mutant embryos. Cellularizing embryos were hybridized with each of the indicated digoxigenin-labeled antisense RNA probes and stained to visualize the gene expression patterns. Embryos are oriented with the anterior to the left and the dorsal up. (a and b) *snail* (*sna*) expression in wild-type (a) and strongly ventralized embryos (b) derived from *Toll^{10B}/+* females. (a) *sna* is expressed in a ventral stripe, the presumptive mesoderm of wild-type embryos. *snail* expression is similar to that of *twist*, which we use in our experiments. (b) *sna* is ubiquitously expressed in strongly ventralized embryos. (c and d) *short gastrulation* (*sog*) expression in wild-type (c) and lateralized embryos derived from *Toll^{rm9/rm10}* females (d). *sog* is expressed in lateral stripes in response to even low levels of nuclear Dorsal in wild-type embryos (c). *sog* is ubiquitously expressed in the lateralized embryos (d). (e and f) *decapentaplegic* (*dpp*) expression in wild-type (e) and completely dorsalized embryos derived from *pipe^{-}/pipe^{-}* females. Loss-of-function mutations in *pipe* and *dorsal* are indistinguishable. *dpp* is confined to a dorsal stripe in wild-type embryos (e), but is uniformly expressed in completely dorsalized embryos (f). Modified from Stathopoulos et al., 2002. Copyright (2002), with permission from Elsevier.

EXPECTED RESULTS AND DISCUSSION

EXPERIMENT 1. CHANGING THE DOSE OF *bicoid* CAUSES SHIFTS IN THE ANLAGEN OF THE AP AXIS

The absence of *bcd* causes a loss of all head and thoracic structures (Frohnhöfer and Nüsslein-Volhard, 1986). This is immediately evident from the cuticle secreted by *bcd* mutant embryos (Figure 19.1e). They lack the head skeleton, all thorax and some anterior abdominal segments. However, the phenotype is more complex. The anterior pole of *bcd* embryos harbours telson structures normally found only at the posterior pole. The cuticles display malformed anterior filzkörper and anal pads. The gastrulation movements reveal not only the absence of a head fold, but also the formation of a posterior plate at the anterior end (Figure 19.1b). The anteriormost *ftz* stripe is not stripe 1, but an anterior duplication of stripe 8 (Figure 19.4). Thus, in addition to its function for head and thorax specification *bicoid* is also required to determine anterior as opposed to posterior terminal structures, the development of which appears to be the ground state in the absence of *bcd*.

Bcd protein is distributed in the embryo in a steep concentration gradient, taking its origin from the anteriorly localized *bcd* mRNA (Figure 19.3). Driever and Nüsslein-Volhard (1988*b*) showed that increasing the number of *bcd*$^+$ copies leads to increased amounts of Bcd protein at the anterior tip and to an expansion of the Bcd gradient towards the posterior. This correlates with a posterior shift of the three landmarks (posterior *hb* border, *ftz* stripe 1, and head fold) which we have analyzed (for *hb* see also Struhl et al., 1989). Thus, a close correlation exists between the local concentrations of Bcd protein and the development of certain anterior cell fates. These data imply that the absolute concentration of Bcd determines at least one position along the AP axis of the embryo. Promoter studies have shown that *hb* is a direct target gene of Bcd (Driever and Nüsslein-Volhard, 1989; Driever et al., 1989; Struhl et al., 1989). In *bcd* null embryos, *hb* is not expressed and the shift of the posterior *hb* border is more sensitive to a change in Bcd amounts than that of the first *ftz* stripe and the head fold.

EXPERIMENT 2. A GRADIENT OF DORSAL DETERMINES THE CELL FATES ALONG THE DV AXIS

Although the loss of *bcd* affects mainly one part of the AP axis, i.e. the anterior half, the loss of *dl* affects the entire DV axis. The *dl* minus embryos lack all dorsoventral polarity (Nüsslein-Volhard, 1979*a*). The cuticle consists of a long irregular tube (Figure 19.1f) which has only dorsal structures (dorsal hairs, labrum at the anterior, and occasionally spiracles at the posterior end). The gastrulation movements are symmetric with regard to the dorsoventral axis (Figure 19.1c). The ventral furrow is absent and dorsal folds encircle the embryo. The deep folds visible in later embryos are presumably the result of a symmetric germband elongation. There is no germband shortening normally seen at the end of wt embryogenesis. This may explain the elongated tube-like appearance of the cuticle. Both *twi* and *sog* expression are absent and *dpp* is expressed uniformly around the circumference (Figure 19.6f). This phenotype was designated as D0 (completely dorsalized; Anderson et al., 1985). The loss of one copy of *dl* (at 29°C; Nüsslein-Volhard, 1979*b*; Nüsslein-Volhard et al., 1980) leads to a reduction in size or complete deletion of the ventralmost region, the mesoderm (D2; Figure 19.5). All cuticular structures are present although the embryos appear twisted and elongated. During gastrulation, germband extension occurs in a polar fashion; i.e., the posterior plate moves to the dorsal side and then toward the anterior. However, the ventral furrow is reduced in size or absent (Nüsslein-Volhard et al., 1980). Correspondingly, *twi* expression in the trunk region is reduced or absent, while instead, *sog* is found at the ventral side. Thus, coordinated fate shifts have taken place: the reduction of mesoderm is accompanied by a shift of the neuroectoderm towards the ventral side. Together these phenotypes suggest that high levels of *dl* specify the mesoderm. The phenotype of *dl*^{2}/*Df* is similar to that of *dl*^{1}/+, although the cuticle pattern does not show ventral denticles, suggesting that both the mesoderm and the ventrolateral anlagen of the neuroectoderm are deleted. The presence of Fk, sense organs of the head, and a clearly polarized gastrulation demonstrate that the embryos consist of dorsolateral and dorsal structures. Compared to *dl*^{1}/+ the *sog* domain is smaller and the *dpp* domain expanded. This phenotype (D1) suggests that intermediate levels of Dorsal specify the neuroectoderm and, if compared to D0, that low levels of Dorsal specify dorsolateral structures. Mutations that lead to even distribution of Dorsal protein demonstrate that the different fates along the DV axis are specified in an autonomous manner. The Toll receptor mutations Tl^{rm9} and Tl^{rm10} cause uniform low-level receptor activation accompanied by intermediate levels of nuclear Dorsal protein around the circumference. The cuticles secreted by Tl^{rm9}/Tl^{rm10} mutant embryos show ventral denticles all around the circumference (Anderson et al., 1985). The embryos gastrulate in an apolar fashion, like dl minus embryos, but the headfold which forms mainly laterally in wt is very pronounced at dorsal and ventral sides. Finally, these embryos express neither twi nor dpp (Roth et al., 1989). However, sog is uniformly present throughout the trunk region (Stathopoulos et al., 2002; Figure 19.6d). Thus, intermediate levels of Dorsal autonomously specify neuroectoderm. The dominant $Toll$ allele Tl^{10b} leads to high levels of Dorsal in all embryonic nuclei (Figure 19.5; Roth et al., 1989). The embryos have almost no cuticle, the gastrulation is frequently apolar, and $twist$ is expressed around the entire circumference. These embryos show that high levels of Dorsal autonomously specify the mesoderm.

TIME REQUIRED FOR THE EXPERIMENTS

The crosses for generating the maternal genotypes have to be set up approximately two weeks in advance. Virgin collection and setting up the final crosses may take about one week of approximately one hour of work in the morning and in the late afternoon (depending on prior experience with fly work and the number of people involved). Once the egg collection chambers are filled with flies, all the described experiments can be conducted within a week. The timing of individual steps has been mentioned above (see Outline of the Experiments). The work plan should be arranged such that the students have plenty of time to watch live embryos (about two afternoons) and analyze the cuticle preparations and the stained embryos. The experiments can be scaled up or down depending on the number of students in the course. If groups of two students each set up two crosses and take care of two egg-collection chambers, and if the groups later exchange their plates and their microscopic slides for analysis with other groups, the experiments as outlined above produce enough material for 14 students.

TEACHING CONCEPTS

Maternal and zygotic genes. The deposited *Drosophila* egg contains a high degree of spatial information for the establishment of the body axes. This spatial information is provided by localized mRNAs and localized signals. The mRNAs, their localization machinery, the signals, and the components of the corresponding signaling pathways are encoded by genes which are active during oogenesis and thus dependant on the maternal genome. These genes have been identified in mutagenesis screens looking for maternal-effect mutations. The embryonic phenotype of such mutations depends only on the genotype of the females producing the eggs. Bicoid and Dorsal are examples of maternal-effect genes. During early development the products of the maternal-effect genes regulate the expression of embryonic genes. Since the embryonic genotype is the same as that of the fertilized egg, the zygote, they are also called zygotic genes. *hb, ftz, dpp, sog,* and *twi* are examples of zygotic embryonic patterning genes.

Morphogen. Morphogens are distributed in concentration gradients and evoke different cell fates at different concentrations. Therefore, they provide an elegant mechanism which dictates cell fate as a function of physical distance. To distinguish morphogens from simple inducing factors the term is usually limited to cases in which the graded substance specifies at least two different cellular behaviors depending on two concentration thresholds. The best way to test whether a molecule acts a morphogen is by generating an even distribution of the molecule at different concentration values. This should lead to expression of one cell type for a given concentration level.

ADDITIONAL INFORMATION

EXPERIMENT 1. CHANGING THE DOSE OF *bicoid* CAUSES SHIFTS IN THE ANLAGEN OF THE AP AXIS

Size regulation. Embryos with one copy of bcd^+ show a reduction and embryos with mutliple copies an expansion of the presumptive head region, which is compensated by respective expansions and compressions of more posterior anlagen. However, these embryos develop into relatively normal larvae and adults (Frohnhöfer and Nüsslein-Volhard, 1986; Busturia and Lawrence, 1994). It has been shown that cell death plays a major role in correcting the altered size of the anlagen. Expanded regions show an increase and compressed regions a decrease of cell death during postgastrulation development (Namba et al., 1997).

Molecular mechanisms of Bicoid action. A gradient-affinity model was proposed to explain Bicoid action (Driever et al., 1989). The model has two assumptions. First, the sharp on–off switch for target gene transcription (e.g., Figure 19.3c, posterior *hunchback* border) results from cooperative gene activation by Bicoid. Second, target genes with different affinity binding sites for Bicoid are expressed at different positions along the AP axis. Low-affinity binding sites lead to more anterior, high-affinity binding sites to more posterior expression domains. The first aspect of the model, the cooperative binding of Bicoid to cis-regulatory regions of target genes, has been experimentally verified (Ma et al., 1996; Burz et al., 1998). The second aspect of the model, though still controversial (Burz et al., 1998), is supported by molecular studies on the *bcd* target gene *orthodenticle* (*otd*), which is expressed in a narrow anterior domain of high Bcd concentrations (Finkelstein and Perrimon, 1990; Gao and Finkelstein, 1998). As predicted by the model, the relevant regulatory regions of *otd* contain low-affinity binding sites for Bicoid (Gao and Finkelstein, 1998).

In a recent paper, precise measurements of the Bicoid gradient and the *hunchback* expression domain of a single embryo have been described (Houchmandzadeh, Wieschaus, and Leibler, 2002). These findings have lead to a surprising result. The Bicoid gradient displayed high embryo-to-embryo variability while the border of the *hunchback* domain was remarkably precise. The genes known to interact with *hunchback* seem not to be responsible for this spatial precision.

bicoid and caudal. Surprisingly, the Bicoid homeodomain, which has been shown to act as a DNA binding motif, can also bind RNA (for review, see Ephrussi and St. Johnston, 2004). Bicoid binds to the 3′ UTR (untranslated region) of *caudal* mRNA and prevents the assembly of the 5′cap complex. This blocks the initiation of *caudal* mRNA translation (Niessing, Blanke, and Jäckle, 2002). Maternal *caudal* mRNA is evenly distributed, but translational repression by Bicoid leads to the formation of a Caudal protein gradient, which decreases from posterior to anterior. Caudal itself is a homeodomain transcription factor that activates posterior gap genes like *knirps* and *gaint* (for review see Rivera-Pomar and Jäckle, 1996).

The evolution of the bicoid gene. Although Bicoid represents the paradigm of a morphogen, it appears to be a recently acquired gene in flies. It has not been found in insects outside the higher Dipterans, and phylogenetic reconstruction indicates that it is a derived Hox class 3 gene (Stauber, Jäckle, and Schmidt-Ott, 1999; Stauber, Taubert, and Schmidt-Ott, 2000; Stauber, Prell, and Schmidt-Ott, 2002). *hunchback* (*hb*) on the other hand is phylogenetically much more conserved. It can substitute for *bcd* in the thorax and abdomen (Wimmer et al., 2000), and synergy between *hb* and *bcd* is required to execute all *bcd* functions (Simpson-Brose, Treisman, and Desplan, 1994). Thus, in other insects *hb*, and/or other head gap genes like *orthodenticle*, might fulfil the function which *bcd* has acquired in flies (Schröder, 2003).

EXPERIMENT 2. A GRADIENT OF DORSAL DETERMINES THE CELL FATES ALONG THE DV AXIS

Molecular mechanisms of Dorsal action. The DNA binding sites for Dorsal protein have been determined within the cis-regulatory regions of a large number of Dorsal target genes (for review see Stathopoulos and Levine, 2002). It has been shown that Dorsal action strongly depends on the context of its binding sites. Thus, the close proximity of binding sites for co-activators or co-repressors determines whether Dorsal activates (e.g., *twist* and *short gastrulation*) or represses its target genes (e.g., *decapentaplegic* and *zerknüllt*; Jiang, Zhon, and Levine, 1993; Kirov et al., 1993). In addition, cofactors are crucial for understanding the concentration-dependent action of Dorsal. For example, only a combination of high-affinity Dorsal with basic helix-loop-helix transcription factor binding sites allows target gene activation by very low levels of nuclear Dorsal protein (Jiang and Levine, 1993).

Evolution of the Toll-rel/NF-κB pathway. Dorsal is a member of the rel/NF-κB family of transcriptions factors which are found throughout the animal kingdom (Friedman and Hughes, 2002). Also the Toll receptors form a large family with multiple members in insects and vertebrates. Both insects and vertebrates use the Toll-rel/NF-κB pathway to control their innate immune response to a variety of different pathogens (Hoffmann et al., 1999). However, there is no comparable conservation of the pathway with regard to axis formation because vertebrates use largely different mechanisms to set up their dorsoventral axis (but see also Armstrong et al., 1998 and Prothmann, Armstrong, and Rupp, 2000). Thus, pathogen defense might be the most ancestral function of the pathway with axis formation being a later co-option (for an alternative viewpoint, see Friedman and Hughes, 2002). We do not know when this co-option occurred. However, within the insects an involvement of the Toll-rel/NF-κB pathway in dorsoventral axis formation appears to be quite old because both Dipteran and Coleopteran embryos form a nuclear Dorsal gradient (Chen, Handel, and Roth, 2000).

The development of the morphogen concept. The intracellular Bicoid morphogen gradient represents a special case because its formation depends on the syncytial organisation of the early *Drosophila* embryo. On the contrary, the formation of extracellular morphogen gradients, like that of Spätzle, is not restricted to the pecularities of the early *Drosophila* embryo. In recent years many examples of extracellular morphogens have been described in insects and vertebrates (for review, see Gurdon and Bourillot, 2001; Freeman and Gurdon, 2002). The first direct proof for the existence of a morphogen gradient that organizes patterning of an epithelial layer of cells was presented in two seminal papers from 1996 (Lecuit et al., 1996; Nellen et al., 1996). In these papers the authors used clonal analysis to demonstrate that the secreted growth factor Decapentaplegic forms a morphogen gradient within the *Drosophila* wing imaginal disc, which specifies the anteriorposterior pattern of the wing.

ACKNOWLEDGEMENTS

I am grateful to Patrick Kalscheuer, Martin Technau and Maurjin van der Zee for critical reading of the manuscript. I thank Hans Georg Frohnhöfer for providing the picture of the *bcd* mutant embryo shown in Figure 19.1.e.

REFERENCES

Anderson, K. V., Jürgens, G., and Nüsslein-Volhard, C. (1985). Establishment of dorsal–ventral polarity in the *Drosophila* embryo: Genetic studies on the role of the *Toll* gene product. *Cell*, 42, 779–89.

Armstrong, N. J., Steinbeisser, H., Prothmann, C., DeLotto, R., and Rupp, R. A. (1998). Conserved Spätzle/Toll signalling in dorsoventral patterning of *Xenopus* embryos. *Mech. Dev.*, 71, 99–105.

Berleth, T., Burri, M., Thoma, G., Bopp, D., Richstein, S., Frigerio, G., Noll, M., and Nüsslein-Volhard, C. (1988). The role of localization of bicoid RNA in organizing the anterior pattern of the *Drosophila* embryo. *EMBO J.*, 7, 1749–56.

Burz, D. S., Rivera-Pomar, R., Jäckle, H., and Hanes, S. D. (1998). Cooperative DNA-binding by Bicoid provides a mechanism for threshold-dependent gene activation in the *Drosophila* embryo. *EMBO J.*, 17, 5998–6009.

Busturia, A., and Lawrence, P. A. (1994). Regulation of cell number in *Drosophila*. *Nature*, 370, 561–3.

Casanova, J., Furriols, M., McCormick, C. A., and Struhl, G. (1995). Similarities between trunk and spätzle, putative extracellular ligands specifying body pattern in *Drosophila*. *Genes Dev.*, 9, 2539–44.

Chen, G., Handel, K., and Roth, S. (2000). The maternal NF-kappaB/dorsal gradient of *Tribolium castaneum*: Dynamics of early dorsoventral patterning in a short-germ beetle. *Development*, 127, 5145–56.

Crick, F. (1970). Diffusion in embryogenesis. *Nature*, 5231, 420–2.

Driever, W. (2004). The bicoid morphogen papers (II). Account from Wolfgang Driever. *Cell*, S116, S7–S9.

Driever, W., and Nüsslein-Volhard, C. (1988*a*). The bicoid protein determines position in the *Drosophila* embryo in a concentration-dependent manner. *Cell*, 54, 95–104.

Driever, W., and Nüsslein-Volhard, C. (1988*b*). A gradient of bicoid protein in *Drosophila* embryos. *Cell*, 54, 83–93.

Driever, W., and Nüsslein-Volhard, C. (1989). The Bicoid protein is a positive regulator of *hunchback* transcription in the early *Drosophila* embryo. *Nature*, 337, 138–43.

Driever, W., Thoma, G., and Nüsslein-Volhard, C. (1989). Determination of spatial domains of zygotic gene expression in the *Drosophila* embryo by the affinity of binding sites for the bicoid morphogen. *Nature*, 340, 363–7.

Ephrussi, A., and St. Johnston, D. (2004). Seeing is believing: The bicoid morphogen gradient matures. *Cell*, 116, 143–52.

Finkelstein, R., and Perrimon, N. (1990). The orthodenticle gene is regulated by bicoid and torso and specifies *Drosophila* head development. *Nature*, 346, 485–8.

Freeman, M., and Gurdon, J. B. (2002). Regulatory principles of developmental signaling. *Ann. Rev. Cell Dev. Biol.*, 18, 515–39.

Friedman, R., and Hughes, A. L. (2002). Molecular evolution of the NF-kappaB signaling system. *Immunogenetics*, 53, 964–74.

Frohnhöfer, H. G., and Nüsslein-Volhard, C. (1986). Organisation of anterior pattern in the *Drosophila* embryo by the maternal gene *bicoid*. *Nature*, 324, 120–5.

Frohnhöfer, H. G., and Nüsslein-Volhard, C. (1987). Maternal genes required for the anterior localisation of *bicoid* activity in the embryo of *Drosophila*. *Genes Dev.*, 1, 880–90.

Gao, Q., and Finkelstein, R. (1998). Targeting gene expression to the head: The Drosophila orthodenticle gene is a direct target of the Bicoid morphogen. *Development*, 125, 4185–93.

Gierer, A., and Meinhardt, H. (1972). A theory of biological pattern formation. *Kybernetik*, 12, 30–9.

Greenspan, R. J. (1997). *Fly Pushing. The Theory and Practice of Drosophila Genetics*. New York: Cold Spring Harbor Laboratory Press.

Gurdon, J. B., and Bourillot, P. Y. (2001). Morphogen gradient interpretation. *Nature*, 413, 797–803.

Hashimoto, C., Hudson, K. L., and Anderson, K. V. (1988). The *Toll* gene of *Drosophila*, required for dorsal-ventral embryonic polarity, appears to encode a transmembrane protein. *Cell*, 52, 269–79.

Hashimoto, C., Gerttula, S., and Anderson, K. V. (1991). Plasma membrane localization of the Toll protein in the syncytial *Drosophila* embryo: Importance of transmembrane signaling for dorsal–ventral pattern formation. *Development*, 111, 1021–28.

Hiromi, Y., Kuroiwa, A., and Gehring, W. J. (1985). Control elements of the *Drosophila* segmentation gene *fushi tarazu*. *Cell*, 43, 603–13.

Hoffmann, J. A., Kafatos, F. C., Janeway, C. A., and Ezekowitz, R. A. (1999). Phylogenetic perspective in innate immunity. *Science*, 284, 1313–8.

Houchmandzadeh, B., Wieschaus, E., and Leibler, S. (2002). Establishment of developmental precision and proportions in the early *Drosophila* embryo. *Nature*, 415, 798–802.

Jackson, P. D., and Hoffmann, F. M. (1994). Embryonic expression patterns of the *Drosophila decapentaplegic* gene: Separate regulatory elements control blastoderm expression and lateral ectodermal expression. *Dev. Dyn.*, 199, 28–44.

Jiang, J., Kosman, D., Ip, Y. T., and Levine, M. (1991). The dorsal morphogen gradient regulates the mesoderm determinant *twist* in early *Drosophila* embryos. *Genes Dev.*, 5, 1881–91.

Jiang, J., and Levine, M. (1993). Binding affinities and cooperative interactions with bHLH activators delimit threshold responses to the dorsal gradient morphogen. *Cell*, 72, 741–52.

Jiang, J., Cai, H., Zhou, Q., and Levine, M. (1993). Conversion of a dorsal-dependent silencer into an enhancer: Evidence for dorsal corepressors. *EMBO J.*, 12, 3201–9.

Kirov, N., Zhelnin, L., Shah, J., and Rushlow, C. (1993). Conversion of a silencer into an enhancer: Evidence for a co-repressor in dorsal-mediated repression in *Drosophila*. *EMBO J.*, 12, 3193–9.

Lecuit, T., Brook, W. J., Ng, M., Calleja, M., Sun, H., and Cohen, S. M. (1996). Two distinct mechanisms for long-range patterning by Decapentaplegic in the *Drosophila* wing. *Nature*, 381, 387–93.

Lewis, J., Slack, J. M., and Wolpert, L. (1977). Thresholds in development. *J. Theor. Biol.*, 65, 579–90.

Ma, X., Yuan, D., Diepold, K., Scarborough, T., and Ma, J. (1996). The *Drosophila* morphogenetic protein Bicoid binds DNA cooperatively. *Development*, 122, 1195–1206.

Markstein, M., Markstein, P., Markstein, V., and Levine, M. S. (2002). Genome-wide analysis of clustered Dorsal binding sites identifies putative target genes in the *Drosophila* embryo. *Proc. Natl. Acad. Sci. USA*, 99, 763–8.

Meinhardt, H. (1982). *Models of Biological Pattern Formation*. London: Academic Press.

Morisato, D., and Anderson, K. V. (1994). The *spätzle* gene encodes a component of the extracellular signaling pathway establishing the dorsal–ventral pattern of the *Drosophila* embryo. *Cell*, 76, 677–88.

Morisato, D., and Anderson, K. V. (1995). Signaling pathways that establish the dorsal–ventral pattern of the *Drosophila* embryo. *Ann. Rev. Genet.*, 29, 371–99.

Namba, R., Pazdera, T. M., Cerrone, R. L., and Minden, J. S. (1997). *Drosophila* embryonic pattern repair: How embryos respond to bicoid dosage alteration. *Development*, 124, 1393–403.

Nellen, D., Burke, R., Struhl, G., and Basler, K. (1996). Direct and long-range action of a DPP morphogen gradient. *Cell*, 85, 357–68.

Niessing, D., Blanke, S., and Jäckle, H. (2002). Bicoid associates with the 5'-cap-bound complex of *caudal* mRNA and represses translation. *Genes Dev.*, 16, 2576–82.

Nüsslein-Volhard, C. (1979*a*). Maternal effect mutations that alter the spatial coordinates of the embryo of *Drosophila melanogaster*. In *Determinants of Spatial Organisation*, eds. I. Koenigsberg and S. Subtelney, pp. 185–211. New York: Academic Press.

Nüsslein-Volhard, C. (1979*b*). Pattern mutants in *Drosophila* embryogenesis. In *Cell Lineage, Stem Cells and Cell Determination*, ed. N. Le Douarin, pp. 69–82. New York: North-Holland.

Nüsslein-Volhard, C. (2004). The bicoid morphogen papers (I). Account from CNV. *Cell*, S116, S1–S5.

Nüsslein-Volhard, C., Lohs-Schardin, M., Sander, K., and Cremer, C. (1980). A dorso-ventral shift of embryonic primordia in a new maternal-effect mutant of *Drosophila*. *Nature*, 283, 474–6.

Prothmann, C., Armstrong, N. J., and Rupp, R. A. (2000). The Toll/IL-1 receptor binding protein MyD88 is required for *Xenopus* axis formation. *Mech. Dev.*, 97, 85–92.

Rivera-Pomar, R., and Jäckle, H. (1996). From gradients to stripes in *Drosophila* embryogenesis: Filling in the gaps. *Trends Genet.*, 12, 478–83.

Roth, S., Stein, D., and Nüsslein-Volhard, C. (1989). A gradient of nuclear localization of the dorsal protein determines dorsoventral pattern in the *Drosophila* embryo. *Cell*, 59, 1189–202.

Rushlow, C. A., Han, K., Manley, J. L., and Levine, M. (1989). The graded distribution of the dorsal morphogen is initiated by selective nuclear transport in *Drosophila*. *Cell*, 59, 1165–77.

Sander, K. (1976). Specification of the basic body pattern in insect embryogenesis. *Adv. Insect Physiol.*, 12, 125–238.

Schneider, D. S., Hudson, K. L., Lin, T. Y., and Anderson, K. V. (1991). Dominant and recessive mutations define functional domains of Toll, a transmembrane protein required for dorsal-ventral polarity in the *Drosophila* embryo. *Genes Dev.*, 5, 797–807.

Schneider, D. S., Jin, Y., Morisato, D., and Anderson, K. V. (1994). A processed form of the Spätzle protein defines dorsal-ventral polarity in the *Drosophila* embryo. *Development*, 120, 1243–50.

Schröder, R. (2003). The genes orthodenticle and hunchback substitute for bicoid in the beetle Tribolium. *Nature*, 422, 621–5.

Simpson-Brose, M., Treisman, J., and Desplan, C. (1994). Synergy between the hunchback and bicoid morphogens is required for anterior patterning in *Drosophila*. *Cell*, 78, 855–65.

St. Johnston, D., Driever, W., Berleth, T., Richstein, S., and Nüsslein-Volhard, C. (1989). Multiple steps in the localization of bicoid RNA to the anterior pole of the *Drosophila* oocyte. *Development*, 107, 13–9.

St. Johnston, D., and Nüsslein-Volhard, C. (1992). The origin of pattern and polarity in the *Drosophila* embryo. *Cell*, 68, 201–19.

Stathopoulos, A., and Levine, M. (2002). Dorsal gradient networks in the *Drosophila* embryo. *Dev. Biol.*, 246, 57–67.

Stathopoulos, A., Van Drenth, M., Erives, A., Markstein, M., and Levine, M. (2002). Whole-genome analysis of dorsal-ventral patterning in the *Drosophila* embryo. *Cell*, 111, 687–701.

Stauber, M., Jäckle, H., and Schmidt-Ott, U. (1999). The anterior determinant bicoid of *Drosophila* is a derived *Hox* class 3 gene. *Proc. Natl. Acad. Sci. USA*, 96, 3786–9.

Stauber, M., Taubert, H., and Schmidt-Ott, U. (2000). Function of *bicoid* and *hunchback* homologs in the basal cyclorrhaphan fly *Megaselia* (Phoridae). *Proc. Natl. Acad. Sci. USA*, 97, 10844–9.

Stauber, M., Prell, A., and Schmidt-Ott, U. (2002). A single *Hox3* gene with composite *bicoid* and *zerknüllt* expression characteristics in non-Cyclorrhaphan flies. *Proc. Natl. Acad. Sci. USA*, 99, 274–9.

Steward, R. (1987). Dorsal, an embryonic polarity gene in *Drosophila*, is homologous to the vertebrate proto-oncogene, *c-rel. Science*, 238, 692–4.

Steward, R. (1989). Relocalization of the dorsal protein from the cytoplasm to the nucleus correlates with its function. *Cell*, 59, 1179–88.

Steward, R., and Nüsslein-Volhard, C. (1986). The genetics of the dorsal-Bicaudal-D region of *Drosophila melanogaster*. *Genetics*, 113, 665–78.

Steward, R., Zusman, S. B., Huang, L. H., and Schedl, P. (1988). The dorsal protein is distributed in a gradient in early *Drosophila* embryos. *Cell*, 55, 487–95.

Struhl, G., Struhl, K., and Macdonald, P. M. (1989). The gradient morphogen *bicoid* is a concentration-dependent transcriptional activator. *Cell*, 57, 1259–73.

Tickle, C., Summerbell, D., and Wolpert, L. (1975). Positional Signaling and specification of digits in chick limb morphogenesis. *Nature*, 254, 199–202.

Turing, A. M. (1953/1990). The chemical basis of morphogenesis. *Bull. Math. Biol.*, 52, 153–97.

Wieschaus, E., and Nüsslein-Volhard, C. (1998). Looking at embryos. In *Drosophila: A Practical Approach*, ed. D. B. Roberts, pp. 199–228. Oxford: IRL Press.

Wimmer, E. A., Carleton, A., Harjes, P., Turner, T., and Desplan, C. (2000). Bicoid-independent formation of thoracic segments in *Drosophila*. *Science*, 287, 2476–9.

Wolpert, L. (1969). Positional information and the spatial pattern of cellular differentiation. *J. Theor. Biol.*, 25, 1–47.

Wolpert, L. (1971). Positional information and pattern formation. *Curr. Top. Dev. Biol.*, 6, 183–224.

Wolpert, L. (1996). One hundred years of positional information. *Trends Genet.*, 12, 359–64.

20 Significance of the temporal modulation of Hox gene expression on segment morphology

J. Castelli-Gair Hombría

OBJECTIVE OF THE EXPERIMENT Hox genes control the morphology of segments in animals. During development, Hox genes are expressed in defined regions, with a specific temporal–spatial dynamic expression in each segment. The purpose of these experiments is to observe the evolution of the temporal expression of some Hox genes and to manipulate their temporal expression to analyse how this affects segment morphology in *Drosophila*.

DEGREE OF DIFFICULTY Antibody staining is time consuming. RNA in situ hybridisation involves many washing steps, and the probe is susceptible to degradation.

The study of dynamic patterns of expression in embryo populations of mixed ages can be daunting to the nonexpert. In addition, the gastrulation movements and the variable orientations of the embryos in the slide make the identification of different stages of embryogenesis difficult.

INTRODUCTION

In the first two hours of embryogenesis, the trunk of *Drosophila* is subdivided into identical segments that become morphologically different by the action of the Hox genes. Hox genes encode transcription factors that activate or repress downstream genes that are responsible for controlling the morphogenesis of the structures formed in each segment.

The trunk consists of three thoracic (T1–T3) and eight major abdominal segments (A1–A8). Some distinguishing features of the segments are the development of legs in the thorax, the formation of ventral denticle belts of different widths in the epidermis, and the formation of the posterior spiracles in A8 (Figure 20.1). In *Drosophila*, the legs are well developed only in the adult. In the embryo, the leg primordia can be detected using probes for leg-specific genes such as *Dll* (Cohen, 1990). The leg primordia invaginate to form the leg imaginal discs that grow inside the larva until they evaginate during metamorphosis. The only visible remnant of the leg in the cuticle of the larva is

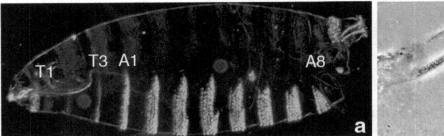
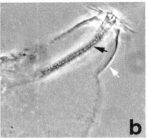

Figure 20.1. Wild-type cuticle. (a) Dark-field image of a wild-type larva. Note the different morphology of the ventral denticle belts in thorax (T) and abdomen (A) and the presence in A8 of the posterior spiracles. (b) A close-up of a posterior spiracle seen in phase contrast. The black arrow points to the filzkörper and the white arrow to the protruding stigmatophore.

the Keilin's organ, a three-haired external sensory organ. The posterior spiracles are a pair of respiratory organs connecting the trachea to the outside. Each spiracle consists of a tube with a refractile filter (the filzkörper) located in a protruding structure that acts as a respiratory "periscope."

There are eight Hox genes in *Drosophila* whose function is to control what structure is formed in each segment. For example, the Hox gene *Sex combs reduced* (*Scr*) is responsible for salivary gland formation in the labial segment (Panzer, Weigel, and Beckendorf, 1992); *Ultrabithorax* (*Ubx*) and *abdominal-A* (*abd-A*) suppress the formation of leg imaginal discs (and Keilin's organs) in the abdomen (Vachon et al., 1992); and *Abdominal-B* (*Abd-B*) controls the formation of spiracles in A8 (Hu and Castelli-Gair, 1999) (see also Chapter 11). It was a surprise when molecular studies showed there were more segments than Hox genes. Especially remarkable is the case of *Ubx* which specifies the third thoracic segment (with legs) and the first abdominal segment (without legs), and the case of *Abd-B* which specifies the A4–A7 segments (without spiracles) and A8 (with spiracles). Several hypotheses were proposed to explain how a single Hox gene can specify two different morphologies:

1. The HOX proteins interact with other proteins that modulate their function. These proteins would be present only in some of the segments where the Hox gene is expressed modifying the outcome of its expression.
2. Different levels of expression in different segments could lead to the modulation of different subsets of downstream targets. Thus, genes with low affinity for the HOX protein would be activated only in segments with higher levels of protein expression, while genes with high affinity for HOX proteins would be activated also in segments with low amounts of HOX proteins.
3. Differential temporal and spatial Hox expression between segments would account for the activation of different targets.

Although all three mechanisms act to some extent to specify segment morphology, the experiments in this practical are designed to study how the differential temporal expression of HOX proteins affects segment morphology. The students will observe the expression of some HOX proteins and manipulate their temporal expression to study the effects of HOX expression on a direct target and on segment morphology.

MATERIALS AND METHODS

EQUIPMENT AND MATERIALS

Per student. Collection of eggs requires egg-laying cages and baskets (see Chapter 11) and the preparation of agar–fruit juice plates and baking yeast.

EMBRYO FIXATION
Shaker
Small paint brushes
Glass vials (see Chapter 11)

STAININGS
1.5-ml Eppendorf tubes
Micropipettes (200 μl, 20 μl) with tips
Watch glasses
Pasteur pipettes with teats
Microscope slides and coverslips
Observing the antibody staining development requires a dissecting microscope
Cuticle preparations require a 60°C hot plate or oven
Observation of the results requires a compound microscope
Induction of heat shocks requires a 37°C incubator
55°C Incubator for RNA in situ

Per practical group
RNA PROBE PREPARATION
Bacterial shaker incubator
Large centrifuge
Qiagen plasmid midi kit [Qiagen]

Biological material. Wild-type flies (Oregon R is commonly used); HS-UBX (González-Reyes and Morata, 1990; Mann and Hogness, 1990) [from the Bloomington Stock Center, University of Indiana http://flystocks.bio.indiana.edu/]; HS-ABD-Bm (Kuziora, 1993; Lamka, Boulet, and Sakonju, 1992).

Dll probe can be obtained from Dr. Stephen Cohen (current address can be found in http://flybase.bio.indiana.edu search under "people"). Monoclonal antibodies: anti-SCR 6H4 (Glicksman and Brower, 1988), anti-ABD-B 1A2E9 (Celniker, Keelan, and Lewis, 1989) [from Developmental Studies Hybridoma Bank, University of Iowa http://www.uiowa.edu/~dshbwww/], anti-UBX FP3.38 (White and Wilcox, 1984) [For Rob White's address search under "people" in http://flybase.bio.indiana.edu/], biotinilated anti-mouse antibody [Vector laboratories].

REAGENTS

Commercial bleach
Methanol
Ethanol
PBS
Hydrogen peroxide
Vectastain ABC Elite standard kit and Normal goat serum [Vector laboratories]
Bovine Serum Albumin
Tween 20
Glycerol
Durcupan

PREVIOUS TASKS FOR STAFF

Solutions. The following solutions should be prepared in advance:

Hoyer's medium (see Chapter 11)
$10 \times$ BBS
(a) Tris   12.1 g
(b) NaCl   32.2 g
(c) KCl   29.8 g
(d) $MgCl_2$ 14.23 g
(e) $CaCl_2$ 7.35 g
(f) Glucose   171 g
(g) Sucrose   39.6 g
(h) Add water up to a litre and adjust pH to 7.4
(i) Sterilize by filtering or aliquot and freeze
BBT: BBS + 0.1% BSA + 0.1% Tween 20
PBT: PBS + 0.1% Tween 20
DAB: 3,3'-diaminobenzidine stock solution 10 mg/ml [Sigma]
1% $NiCl_2$ 1% $CoCl_2$ solution
Permanent embryo mounting medium (optional) [Durcupan ACM Fluka]:
   Mix at room temperature:
(a) Resin A   54 g
(b) Hardner B   44.5 g
(c) Accelerator C   2.5 g
(d) Plasticiser D   11 g
(e) Shake vigorously for 5 minutes, aliquot, and store at −20°C.
Hybridisation solution (hyb)
(a) 50% formamide (25 ml)
(b) $5 \times$ SSC (12.5 ml of $20 \times$ SSC)
(c) $5 \times$ Denharts solution (2.5 ml of $100 \times$ DH)
(d) 0.1% Tween 20 (50 μl)
(e) 100 μl tRNA (or yeast RNA) (500 μl of a 50 mg/ml stock solution)
(f) H_2O up to 50 ml

Alkaline phosphatase buffer (make fresh)

(a) 100 mM NaCl

(b) 50 mM $MgCl_2$

(c) 100 mM Tris pH 9.5

(d) 0.1% Tween 20

Siliconise watch glass and Pasteur pipettes by rinsing them with Sigmacote [Sigma].

Proteinase K stock 20 mg/ml.

Glycine 2 mg/ml in PBT (these last two can be aliquoted and stored at $-20°C$)

PROCEDURES

Fixation of embryos for antibody staining and RNA in situ

1. Collect eggs in agar–juice plates (100 flies per cage). Make a 16-hour and a 6-hour collection to obtain all embryonic stages.
2. Transfer eggs with brush to a basket.
3. Dechorionate eggs, submerging basket in 50% commercial bleach for 2–5 minutes (you can check dechorionation under the dissecting scope: the two long anterior appendixes will disappear when the chorion has dissolved).
4. Wash away bleach with tap water.
5. Transfer the eggs with brush to a glass vial with 4 ml equal parts of heptane/fixative [Formaldehyde 5% in PBS].
6. Shake 40 minutes.
7. Remove fixative (lower phase) and replace it with methanol.
8. Shake well. (Do not worry, fixed embryos are very tough.) Devitellinised embryos sink; ignore those remaining on the interphase.
9. Transfer devitellinised embryos to an Eppendorf tube using a Pasteur pipette. Tap the tube to help embryos sink.
10. Discard methanol (leaving some covering the embryos).
11. Wash embryos with clean methanol (to get rid of heptane traces).
12. Fixed embryos can be stored in methanol at $-20°C$ for several months.

Probe for RNA in situ hybridisation. The secret of a good in situ is a good probe.

1. Isolate plasmid with Qiagen midiprep kit following manufacturer's instructions.
2. Label sense and anti-sense probe with a nonradioactive method (Roche Digoxigenin RNA labelling kit (SP6/T7)). There is no need to treat solutions to inactivate RNAses as the kit provides RNAse inactivating agents.
3. Probes stored at $-20°C$ can last for years.

Protocol for antibody staining

STAINING (DAY 1)

1. Rehydrate in BBT 50 μl of fixed embryos per experiment and wash in fresh BBT for 15 minutes.
2. Wash 15 minutes in BBT +3% goat serum.
3. Discard BBT (leaving a small amount over the embryos to avoid desiccation).
4. Add primary antibody in 100 μl volume of BBT -3% goat serum.
5. Incubate at 4°C overnight.

DAY 2

1. Wash 4 times for 10 minutes in BBT.
2. Wash 15 minutes in BBT + 3% goat serum.
3. Discard wash and add biotinilated secondary antibody in 100 μl final volume of BBT + 3% goat serum.
4. Incubate one hour at room temperature.
5. Wash 4 times for 10 minutes in PBS. (Separately mix at room temperature Vectastain Elite A + B reagents for at least 30 minutes: two drops of solution A and two of B in 5 ml of PBS. It is stable at least for a week.)
6. Incubate 30 minutes with 200 μl of the A + B solution.
7. Wash 4 × 10 minutes in PBS.
8. Transfer to a watch glass with a Pasteur pipette in approximately 400 μl of PBS. Add 100 μl of DAB (0.5 mg/ml), (1 μl of 1% $NiCl_2$ 1% $CoCl_2$) to enhance and make the reaction colour grey instead of the otherwise brown colour) and finally add 6 μl of 3% hydrogen peroxide. [WARNING! DAB IS A CARCINOGEN. CONTAMINATED TIPS, EPPENDORFS, AND SOLUTIONS SHOULD BE TREATED WITH BLEACH AND DISPOSED OF PROPERLY.] Staining will appear in 2–10 minutes.
9. Stop the reaction by washing twice with BBT.
10. Wash in PBS.
11. Embryos can be mounted in glycerol for observation. Alternatively, dehydrate by successive washes in 50, 75, 96, and 100% ethanol. (The embryos can be stored at 4°C for several days before mounting.)
12. Cut the end of a yellow tip. Suck the embryos into the tip with micropipette. Let embryos sink until they form a drop outside the tip's mouth and let them fall on a drop of Durcupan in the slide. Put a coverslip on the slide and the embryos are ready for observation. [BEWARE NOT TO GET FRESH DURCUPAN ON THE MICROSCOPE OBJECTIVES!]

RNA in situ hybridisation (Modified from Tautz and Pfeifle, 1989)
DAY 1. All steps in Eppendorf tube:

1. Rehydrate 50 μl of embryos, washing twice in PBT.
2. Fix 30 minutes in 6% paraformaldehyde.
3. Wash 3 times for 2 minutes with PBT.
4. Proteinase treatment (critical step for morphology and permeability). Add 1 μl of Proteinase K from a 20 mg/ml frozen batch to 1 ml of PBT. Shake gently 1–2 minutes by hand (incubation times can vary between proteinase batches). Overdoing destroys the embryo.
5. Stop *immediately*. Wash 2 times for 2 minutes with 2 mg/ml glycine in PBT.
6. Wash 2 times with PBT.
7. Fix for 20 minutes in 6% paraformaldehyde.
8. Wash 4 times 2 minutes in PBT.
9. Wash 10 minutes in 100 μl of equal parts of hybridisation solution and PBT (pipetting up and down gently mixes solution with embryos).
10. Wash 10 minutes in 100 μl of hybridisation solution.

11. Wash 1 hour in hybridisation solution at 55°C.
12. Add probe in 100 μl of hybridisation solution (before adding the probe, boil it for 5 minutes to get rid of any secondary structure. Move immediately into salted ice).
13. Incubate overnight at 55°C.

DAY 2. All steps in Eppendorf tube:

1. Wash at 55°C in 100 μl volume:
   2 times for 20 minutes in hybridisation solution (hyb) and one time in the following:
   20 minutes 8 hyb/2 PBT
   20 minutes 6 hyb/4 PBT
   20 minutes 4 hyb/6 PBT
   20 minutes PBT
2. Wash at room temperature for 20 minutes in PBT.
3. Incubate for 1 hour with anti-DIG alkaline phosphatase conjugated antibody 1/2,000 in PBT (no need to pre-adsorb antibody).
4. Wash 4 times for 10 minutes in PBT.
5. Wash in alkaline phosphatase buffer 3 times for 5 minutes.
6. Transfer to a siliconised watch glass *with a siliconised glass pipette (the embryos stick to glass at this stage)* and add to the last wash 4.5 μl NBT and 3.5 μl X-phosphate in 1 ml of PBT.
7. Develop in the dark. Staining develops in 10 minutes to 2 hours; stop before background develops, so that most of the embryo looks clear on a white background.
8. Stop reaction by washing 2 times for 5 minutes in ice-cold PBT (if not cold, a filthy precipitate will form). Mount in glycerol, or
9. Dehydrate by washing in 50, 75, 96, and 100% ethanol. The embryos can be stored at 4°C in ethanol for several days before mounting in Durcupan.

OUTLINE OF THE EXPERIMENTS

EXPERIMENT 1. ANALYSIS OF THE EXPRESSION OF THREE HOX GENES

In this experiment the students will analyse the expression of the *Scr, Ubx*, and *Abd-B* genes.

Collection and embryo fixation. The technique is similar to those described in Chapter 11.

1. Place about 100 flies in an egg-laying cage with an agar juice plate with a dub of fresh yeast in the centre. The cages should preferably be maintained at 25°C but can be left at room temperature.
2. To collect all stages of embryogenesis two egg collections should be performed:
   (a) An overnight egg collection (yields embryos up to 16 hours).
   (b) An eight-hour collection (enriches early stages of Hox expression).
3. To induce maximum egg laying, keep the flies in the cage for two days, changing the plates twice daily prior to the egg collection.
4. Fix embryos as described in Materials and Methods.

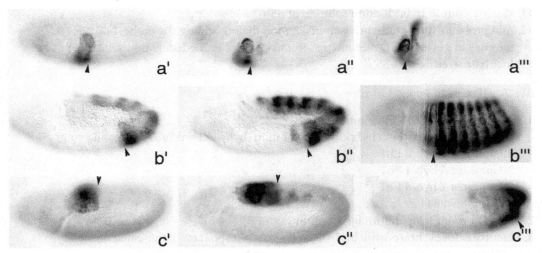

Figure 20.2. SCR, UBX and ABD-B protein expression patterns. Row (a) shows SCR expression, row (b) UBX expression, and row (c) ABD-B expression. Embryos of progressively older stages are shown with early stages to the left. The first two columns show embryos at the extended germ band stage when the embryo is folded on itself. This is easier to visualise in the UBX stained embryos (row b). All embryos are oriented with the head to the left and the ventral side down. Arrowheads in a row mark equivalent positions in the embryo to facilitate the identification of new areas of HOX expression.

5. Fixed embryos can be stored in methanol at −20°C.
6. Stain embryos with antibody.

Data recording. The protein expression patterns (Figure 20.2) can be observed under the compound microscope. If mounted on Durcupan, the embryos can be stored and used as examples for future students.

EXPERIMENT 2. ANALYSIS OF THE EFFECTS OF TEMPORAL REGULATION OF UBX AND ABD-B ON CUTICLE DEVELOPMENT

In the previous experiment, the students will have observed that not all segments simultaneously express the HOX proteins. In this experiment, students will modulate the temporal HOX expression and observe the defects this causes. Flies carrying *Ubx* or *Abd-B* genes under the control of heat-inducible promoters (HS) will be used to induce timed HOX expression. The following heat shock regime induces ubiquitous activation of UBX or ABD-B at the stages when endogenous HOX expression is more dynamic.

Heat shock protocol (Regime 1, Figure 20.3)
1. Place 100 young flies of each genotype in egg-laying cages (keep at 25°C).
2. Change plates every hour for five hours, keeping the plates and cages at 25°C (discard first plate).
3. After the last plate has been changed, allow 3.5 hours of development.
4. Transfer the four plates to a 37°C incubator for 1 hour.

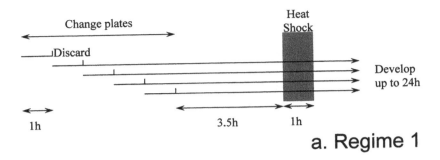

a. Regime 1

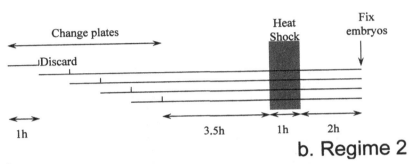

b. Regime 2

Figure 20.3. Heat shock regimes for Experiments 2 and 3.

5. Return plates to 25°C until eggs are 24 (+ 4) hours old.
6. Make cuticle preparations as described in Chapter 11.

Data recording. To interpret the stages of embryogenesis an atlas of *Drosophila* development (Campos-Ortega and Hartenstein, 1997) as well as some Web pages (http://flymove.uni-muenster.de/) can be of great help.

Cuticles will be observed in a compound microscope using phase contrast or dark field. WT embryos will serve as control for defects arising from heat shock induction. Ectopic UBX causes denticle belt transformations (Figure 20.4) and sensory organ alterations (partial or total absence of thoracic Keilin's organs). The three hairs of the Keilin's organs are difficult to see but it is an important element to study. Ectopic ABD-B generates ectopic filzkörper and abnormal denticle belts. Ectopic expression of both HOX proteins results in head-involution defects.

EXPERIMENT 3. ANALYSIS OF THE EXPRESSION OF UBX AND ONE OF ITS DOWNSTREAM TARGETS

The *Distal-less* (*Dll*) gene is required for the formation of legs and Keilin's organs (Cohen and Jürgens, 1989). UBX and ABD-A prevent formation of legs in the abdomen by repressing *Dll* but UBX does not repress *Dll* in T3 (see discussion).

In this practical the students will observe how ectopic UBX induced at 4 hours of development represses *Dll* in the leg primordia, while at 8 hours it does not.

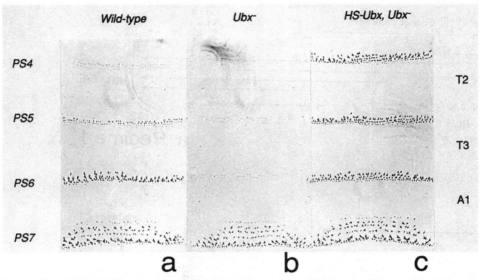

Figure 20.4. Effect of UBX expression in *Drosophila* larval cuticles. (a) Close-up of the ventral cuticles of a wild-type embryo; (b) a *Ubx* embryo; and (c) a *Ubx* mutant embryo where UBX is expressed ubiquitously from a *HS-UBX* transgene. Note the small size of the denticles in the thoracic segments in the wild-type (a). The abdominal segments have bigger denticles and the shape of the denticle is different in A1 and A2. In the *Ubx* mutant the denticles in A1 resemble those of a thoracic segment (b). After ectopic UBX expression the thoracic denticles resemble the A1 belt. Panels a–c are phase contrast images.

Heat shock protocol (Regime 2, Figure 20.3)
1. Place 100 flies of each genotype in egg-laying cages (keep at 25°C).
2. Change plates every hour for five hours keeping the plates and cages at 25°C (discard first plate).
3. After the last plate has been changed allow 3.5 hours of development.
4. Transfer the four plates to a 37°C incubator for 1 hour.
5. Return plates to 25°C for 2 hours.
6. Fix embryos (Materials and Methods).
7. Proceed with RNA in situ or store in methanol at −20°C.

RNA whole-mount in situ protocol. Probe and embryo fixation (for a shorter practical these should be made in advance by staff): see Materials and Methods.

Data recording. As in Experiment 1.

EXPECTED RESULTS

EXPERIMENT 1. ANALYSIS OF THE EXPRESSION OF THREE HOX GENES
Embryos at early stages of development do not express HOX proteins. HOX expression does not develop in all segments simultaneously (Figure 20.2). SCR appears first in the labial segment and later in T1. UBX expression develops first in A1, then in posterior

segments (A2–A7), and last in the T3 segment. ABD-B expression develops first in A8–A9 and later in more anterior segments. This modulation is very transient but, as Experiments 2 and 3 show, has an important effect on the animal's morphology. Observation at high magnification (40 × objective) will reveal the protein localised to the nucleus, giving the appearance of being expressed in closely packed balls. Careful observation will also reveal that the HOX protein is expressed at many stages not in segmental units, but in parasegmental units that encompass the posterior part of one segment and the anterior part of the next one (Martínez-Arias and Lawrence, 1985). This can be observed clearly by analysing HOX protein expression compared to the segmental grooves formed after the retraction of the germ band. This out-of-register expression is due to Hox gene activation by the segmentation genes that are expressed in parasegments.

EXPERIMENT 2. ANALYSIS OF THE EFFECTS OF TEMPORAL REGULATION OF UBX AND ABD-B ON CUTICLE MORPHOLOGY

WT controls should develop normally, with some exceptions in which the deletion of segments is observed. HS-UBX embryos should have Keilin's organs missing totally or partially when the heat shock has been induced at 4 hours but not at 8 hours. HS-ABD-Bm embryos should form filzkörper anterior to A8 when the heat shock is induced at 4 hours but not at 8 hours. Expect intermediate effects at intermediate ages.

EXPERIMENT 3. ANALYSIS OF THE EXPRESSION OF UBX AND ONE OF ITS DOWNSTREAM TARGETS

Dll RNA expression in the leg primordium should be reduced or absent when the heat shock is induced at 4 hours but normal when induced at 8 hours. *Dll* expression in the head serves as an internal control for staining quality. If the heat shock lines are not homozygous, 25% of the embryos will have normal *Dll* expression. Only the anti-sense probe should result in signal; the sense probe acts as a negative control.

DISCUSSION

Hox genes encode transcription factor molecules that regulate downstream target genes. Because each Hox gene has a different set of targets and they are expressed in different segments, they are indirectly conferring segmental diversity. Paradoxically, there are more segments than Hox genes. We have observed that Hox genes are not expressed identically all along their domain, but that the expression in each segment has a unique temporal–spatial modulation. Heat shock induction of *Abd-B* and *Ubx* shows that the cells respond differently to Hox expression at each stage. At 4 hours of development ABD-B induces the formation of posterior spiracles, while at 7–8 hours it modulates only the cuticle shape and the sensory organs (Castelli-Gair, 1998). Similarly, at 4 hours UBX represses the formation of leg primordia, while at 7–8 hours it does not (Castelli-Gair and Akam, 1995). This last case is particularly interesting because in the wild-type fly UBX contributes to the morphological diversification of the T3 leg. The explanation for the differential responsiveness of the legs to UBX can be found at the *Dll*

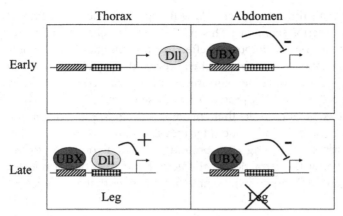

Figure 20.5. Hypothetical scheme of UBX and *Dll* interactions in the leg primordium. View of *Dll* in a thoracic and an abdominal segment at early (top) and late (bottom) stages. The boxes represent the early enhancer (striped) or late *Dll* leg enhancer (checkered). At early stages absence of UBX protein (dark oval) in the thorax allows *Dll* transcription leading to the formation of DLL protein (light shaded oval) in T3. Presence of UBX in the abdomen at early stages blocks *Dll* transcription. Late expression of UBX in the thorax cannot prevent *Dll* transcription as it is now controlled by DLL binding to the late enhancer. This situation allows UBX to have two different functions in T3 and A1.

regulatory regions (Figure 20.5). The non-coding cis regulatory regions of a gene bind regulatory proteins that influence its transcription. The dissection of the *Dll* regulatory region has resulted in the localisation of two enhancer elements responsible for leg activation (Vachon et al., 1992). One is an early leg enhancer that functions only transiently, and the other is a late enhancer that maintains *Dll* expression. UBX binds only the early enhancer. If the UBX binding site is deleted, the *Dll* leg enhancer becomes expressed in all abdominal segments (Vachon et al., 1992). The late enhancer, on the contrary, is not responsive to UBX repression but its expression requires DLL protein (Castelli-Gair and Akam, 1995). The use of these two enhancers explains the temporal sensitivity of *Dll* to UBX (Figure 20.5). Early UBX is expressed exclusively in the abdomen where it represses *Dll* expression. In the thorax, the absence of UBX expression allows *Dll* activation from the early enhancer. At late stages, when UBX is expressed in the T3 segment, UBX binding of the early enhancer does not block *Dll* transcription, as *Dll* has entered an auto-regulatory phase controlled by DLL binding to the UBX-independent late enhancer. As a result, UBX protein can now influence leg patterning.

Experiments 1 and 2 show that ABD-B behaves in a similar way (Castelli-Gair, 1998) and experiments with vertebrate Hox genes suggest that modulation of temporal expression is a conserved mechanism (Zakany et al., 2001).

TIME REQUIRED FOR THE EXPERIMENTS

If the probe and the fixed embryos are provided by staff, this exercise could take three to four days. Antibody staining of embryos could be done in Days 1 and 2. Heat shocks can be induced in Day 1 and cuticle preparations mounted in Day 2. RNA in situ can be done in Days 2 and 3. Observation of antibody staining and cuticle preparations can be done on Day 3. In situ can be analysed on Day 3. An extra day can be given to analyse all data.

POTENTIAL SOURCES OF FAILURE

To prevent desiccation and embryo loss, do not remove all liquid during washes. The protocols already account for the liquid left to cover the embryos.

➤ Fixation: embryos sink and do not stay in the interphase (the embryos are not correctly dechorionated). The heptane–methanol phases do not separate (bleach carried over).

➤ Rehydration steps: the embryos stick to the Eppendorf walls (heptane carried over, wash with methanol).

➤ RNA in situ: no signal observed (make a new probe). Sticky and "mashed" embryos (too much Proteinase K). Black precipitate forming when stopping the staining reaction (use ice-cold PBT for the washes). Background: too much probe. Typical areas for artifactual staining are the trachea and spiracle lumen.

➤ Flies lay few eggs: flies should be young (three to fifteen days old). Start changing plates two days before the experiment so that the flies adapt to the disruption. Ideally, use four times more females than males. Keep cages clean.

TEACHING CONCEPTS

Regulatory protein. Transcriptional activators or repressors of other genes.

Antibody and in situ RNA stainings. These techniques allow the observation of regulatory proteins as they control development. Prior to their introduction researchers had to guess the gene's function from phenotypes detectable hours after the function had occurred.

Regulation of segment diversification by Hox expression. Each segment develops a characteristic morphology due to the expression of a different HOX protein. The effect of a Hox gene on its targets changes dramatically in time. Thus, subtle modifications of Hox spatial and temporal expression can have major effects on the morphology of the segment. This has wide evolutionary implications (Castelli-Gair, 1998).

Inducible promoters. Artificial activation of gene expression complements the study of mutants, helping us to understand the function of a gene.

REFERENCES

Campos-Ortega, J. A., and Hartenstein, V. (eds.) (1997). *The Embryonic Development of Drosophila melanogaster*, 2nd ed. Berlin: Springer-Verlag.

Castelli-Gair, J. (1998). Implications of the spatial and temporal regulation of *Hox* genes on development and evolution. *Int. J. Dev. Biol.*, 42, 437–44.

Castelli-Gair, J., and Akam, M. (1995). How the Hox gene *Ultrabithorax* specifies two different segments: The significance of spatial and temporal regulation within metameres. *Development,* 121, 2973–82.

Celniker, S. E., Keelan, D. J., and Lewis, E. B. (1989). The molecular genetics of the bithorax complex of *Drosophila*: Characterization of the products of the *Abdominal-B* domain. *Genes Dev.,* 3, 1424–36.

Cohen, S. M. (1990). Specification of limb development in the *Drosophila* embryo by positional cues from segmentation genes. *Nature,* 343, 173–7.

Cohen, S. M., and Jürgens, G. (1989). Proximal-distal pattern fomation in *Drosophila*: Cell autonomous requirement for *Distal-less* gene activity in limb development. *EMBO J.,* 8, 2045–55.

Glicksman, M. A., and Brower, D. L. (1988). Expression of the Sex combs reduced protein in *Drosophila* larvae. *Dev. Biol.,* 127, 113–8.

González-Reyes, A., and Morata, G. (1990). The developmental effect of overexpressing a *Ubx* product in *Drosophila* embryos is dependent on its interactions with other homeotic products. *Cell,* 61, 515–22.

Hu, N., and Castelli-Gair, J. (1999). Study of the posterior spiracles of *Drosophila* as a model to understand the genetic and cellular mechanisms controlling morphogenesis. *Dev. Biol.,* 214, 197–210.

Kuziora, M. A. (1993). *Abdominal-B* protein isoforms exhibit distinct cuticular transformations and regulatory activities when ectopically expressed in *Drosophila* embryos. *Mech. Dev.,* 42, 125–37.

Lamka, M. L., Boulet, A. M., and Sakonju, S. (1992). Ectopic expression of UBX and ABD-B proteins during *Drosophila* embryogenesis: Competition, not a functional hierarchy, explains phenotypic suppression. *Development,* 116, 841–54.

Mann, R. S., and Hogness, D. S. (1990). Functional dissection of Ultrabithorax proteins in *D. melanogaster. Cell,* 60, 597–610.

Martínez-Arias, A., and Lawrence, P. A. (1985). Parasegments and compartments in the *Drosophila* embryo. *Nature,* 313, 639–42.

Panzer, S., Weigel, D., and Beckendorf, S. K. (1992). Organogenesis in *Drosophila melanogaster.* Embryonic salivary gland determination is controlled by homeotic and dorsoventral patterning genes. *Development,* 114, 49–57.

Tautz, D., and Pfeifle, C. (1989). A non-radioactive in situ hybridisation method for the localisation of specific RNAs in *Drosophila* embryos reveals translational control of the segmentation gene *hunchback. Chromosoma,* 98, 81–5.

Vachon, G., Cohen, B., Pfeifle, C., McGuffin, M. E., Botas, J., and Cohen, S. M. (1992). Homeotic genes of the bithorax complex repress limb development in the abdomen of the *Drosophila* embryo through target gene *Distal-less. Cell,* 71, 437–50.

White, R. A. H., and Wilcox, M. (1984). Protein products of the bithorax complex in *Drosophila. Cell,* 39, 163–71.

Zakany, J., Kmita, M., Alarcon, P., de la Pompa, J. L., and Duboule, D. (2001). Localized and transient transcription of Hox genes suggests a link between patterning and the segmentation clock. *Cell,* 106, 207–17.

21 The UAS/GAL4 system for tissue-specific analysis of EGFR gene function in *Drosophila melanogaster*

J. B. Duffy and N. Perrimon

OBJECTIVE OF THE EXPERIMENT Genetic methodologies have historically provided an important means for unraveling the mysteries of development. The UAS/GAL4 system in *Drosophila melanogaster* provides one such example. This bipartite genetic system, based on the yeast transcriptional activator GAL4, was constructed in 1993 as a means to direct gene expression *in vivo* (Brand and Perrimon, 1993). The system is based on the ability of GAL4 to bind to an Upstream Activating Sequence (UAS) element and stimulate transcription of the associated gene (responder). By placing GAL4 under the control of tissue-specific regulatory elements, transcription of the responder can be directed in a similar pattern. Because expression of the responder is dependent on the presence of GAL4, the absence of GAL4 renders the responder gene transcriptionally silent. With a simple mating scheme, GAL4 and the responder can be combined, resulting in expression of the responder (Figure 21.1). In this chapter, we describe experiments that explore the utility of the UAS/GAL4 technique in the study of post-embryonic development in the fly.

As our knowledge of development has increased, we have come to realize that many genes are often required at multiple stages throughout the life cycle. While providing an example of developmental efficiency, it simultaneously presents an experimental hurdle that must be overcome in order to assess the role of such genes in any one spatial or temporal context. This is evident in the case of developmental processes involving cellular communication, which rely heavily upon conserved signal transduction modules. Mutation of any of their numerous molecular elements results in pleiotropic effects and embryonic lethality, thus making the nature of their role in any one process difficult to discern. This presents a hurdle to those studying post-embryonic development, because the embryonic roles of such molecules hinder their identification in simple screens for mutations affecting the tissue or process of interest. While genetic tools have been created to overcome problems like those presented earlier, over the past decade the GAL4 system has opened many new doors for the analysis of gene function *in vivo*. When coupled with the appropriate responders, this system can be

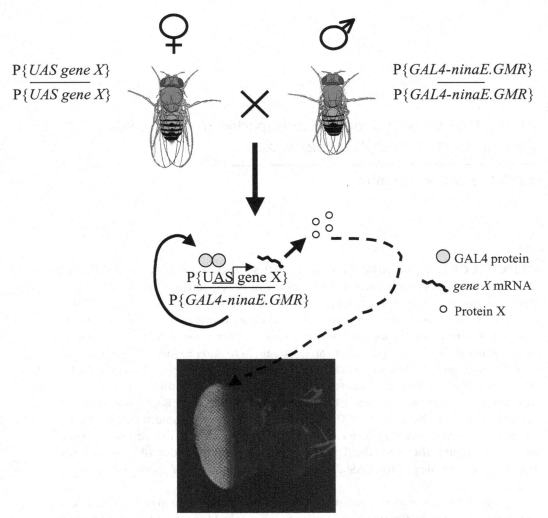

Figure 21.1. The bipartite UAS/GAL4 system in *Drosophila*. Females carrying a UAS responder (*UAS-Gene X*) are mated to males carrying the eye-specific GAL4 driver P{*GAL4-ninaE.GMR*}. The F1 contain both elements of the system resulting in the eye-specific expression of *Gene X*, depicted in the eye imaginal disc as white staining.

used to help elucidate gene function through tissue-specific loss-of-function (LOF) and gain-of-function (GOF) studies. This has been particularly useful for analysis of the *in vivo* roles of pathways involved in cell-cell communication, such as the Epidermal Growth Factor Receptor (EGFR) pathway.

The objective of the experiments outlined in this chapter is to demonstrate the utility of the UAS/GAL4 system in studying gene function in a tissue-specific fashion. Using the UAS/GAL4 system, we will analyze the phenotypic effects of decreasing or increasing EGFR activity in two distinct developmental processes: formation of the adult compound eye and organization of the dorsal–ventral (DV) axis during oogenesis. Finally, we will combine the UAS/GAL4 and FLP/FRT systems to perform clonal analysis in the ovary of two lethal mutations that affect EGFR signaling.

DEGREE OF DIFFICULTY The experiments in this chapter require the ability to distinguish the sexes properly (male and female), carry out mating schemes, recognize dominant markers to identify the appropriate genotypes, and finally to analyze chorion (eggshell) phenotypes. All of these experiments can be performed with the aid of a dissecting microscope and standard methods for culturing *Drosophila*.

INTRODUCTION

The disruption of cellular communication is one of the hallmarks of cellular transformation and the progression of cancer. A molecular understanding of the contribution of EGFR signaling to these processes was initiated by identification of the rat *neu* (*ErbB-2*) oncogene as one of the first oncogenic receptor tyrosine kinases (RTK). This was followed by the characterization of four members of this family of RTKs in humans variously termed the EGFR or ErbB family. Since its identification, misregulation of the EGFR family has been implicated in most cellular aspects of oncogenesis, including immortalization, proliferation, migration, and chemoresistance. Structurally, the EGFR is composed of an extracellular ligand-binding domain, a transmembrane region, and a cytoplasmic tyrosine kinase domain (Schlessinger, 2000). Upon ligand binding, dimerization of receptor monomers occurs, leading to trans-phosphorylation of cytoplasmic tyrosines. These phosphorylated tyrosines then serve as molecular docking sites for adaptor proteins, resulting in the activation of cytoplasmic signaling cascades.

In *Drosophila*, null mutations in the EGFR result in embryonic lethality and an appropriately termed "faint little ball" phenotype, reflecting the overwhelming number of cellular decisions the receptor participates in and thus the poor differentiation of the cuticle (Price, Clifford, and Schupbach, 1989; Schejter and Shilo, 1989). During embryogenesis, EGFR is involved in the establishment of ventral cell fates, survival of the amnioserosa and ventral ectodermal cells, CNS development, production of embryonic cuticle, segmentation, and germ band retraction. Initially, insight into the receptor's role in post-embryonic processes was gained through the isolation of rare hypomorphic LOF alleles. For example, a role for the receptor in DV axis formation was discovered upon the isolation of the hypomorphic allele *torpedo*^{QY1} (Schupbach, 1987). This viable allele resulted in the chorion's displaying a ventralized phenotype, demonstrating a tissue-specific requirement for the receptor within the ovary. Likewise, a role for EGFR in eye development was uncovered with the identification of the viable hypermorphic, GOF *Ellipse* allele of the receptor, which inhibits eye morphogenesis (Baker and Rubin, 1989).

In both tissues, restricted expression of ligands delimits receptor activation to distinct cellular populations. In the eye, coordinated expression of EGFR ligands results in a highly dynamic pattern of receptor activation, which regulates the specification of photoreceptors, prevents cell death, and promotes cell proliferation. In the oocyte, asymmetric accumulation of transcripts for the ligand Gurken (Grk) leads to spatially restricted EGFR activation in the overlying follicular epithelium or follicle cells. This results in the specification of dorsal fates and the production of two dorsal respiratory appendages on the chorion by the follicle cells. The experiments in this chapter are designed to alter EGFR signaling in the eye and ovary and explore the UAS/GAL4 system as a more versatile alternative to the recovery of tissue-specific alleles in addressing gene function throughout development.

MATERIALS AND METHODS

EQUIPMENT AND MATERIALS

Standard materials and equipment for maintenance of *Drosophila* are described in Chapter 14 and at http://flystocks.bio.indiana.edu/working-with-docs.htm. In addition, the preparation of the chorions requires egg-laying cages and juice–agar plates.

1. Egg-laying cages are made using 50-ml plastic conical tubes that have ~10 holes made with a 22-gauge needle.
2. A small amount of liquified juice–agar mixture is poured into the caps and allowed to harden.
3. Flies of the desired genotype are then anesthetized and placed into the cage, which is sealed by a juice agar cap.

Biological material. The *Drosophila* stocks necessary for the experiments described in this chapter and sources for obtaining them are listed below:

MISEXPRESSION EXPERIMENTS

P{*GAL4-ninaE.GMR*} commonly referred to as *GMR-GAL4*	Bloomington Stock Center (http://flystocks.bio.indiana.edu/)
P{*CY2-GAL4*}	Bloomington Stock Center
P{*UAS-λtop*}	Duffy Lab (http://sunflower.bio.indiana.edu/~duff/lab.html)
P{*UAS-DNEGFR*}	Bloomington Stock Center
P{*UAS-GFP*}	Bloomington Stock Center

DIRECTED MOSAIC EXPERIMENTS

y w nej[131] P{*FRT(w[hs])*} *101/FM7*	Duffy Lab
y w P{*FRT(w[hs])*} *101*; P{*en2.4-GAL4*}*e22c* P{*UAS-FLP1.D*}*JD1/CyO*	Bloomington Stock Center
w; bwk[hab] P{*neoFRT*}*82B/TM8*	Duffy Lab
w; P{*en2.4-GAL4*}*e22c* P{*UAS-FLP1.D*} *JD1/CyO*; P{*neoFRT*}*82B*	Bloomington Stock Center

A description of Balancer chromosomes is available at http://flystocks.bio.indiana.edu/working-with-docs.htm. The dominant markers associated with the Balancers used in this chapter are as follows:

Dominant marker	Phenotype	Balancer
Bar	reduction in size of eye and presence of anterior indentation	*FM7*
Curly	wings curled up at edges	*CyO*
Stubble	short bristles	*TM8*

PROCEDURES

For information on distinguishing females from males and on virgin collection refer to Chapter 11. Chorions can be observed directly on the egg-laying plate using a dissecting microscope, or alternatively mounted in lacto-hoyers and observed by dark-field microscopy.

OUTLINE OF THE EXPERIMENTS

EXPERIMENT 1. DEFINING THE TISSUE-SPECIFIC EFFECTS OF ALTERED EGFR SIGNALING DURING EYE DEVELOPMENT

The aim of this section is to demonstrate that EGFR mediated cell–cell communication is essential for formation of the adult eye. The phenotypic consequences of decreasing, as well as increasing, receptor activity will be examined in the adult compound eye. EGFR signaling will be altered using the P{*GAL4-ninaE.GMR*} driver to express two different variants of the receptor, P{*UAS-DNEGFR*} and P{*UAS-λtop*} during eye development. Eye-specific regulatory elements target GAL4 expression posterior to the morphogenetic furrow in the developing eye imaginal disc, which consequently directs expression of the responders. The first responder, P{*UAS-DNEGFR*}, decreases wild-type receptor signaling, whereas the second increases receptor signaling. The first variant encodes a dominant negative version of the receptor that is capable of binding ligand, but unable to activate downstream signaling due to deletion of the cytoplasmic tyrosine kinase domain (Freeman, 1996). Overexpression of this dominant negative receptor sequesters ligand in inactive complexes. Consequently this variant prevents ligand association with, and activation of, the wild-type EGFR. In contrast, the second variant, P{*UAS-λtop*}, increases receptor signaling bacause it encodes a constitutively active form of the receptor (Queenan and Schupbach, 1997). To create this version of the receptor, the extracellular domain was replaced with a dimerization domain from the bacteriophage lambda repressor, cI. This triggers ligand-independent dimerization, resulting in high levels of receptor signaling.

Protocol. The same crossing scheme will be used for all crosses in this section and is diagrammed in the table below.

Virgin females		Males	
P{GAL4-ninaE.GMR}	×	P{UAS-DNEGFR}	**Cross A**
"	×	P{UAS-λtop}	**Cross B**
"	×	P{UAS-GFP}	**Cross C**

1. Cross 5–10 virgin females from the P{GAL4-ninaE.GMR} strain to 5–7 males from the indicated strains for Crosses A–C.
2. Maintain the cross at 25°C and transfer the parental generation to a new tube after 4–5 days.
3. After an additional 4–5 days discard the parental generation from the second tube.
4. Once the F1 have eclosed (after ~10 days at 25°C), anesthetize them and examine the compound eye for defects.

For each of the crosses indicated (A–C), the F1 progeny will have one copy of the eye-specific GAL4 driver P{GAL4-ninaE.GMR} and the indicated responder. Progeny from Cross A will experience a decrease in EGFR activity during eye development as a result of misexpression of a dominant negative variant of the receptor P{UAS-DNEGFR}. In contrast, progeny from Cross B will experience an increase in EGFR activity through misexpression of a constitutively active receptor P{UAS-λtop}. Finally, progeny from Cross C represent normal levels of EGFR signaling during eye development and provide an important control and reference phenotype.

Data recording. The *Drosophila* compound eye is composed of approximately 800 ommatidia in an ordered array (Figure 21.2). For each cross, the phenotype of the compound eye should be examined for any disruptions, including ommatidial loss or fusion and changes to the overall architecture or size of the eye (Figure 21.2). Comparisons should then be drawn between the progeny from all crosses.

EXPERIMENT 2. DEFINING THE TISSUE-SPECIFIC EFFECTS OF ALTERED EGFR SIGNALING DURING OOGENESIS

The aim of this section is to demonstrate that EGFR mediated cell–cell communication is required during oogenesis for the formation of the DV axis. The phenotypic consequences of decreasing, as well as increasing, receptor activity in the ovarian follicle cells will be examined. As in Experiment 1, EGFR signaling will be altered using the two receptor variants, P{UAS-DNEGFR} and P{UAS-λtop}. However, in this experiment responder expression will occur in the somatic follicle cells during stages 8–10 due to the P{CY2-GAL4} driver.

Protocol. The same scheme will be used for all crosses in Experiment 2 and is diagrammed in the following table.

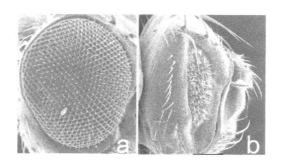

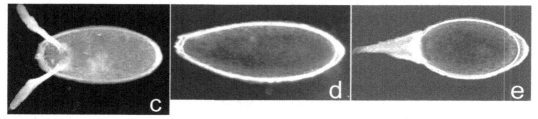

Figure 21.2. GAL4 directed alterations in EGFR activity. Panels (a) and (b) are scanning electron micrographs of wild-type and P{*GAL4-ninaE.GMR*}/P{*UAS-DNEGFR*} eyes, respectively. Panels (c), (d), and (e) represent dark-field micrographs of a wild-type, ventralized P{*CY2-GAL4*}/P{*UAS-DNEGFR*}, and dorsalized P{*CY2-GAL4*}/P{*UAS-λtop*} chorion, respectively.

Virgin females		Males	
P{*CY2-GAL4*}	×	P{*UAS-DNEGFR*}	**Cross A**
"	×	P{*UAS-λtop*}	**Cross B**
"	×	P{*UAS-GFP*}	**Cross C**

1. Cross 5–10 virgin females from the P{*CY2-GAL4*} strain to 5–7 males from the indicated strains for Crosses A–C.
2. Maintain the crosses at 20°C and transfer the parental generation to a new tube after 5–8 days. (20°C should be used in this experiment to reduce lethality caused by CY2-GAL4 driven responder expression during the larval stages.)
3. After an additional 5–8 days discard the parental generation from the second tube.
4. Once the F1 have eclosed (after ∼10 days), anesthetize, select 5–10 females, and place them along with 3–4 males into a fresh vial at 25°C.
5. After 2 days at 25°C in a vial, transfer these progeny into an egg-laying chamber at 25°C overnight. Be sure to record the appropriate genotype on both chamber and the plate.
6. The next day remove the first plate for examination, placing a fresh juice plate on the chamber for a second overnight collection.

For each of the crosses indicated (A–C), the F1 progeny will have one copy of the follicle cell GAL4 driver P{*CY2-GAL4*} and the indicated responder, resulting in the corresponding effects on EGFR signaling in the follicle cells.

Data recording. For each cross, examine the chorions deposited on the first juice–agar plate at the highest magnification available on the dissecting microscope. Comparisons of the chorion phenotypes should then be drawn between F1 females of the appropriate genotypes from all crosses (Figure 21.2). For a description of oogenesis, see Spradling, 1993.

EXPERIMENT 3. REVEALING TISSUE-SPECIFIC EFFECTS OF VITAL LOCI USING DIRECTED MOSAICS

The aim of this section is to use the directed mosaic system to characterize genes that are required in the follicle cells for the formation of the dorsal–ventral axis. Because many components of the EGFR signaling pathway are essential for viability, mutations in these genes result in lethality, thereby hindering an analysis of their development roles at other stages of the life cycle. As a means to overcome this, the experiments in this chapter combine the UAS/GAL4 system with the FLP/FRT system. The FLP/FRT system represents another instance of *Drosophila* research taking advantage of yeast biology (Golic and Lindquist, 1989). In this system, site-specific recombination at FRT sequences is induced with the yeast FLP recombinase. By combining the two systems, clones homozygous for the specified mutation can be induced in the somatic follicle cells of the adult ovary (Figure 21.3) (Duffy, Harrison, and Perrimon, 1998). This allows for a direct analysis of the function in DV patterning of the gene under study. For a more extensive description of the FLP/FRT system, refer to Chapter 15 and Figure 21.3.

Protocol. The schemes for clonal analysis in follicle cells with two distinct lethal mutations are diagrammed below. The lethality associated with these mutations in *nejire* (*nej*) and *bullwinkle* (*bwk*) prevents a straightforward analysis of the function of these genes in the ovary. The clonal analysis presented below overcomes this problem, allowing their roles of EGFR signaling in the follicle cells to be uncovered.

CROSS A – CLONAL ANALYSIS OF *nej*

Virgin females		Males
y w nej¹³¹ P{FRT(w[hs])}101/FM7	×	*y w P{FRT(w[hs])}101; P{en2.4-GAL4}e22c P{UAS-FLP1.D}JD1/CyO*

Progeny:

I. *y w nej¹³¹ P{FRT (w[hs])}101/y w P{FRT (w[hs])}101; P{en2.4-GAL4}e22c P{UAS-FLP1.D}JD1/ +*
   * Phenotype = non–bar-eyed, non–curly-winged females.

II. *y w nej¹³¹ P{FRT (w[hs])}101/y w P{FRT(w[hs])}101; CyO/ +* Phenotype = non–bar-eyed, curly-winged females.

III. *FM7/y w P{FRT(w[hs])}101; P{en2.4-GAL4}e22c P{UAS-FLP1.D}JD1/+*
   Phenotype = bar-eyed, non–curly-winged females.

IV. *FM7/y w P{FRT(w[hs])}101; CyO/+*
   Phenotype = bar-eyed, curly-winged females.

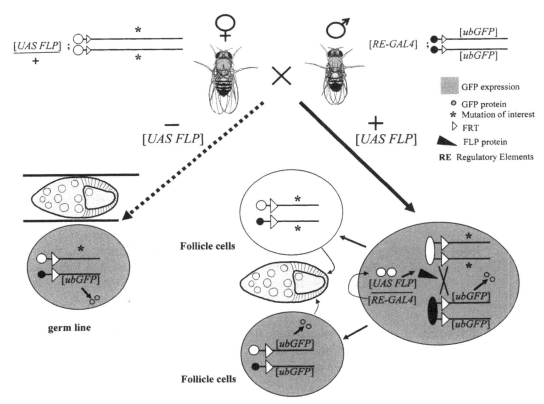

Figure 21.3. Generation of follicular clones. Using the directed mosaic system, follicle cells homozygous for the mutation of interest can be produced. GFP expression serves as a convenient marker to distinguish homozygous mutant cells from wild-type cells. During mitosis, GAL4-directed expression of the FLP recombinase leads to recombination in the follicle cells at the FRT sequences. This results in the production of two daughter cells, each homozygous for the indicated chromosomes. The effects of clones of these mutations can then be assayed on the chorion.

CROSS B – CLONAL ANALYSIS OF *bwk*

Virgin females	Males
w; bwk^hab P{neoFRT}82B/TM8 ×	w; P{en2.4-GAL4}e22c
	P{UAS-FLP1.D}JD1/CyO; P{neoFRT}82B

Progeny:

I. w/ w; P{en2.4-GAL4}e22c P{UAS-FLP1.D}JD 1/+ ; bwk^hab P{neoFRT}82B/ P{neoFRT}82B

* Phenotype = non–curly-winged, non–stubble-bristled females.

II. w/ w; CyO/+ ; bwk^hab P{neoFRT }82B/P{neoFRT}82B

Phenotype = curly-winged, non–stubble-bristled females.

III. w/ w; P{en2.4-GAL4}e22c P{UAS-FLP1.D}JD 1/+; TM8/ P{neoFRT}82B

Phenotype = non–curly-winged, stubble-bristled females.

IV. w/ w; CyO/+ ; bwk^hab P{neoFRT}82B/P{neoFRT}82B

Phenotype = curly-winged, non–stubble-bristled females.

Procedure

1. For each cross mate 5–10 virgin females to 5–7 males from the indicated strains.
2. Maintain the cross at 25°C and transfer the parental generation to a new tube after 4–5 days.
3. After an additional 4–5 days discard the parental generation from the second tube.
4. Once the F1 have eclosed (after ~10 days), anesthetize, select 5–10 females of the correct genotype (class I for both crosses, marked above with *), and place them along with 3–4 males (of any genotype) into a fresh vial at 25°C.
5. After 2–4 days at 25°C in a vial, transfer these progeny into an egg-laying chamber at 25°C overnight. Be sure to record the appropriate genotype on both chamber and the plate.
6. The next day remove the first plate for examination, placing a fresh juice plate on the chamber for a second overnight collection.

For each of the crosses (A and B), F1 progeny of the appropriate genotype will be generating cells homozygous for the mutation indicated within the follicle cells. Thus, the lethality associated with these mutations can be bypassed, allowing one to determine their function in the adult ovary.

Data recording. For each cross, examine the chorions deposited on the first juice–agar plate at the highest magnification available on the dissecting microscope. Compare the chorion phenotypes between crosses and with Figure 21.2.

EXPECTED RESULTS

DISRUPTION OF EGFR SIGNALING IN THE EYE (EXPERIMENT 1)

During eye development, EGFR signaling functions in many cellular decisions, such as the establishment of photoreceptor cell fates, cell survival, and mitotic progression. Misexpression of the dominant negative EGFR reduces EGFR signaling, leading to increased levels of cell death, loss of photoreceptors, and reduced proliferation. Combined, these effects result in dramatic alterations to the highly ordered ommatidial array of the adult eye, including a significant decrease in the overall size. Likewise, alterations in the structure of the adult compound eye will be caused by unregulated activation of the receptor, which triggers inappropriate cellular decisions.

DISRUPTION OF EGFR SIGNALING DURING OOGENESIS (EXPERIMENT 2)

Activation of the EGFR in follicle cells during stages 8–10 (S8–10) of oogenesis specifies dorsal fates, easily observed by the presence of the two dorsal respiratory appendages on the chorion. In the absence of receptor activity ventral fates are specified, resulting in a loss of the respiratory appendages. Therefore, misexpression of P{UAS-DNEGFR} during S8–10 will result in chorions lacking dorsal respiratory appendages. In contrast, because receptor activation in follicle cells specifies dorsal fates, misexpression of P{UAS − λtop} during S8–10 will result in the specification of dorsal fates throughout the epithelium and the production of ectopic respiratory appendages around the entire anterior circumference of the chorion.

CLONAL ANALYSIS OF VITAL LOCI REVEALS THEIR ROLE IN THE FOLLICLE CELLS (EXPERIMENT 3)

In the last experimental section, clonal analysis is used to reveal that the phenotypic consequences of eliminating the function of these two genes in follicle cells are distinct. Both genes encode molecules that alter the transcriptional profile of cells in response to EGFR signaling. The production of follicle cells homozygous for a mutation in *nej* will result in the appearance of a single respiratory appendage, while the mutation in *bwk* will result in the formation of ectopic respiratory appendages around the anterior circumference of the chorion. In addition to demonstrating a tissue-specific role in EGFR signaling for these genes, these phenotypes reveal the distinct requirements for these gene products in the elaboration of the DV axis.

DISCUSSION

The results from Experiments 1 and 2 indicate that wild-type levels of EGFR signaling are essential for development of the compound eye and the DV axis, respectively. Therefore, signaling via the EGFR pathway is utilized repeatedly throughout development to specify a diverse range of cellular fates. Similarly, the results from Experiment 3 demonstrate that the UAS/GAL4 system can be used to identify additional genes that function downstream of the receptor in a tissue-specific fashion during adult development.

TIME REQUIRED FOR THE EXPERIMENT

➤ The largest time expenditure will involve stock maintenance, preparation, virgin collection, and sorting the F1 progeny.
➤ At 25°C the generation time of *Drosophila* is approximately 10–12 days (Experiments 1 and 3), while at 20°C the generation time will take 15–17 days (Experiment 2).
➤ Identification, collection, and phenotypic characterization of the appropriate genotypes once the F1 has eclosed will require 2–3 hours work for 3–4 days.

POTENTIAL SOURCES OF FAILURE

➤ Inclusion of non-virgins in the mating schemes.
➤ Incorrect selection of F1 for analysis.
➤ In Experiment 2 expression of responders during pre-adult stages caused by the P{CY2-GAL4} driver can result in lethality. To minimize this lethality the crosses should be maintained between 18°–20°C to reduce GAL4 activity during embryonic, larval, and pupal stages.

TEACHING CONCEPTS

Tissue-specific gene function. Often, genes are expressed and function in complex and dynamic patterns throughout development, indicative of their ability to function in multiple processes. Mutations in such genes might produce a phenotype representing a single role, a subset or composite of all these roles. To overcome the developmental complexity associated with genes, genetic methods to alter gene or protein activity in a temporal and spatial fashion have been created. Using such methods, like the UAS/GAL4 system, a researcher can achieve a better understanding of the developmental contributions of a gene through tissue-specific analyses.

Complexity of signaling pathways. The experiments performed in this chapter demonstrate two important concepts. First, signaling pathways often function in multiple spatial and temporal contexts because of the importance of cell–cell communication in cell fate specification. Second, under different developmental contexts the same signaling pathway can specify distinct cell fates. For example, EGFR signaling controls photoreceptor fate during eye development and dorsal follicle cell fates during oogenesis.

ALTERNATIVE EXERCISES

UNDERSTANDING THE TECHNOLOGY – LINKING THE METHODS TO THE OBSERVED RESULTS

In Experiment 2, the temperature at which the crosses are maintained is 20°C. This reduces the activity of GAL4 leading to lower levels of responder expression. To demonstrate this directly, repeat Experiment 1 at three different temperatures (18°C, 25°C, and 29°C). Phenotypes should be less severe at 18°C and increase in severity as the temperature is increased.

QUESTIONS FOR FURTHER ANALYSIS

➤ Do the progeny of any of the crosses display defects in tissues other than those discussed? What might be an explanation for such effects? Is it likely that the GAL4 drivers direct expression solely within the tissue of interest and not in any other tissue or at any other developmental stage? How precisely have the expression domains produced by these GAL4 drivers been characterized?

➤ In Experiment 3, is a single chorion phenotype observed from the selected progeny for each of the crosses or is a range of phenotypes observed? How can a range of phenotypes from a single class of progeny be explained? How much mosaicism is induced during oogenesis?

REFERENCES

Baker, N. E., and Rubin, G. M. (1989). Effect on eye development of dominant mutations in *Drosophila* homologue of the EGF receptor. *Nature*, 340, 150–3.

Brand, A., and Perrimon, N. (1993). Targeted gene expression as a means of altering cell fates and generating dominant phenotypes. *Development*, 118, 401–15.

Duffy, J. B., Harrison, D. A., and Perrimon, N. (1998). Identifying loci required for follicular patterning using directed mosaics. *Development*, 125, 2263–71.

Freeman, M. (1996). Reiterative use of the EGF receptor triggers differentiation of all cell types in the *Drosophila* eye. *Cell*, 87, 651–60.

Golic, K. G., and Lindquist, S. (1989). The FLP recombinase of yeast catalyzes site-specific recombination in the *Drosophila* genome. *Cell*, 59, 499–509.

Price, J. V., Clifford, R. J., and Schupbach, T. (1989). The maternal ventralizing locus *torpedo* is allelic to *faint little ball*, an embryonic lethal, and encodes the *Drosophila* EGF receptor homolog. *Cell*, 56, 1085–92.

Queenan, A. M., and Schupbach, T. (1997). Ectopic activation of *torpedo/EGFr*, a *Drosophila* receptor tyrosine kinase, dorsalizes both the eggshell and embryo. *Development*, 124, 3871–80.

Schejter, E. D., and Shilo, B. Z. (1989). The *Drosophila* EGF receptor homolog (DER) gene is allelic to *faint little ball*, a locus essential for embryonic development. *Cell*, 56, 1093–1104.

Schlessinger, J. (2000). Cell signaling by receptor tyrosine kinases. *Cell*, 103, 211–25.

Schupbach, T. (1987). Germ line and soma cooperate during oogenesis to establish the dorsoventral pattern of the egg shell and embryo in *Drosophila melanogaster*. *Cell*, 49, 699–707.

Spradling, A. (1993). Developmental genetics of oogenesis. In *The Development of Drosophila melanogaster*, eds. M. Bate and A. Martínez-Arias, pp. 1–70. New York: Cold Spring Harbor Laboratory Press.

22 Neurogenesis in *Drosophila*: A genetic approach

C. Klämbt and H. Vaessin

OBJECTIVE OF THE EXPERIMENT The objective of these experiments is to understand some of the major principles underlying neural development by applying classic and modern genetic experimental techniques to this problem. In this chapter we will study (1) the neural and ectodermal phenotypes of different mutations, (2) the use of classic loss-of-function and conditional alleles with modern gain-of-function techniques, and (3) genetic interactions that allow the determination of epistatic relationships among genes.

DEGREE OF DIFFICULTY Moderate. The experiments require some experience in the handling of flies and in the preparation of microscope slides for cuticle preparations and whole-mount antibody staining.

INTRODUCTION

The precisely regulated formation of neurons and glial cells is of obvious importance during development of higher metazoan organisms. Work over the last few decades has led to the initially surprising finding that invertebrates and vertebrates utilize very similar molecular mechanisms to elaborate a functional nervous system.

In all animals, neural tissue develops from the ectoderm. This raises the question of how ectodermal cells are initially routed towards the neural fate and how the correct number of neural founder cells is established in a stereotyped pattern. Pioneering work on *Drosophila* neurogenesis has set the stage for efficient molecular dissection of neural development.

THE PROCESS

During the blastoderm stage, positional information is translated to define the neurogenic ectoderm that resides between the mesodermal anlage and the future dorsal ectoderm (Figure 22.1a). In addition, a domain that will give rise to the brain is set aside as the procephalic ectoderm. Neurogenesis occurs only within these neurogenic

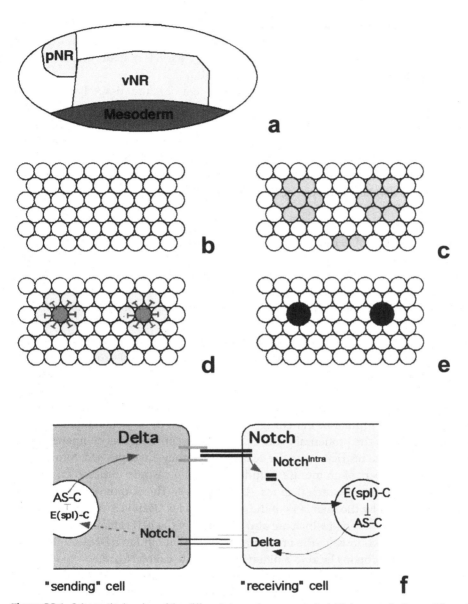

Figure 22.1. Schematic drawing of the different steps of neurogenesis. (a) Fate map of a *Drosophila* embryo. Mesoderm, ventral neurogenic region (vNR) and the procephalic neurogenic region (pNR) are indicated. Dorsal is up anterior to the left. (b–e) Different phases of neural progenitor selection, for details visit Fly*Move* (see p. 284). (f) Molecules involved in mutual/lateral inhibition.

epithelia. Cells initially become larger and individual cells delaminate from the epithelium into the interior of the embryo. These are the so-called neuroblasts; they divide up to 10 times in a stem-cell mode to generate the diversity of neurons and glial cells found in the ventral nerve cord. Within the PNS, individual progenitor cells (sensory organ precursor cells, or SOPs) are selected that again will divide in a stereotyped manner. In contrast to the forming neuroblasts, however, the SOPs initially do not leave the epithelium. Movies and images illustrating the different aspects of neurogenesis can be viewed in Fly*Move* (http://flymove.uni-muenster.de/).

A three-step hypothesis to explain the development of neural tissue has been formulated as a result of numerous experiments on this system. (1) Initially all cells of the neuroectoderm have the ability to acquire neural fate; smaller groups of cells, called proneural clusters, are all able to generate a particular neural stem cell. (2) One cell of this so-called equivalence group is singled out to become the neural stem cell. (3) It simultaneously hinders its neighbors from adopting the primary neural fate so that they take on a secondary, ectodermal, fate (Figures 22.1b–e).

THE GENES

The two main groups of genes involved in the regulation of neurogenesis have been known for quite some time. The first group of genes comprises the so-called proneural genes, which are required to set up neural competence. All proneural genes encode transcriptional regulators of the basic helix-loop-helix (bHLH) family. The most intensely studied proneural genes are members of the *achaete-scute*-complex (AS-C) (see Chapter 23).

Subsequently, individual neural progenitors need to be selected from the pool of cells that initially develop neural competence. This process is controlled by the so-called neurogenic genes. (Note that the name is given to describe the loss-of-function phenotype; thus, neurogenic genes are actually required to suppress neural development, and thus promote epidermal development.) The most well-known member of this gene group is *Notch*. Mutations in neurogenic genes result in very similar phenotypes. This is in contrast to the proneural genes that not only confer neural competence but also impose some fate restrictions to the cells in which they are expressed. Neurogenic genes encode members of an intensely studied signaling cascade centered on the ligand Delta and its receptor Notch (Figures 22.1f and 22.2). The output of this signaling cascade is mediated by the members of the *E(spl)*-complex [*E(spl)*-C], which comprises more than 12 genes, seven of which are also members of the bHLH family of transcriptional regulators. In contrast to the proneural proteins that bind to a DNA-motif called E-box, the E(spl)-C proteins bind to a different target site called N-box.

How do these genes act? Proneural genes encode the first function required for neurogenesis. Their expression conveys the competence to develop as a neural fate. Genetic mosaic studies have demonstrated that proneural genes act cell-autonomously to specify the proneural clusters (Figure 22.1). Within the proneural cluster or equivalence group, the neurogenic genes mediate the process of lateral inhibition. Some of these genes act cell-autonomously (for example, as a transcription factor or receptor), while others function non-autonomously (for example, as a ligand). Again, genetic mosaics demonstrated that for example *Delta* acts non-autonomously whereas *E(spl)* or *Notch* function cell-autonomously. The next obvious question is the order in which these genes act. A first answer to this can be obtained by simple genetic experiments described below.

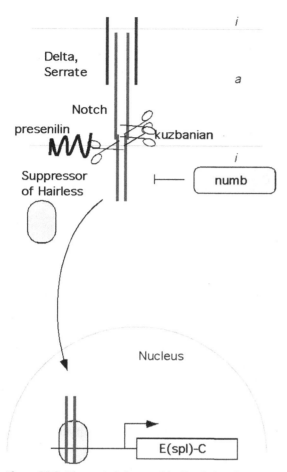

Figure 22.2. The central players of the Notch signaling pathway. **Delta (Dl),** transmembrane protein with 9 EGF-like repeats, binds and activates the Notch receptor. **Serrate (Ser),** transmembrane protein with 11 EGF-like repeats, binds and activates the Notch receptor in non-neural tissues. **Notch (N),** transmembrane protein with 36 EGF-like repeats, receptor of Dl and Ser. A furin type protease cleaves Notch before it reaches the plasma membrane. **kuzbanian (kuz),** an extracellular protease cleaves Notch following binding of Delta. **presenilin (psn),** together with additional proteins such as Nicastrin mediates cleavage of Notch in the membrane. **numb (numb)** counteracts Notch function following cell division. **Suppressor of Hairless (Su(H)),** intracellular protein binds Notch upon activation, translocates to the nucleus. Binds to DNA and acts as a transcription factor. **Enhancer of split-Complex (E(spl)-C),** gene complex comprising 7 bHLH transcription factors. Target of Notch signaling.

MATERIALS AND METHODS

EQUIPMENT AND MATERIALS

Equipment. See Chapter 14.

> Standard size slides, coverslips.
> Scintillation counter tubes to devitillinize the embryos (remove the vitelline membrane).

Table 22.1. Fly stocks

Gene/allele name	Chromosomal location	Bloomington stock #
Oregon R	wild-type	BL-4269
Df(1)AS-C^{B57}	deletion of the region 1B4-1B7on X chromosome	BL-1511
Notch^{55e11} (lof)	3C on X chromosome	BL-3015
Delta^3	92A1 on 3rd chromosome	BL-483
Df(3)E(spl)^{R1}	96F on 3rd chromosome	BL-199
DpE(spl)^+ = Dp(3;3)Su^8	Tandem duplication of the E(spl)-C	BL-3468
UAS fred RNAi	3rd chromosome	from HV
pnr-Gal4	3rd chromosome	BL-3039
neur-lacZ A101, pnr-Gal4	3rd chromosome	from HV
UAS-E(spl)m7	2nd chromosome	from HV

Receptacles for the flies to lay eggs. Any cylindrical receptacle can be used, if it is large enough to include 100–150 flies and it is open at the top or the bottom. A plate with fly food will be installed at the top or the bottom; the flies will lay their eggs on this plate. The plate should fit the open side of the receptacle, and be fixed to it with tape. In this way flies cannot escape and will not be trapped between the receptacle and the plate.

Funnel. Helps to pass the flies from the bottles to the receptacle for laying eggs and vice versa.

Plastic bottle with water. This bottle is used to squirt water onto the plate, helping to remove the eggs from the fly food.

Pasteur pipettes.

Baskets. Small baskets are used to collect the embryos after they are removed from the food. The baskets can be made with 50-ml Falcon tubes, cut about one-third of their length and with a thin gauze "trapped" between the screw top and the tube (see Chapter 11, Figure 11.4).

Pipettemen, and yellow and blue pipette tips.

Microscope with phase-contrast and Normarski optics.

Fly food and maintenance. See Chapter 14.

Cuticle preparation. Egg collection and cuticle preparation; see Chapter 11.

Biological material. Flies can be obtained from the Bloomington Stock Center (URL: http://flystocks.bio.indiana.edu/). The relevant stock numbers are indicated in Table 22.1 or are available from the author (HV).

ANTIBODIES. Rabbit anti-HRP (Jackson Immunoresearch Laboratories). Goat anti-Rabbit-HRP conjugated (Jackson Immunoresearch Laboratories).

PREVIOUS TASKS FOR STAFF

Solutions

Methanol

n-Heptane

Hoyer's medium: This medium is used to observe the embryonic cuticle (it is also used for the adult cuticle). The procedure to make Hoyer's medium is as follows:

> 30 g of arabic gum is added to 50 ml of distilled water, and the solution is stirred overnight until the arabic gum is completely dissolved.
>
> While stirring, 200 g of chloral hydrate is added in small amounts, to prevent the formation of lumps.
>
> Subsequently, 20 ml of glycerol is added.
>
> The mixture is centrifuged until it clarifies and has no particles on it (3 h to overnight at 12,000 *g*).

Note: The use of chloral hydrate may require a special permit. IT IS ESSENTIAL THAT ALL INDIVIDUALS WEAR PROTECTIVE EQUIPMENT SUCH AS GLOVES, LABORA-TORY COAT AND GLASSES AND PERFORM AND FOLLOW THE LOCAL RULES THAT GOVERN THE SAFE HANDLING AND DISPOSAL OF DAB (SEE BELOW) AND CHLO-RAL HYDRATE. Please contact the nearest *Drosophila* lab for help (URL: http://flybase. harvard.edu:7081/people/fbpeople.hform).

Apple and grape juice–agar mix (from DocFrugal Scientific; www.flystuff.com). Possible source for materials and chemicals (except where noted): Fisher Scientific, Inc.

PROCEDURES

Mounting of the embryonic cuticle. Place about 60–70 flies in an egg collection chamber and let the flies lay eggs on apple (or grape) juice plates at room temperature. Change apple juice plates every 24 hours and let the larvae develop for an additional 24–48 hours. Remove larvae from the plate. Collect brown eggs, which will contain the homozygous mutant larvae. Transfer eggs onto a slide, add a drop of 25% bleach, and leave for 5 minutes to remove the chorion. Transfer embryos to a drop of Hoyer's. Cover with coverslip and leave overnight at 65°C.

Immunohistochemical analysis of the embryo

1. Collect embryos into mesh baskets.
2. Dechorionate eggs in 50% bleach for 5 minutes.
3. Rinse with water.
4. Transfer embryos into a glass vial and fix in 5 ml 1 × PBS, 3.7% formaldehyde mixed with 5 ml heptane. Close vial and shake at 300 rpm for 20 minutes.
5. Remove PBS/formaldehyde (bottom layer).
6. Add 5 ml methanol to the vial.
7. Shake vigorously for 30 seconds.
8. Allow phases to separate. Devitellinized embryos sink to bottom.
9. Remove top phase (heptane) and most of bottom phase (methanol).

10. Add 5 ml fresh methanol.
11. Transfer eggs with pipette into microcentrifuge tube.
12. Wash eggs three times with absolute ethanol (embryos can be stored in ethanol at 4°C).
13. Perform 3 half-volume exchanges with H_2O, followed by 1 complete exchange with PBT.
14. Block embryos in 1 ml PBT/BSA (PBT + 1% BSA) 1 hour at RT.
15. Primary antibody: exchange blocking solution with 500 μl fresh PBT/BSA containing a 1:3,000 dilution of rabbit-antiHRP.
16. Incubate 1–2 hours at RT, or overnight at 4°C.
17. Wash 4 times 10 minutes each with PBT/BSA.
18. Secondary antibody: exchange with 500 μl fresh PBT/BSA containing 1:3,000 dilution of goat-anti-Rabbit – HRP; incubate 1 hour at RT.
19. Repeat 17.
20. Exchange with 500 μl 1 mg/ml DAB (diaminobenzidine; adhere to local security regulations when handling DAB). Incubate 3 minutes at RT.
21. Add 500 μl 0.03% H_2O_2 and transfer embryos into a watchglass to observe color development.
22. After sufficient color has developed (15–30 minutes), stop reaction with 4 rapid washes of PBT.
23. Transfer embryos into microcentrifuge tubes and perform 4 half-volume exchanges with 95% ethanol, 2 full-volume exchanges with 100% ethanol, and 1 full-volume exchange with xylenes (solution should be clear; if not, repeat 23).
24. Transfer embryos onto microscope slide. Remove excess xylenes and cover with a drop of Permount mounting medium (Fisher Scientific).
25. Cover with cover slide (24 × 50mm).
26. Incubate slides at 42°C for 1 day. At this time, slides should be stored at RT.

OUTLINE OF THE EXPERIMENTS

EXPERIMENT 1. OBSERVATION OF EMBRYONIC PHENOTYPES

Experimental setup. To begin the experiment, set up overnight collections of embryos of the following genotypes: *Df(1)AS-C, Dl, N, Df(3)E(spl), DpE(spl)*. Change apple juice plates every twelve hours. The first two collections will be used for cuticle preparations. For each plate, count the number of freshly laid eggs and the number of dead embryos when you change the plates. If you are looking at an embryonic lethal phenotype, you expect about 25% dead animals. This number can be higher (can you explain why?) when you are working with certain balancer chromosomes (Greenspan, 1997).

The subsequent egg collections will be used for whole mount staining preparations. Change the collection plate every 12 hours. The "old" plate should remain at 25°C for another three hours (or at 18°C overnight) before fixing and processing the embryos for whole-mount antibody stainings. Alternatively, plates can also be stored [4°C] for a maximum of 48 h before beginning the fixation and processing steps.

Data recording

CUTICLE. On the first egg-collection plates, dead embryos can be easily recognized after two days because of intense melanization. Cuticles will be prepared and will be viewed using phase contrast microscopy.

WHOLE-MOUNT STAINING. To monitor neural development, we will follow expression of the HRP-epitope, a carbohydrate moiety which is expressed on all neuronal membranes and highlights axons as well as dendrites. Within the ventral nerve cord one can easily recognize the segmental commissures and the longitudinal axon tracts that connect the different neuromeres.

One-quarter of the stained embryos will be mutant for the genotype of interest. In a research laboratory setting, one can use balancer chromosomes with dominant markers that confer lacZ or GFP expression, making it easier to identify mutant embryos.

Expected results

CUTICLE. Loss of the *achaete-scute* complex does not lead to alterations in the larval cuticle pattern. In contrast, *Notch* mutant embryos are characterized by a lack of all ventral cuticle, which normally displays a characteristic denticle pattern. The remaining epidermis can be identified as dorsal based on the remains of the trachea and the dorsal hairs.

Loss of *Delta* function gives rise to a similar cuticle phenotype in which the ventral epidermis is missing and only a small piece of dorsal ectoderm can be detected. However, due to its strong maternal component, the *Notch* cuticle phenotype is weaker than the *Delta* phenotype. (Which genetic condition would you analyze to test whether the two mutations can lead to the same phenotype?) Similarly, embryos lacking the *E(spl)*-complex display the more dramatic cuticle phenotype.

When the gene dose of the *E(spl)*-complex is increased by using a chromosomal duplication, no alteration of normal development is observed, indicating that embryos tolerate triplication of this chromosomal region quite well.

WHOLE MOUNT STAINING. Homozygous *Df(1)AS-C* mutants have neural hypoplasia. Note that neurogenesis is not completely blocked. On the other hand, *Notch* mutant embryos are characterized by neural hyperplasia. All cells of the neurogenic ectoderm develop as neural progenitor cells at the expense of epidermal progenitor cells. By using Normarski optics, one can also recognize the lack of ventral epidermis in older embryos. The normally structured PNS is hardly recognizable and only large clumps of neural cells form. Loss of *Delta* or the *E(spl)-C* results in a very similar phenotype. Manipulation of the *E(spl)* copy number does not lead to any abnormal embryonic phenotype.

EXPERIMENT 2. EPISTATIC RELATIONSHIPS

To look for epistatic relationships between individual gene functions, one can use a variety of experimental approaches. The most classic one will be demonstrated in the next experiment. Another approach using the GAL4 system will be introduced in Chapter 23 on adult PNS development.

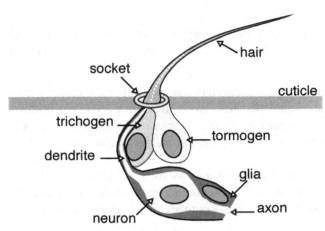

Figure 22.3. Schematic drawing of an external sensory organ.

A simple way to determine epistatic relationships is to increase the copy number of one gene in the mutant background of another gene and to analyze the corresponding phenotype (de la Concha et al., 1988). We will cross *Notch* mutant virgins (why not use males?) to males carrying a *Dp(3,3)E(spl)* chromosome. Eggs will be collected and processed for cuticle preparations.

Data recording. One-quarter of the embryos will be hemizygous for the *Notch* mutation. 50% of the mutant *Notch* embryos will carry an additional copy of the *E(spl)-C*. A larger number of mutant embryos must be analyzed. The extent of the dorsal cuticle shield can be roughly measured in the microscope by judging the % egg length and width.

Expected results. An increase in the *E(spl)-C* copy number weakens the *Notch* mutant phenotype.

EXPERIMENT 3. ADULT PNS DEVELOPMENT

The adult PNS of *Drosophila* consists of a range of sensory organs that provide the fly with environmental information, as well as information regarding body positioning and movement. Adult sensory organs include the compound eye, ocelli, antenna, internal stretch receptors, and many sensory bristles (macrochaeta and microchaeta) that cover the body of the adult fly. Careful examination with a dissection microscope allows identification of the socket, a ring-like structure surrounding the entry point of the hair shaft through the adult cuticle. This socket is formed by a single cell. Below the surface, the hair shaft is connected to the dendrite of a sensory neuron that is surrounded by a glial, or support, cell (Jan and Jan, 1993; Figure 22.3).

Macro- and microchaeta formation initiates in the imaginal discs of larvae and pupae, and involves proneural and neurogenic genes (see also Chapter 23). Several genes have been identified that function in close concert with the Notch signaling pathway. We will utilize one gene, *friend of echinoid* (*fred*; Chandra, Ahmed, and Vaessin, 2003)

P:

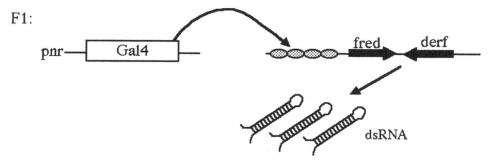

Figure 22.4. Inducible RNAi. The Gal4/UAS system is used for the expression of hairpin dsRNA.

to illustrate how an ectopic expression system can be used to analyze epistatic relation-ships. To analyze the function of *fred* in adult PNS formation, we will use induceable RNA interference (RNAi, see also Chapter 9, Kennerdell and Carthew, 2000) to suppress the expression of the internal *fred* gene. Double-stranded RNA (dsRNA) is expressed via the Gal4/UAS system (see also Chapter 21). To this end, transgenic flies carrying a pUAST construct with a dyad symmetric array of *fred* sequences are crossed to flies carrying a Gal4 construct under the control of a tissue-specific promotor. Transheterozygous progenies, carrying both constructs, will express an RNA that can snap back and thus give rise to a hairpin-loop RNA (Figure 22.4), which mediates the degradation of the corresponding endogenous *fred* mRNA. To determine whether there is an epistatic re-lation between the phenotype associated with the suppression of *fred* gene function and the Notch signaling pathway genes, we will use ectopic expression of genes of the Notch signaling pathway, in parallel with suppression of *fred* gene function. We will use the *pannier* (*pnr*) *Gal4* driver in this experiment. This *Gal4* driver mediates expression in the dorsal part of the wing disc that gives rise to the dorso–central region of the adult thorax, an area rich in sensory organs.

The following crosses are set up at 25°C two to three weeks before data collection:

➤ Cross 1: mate 5 *UAS-fred RNAi/ UAS-fred RNAi* virgin females with 5 *pnr-Gal4/TM6, Tb* males.
➤ Cross 2: mate 5 *UAS-E(spl)m7/ UAS-E(spl)m7; +/+* virgin females with 5 +/+; *pnr-Gal4/TM6, Tb* males.
➤ Cross 3: mate 5 *UAS-E(spl)m7/ UAS-E(spl)m7; UAS-fred RNAi/ UAS-fred RNAi* virgin females with 5 +/+; *pnr-Gal4/TM6, Tb* males.
➤ Cross 4: mate 5 wild-type virgin females with 5 *pnr-Gal4/TM6, Tb* males.
➤ For all crosses, transfer flies into new container when larvae become visible.

Data recording. Isolate F1 flies of the following genotypes for further analysis:

➤ *Cross 1*: identify *UAS-fred RNAi/pnr-Gal4* flies. These flies do not carry the *TM6, Tb* balancer and thus *do not* display the *Tb* phenotype (short, stocky abdomen).

➤ *Cross 2*: identify *UAS-E(spl)m7/+; pnr-Gal4/+* flies. These flies do not carry the *TM6, Tb* balancer.

➤ *Cross 3*: identify *UAS-E(spl)m7/+; UAS-fred RNAi/pnr-Gal4* flies. These flies do not carry the *TM6, Tb* balancer.

➤ *Cross 4*: identify *+/+; pnr-Gal4/+* flies. These flies do not carry the *TM6, Tb* balancer and will serve as controls to account for any potential effects due to Gal4 expression.

Use the dissection microscope to compare and document adult bristle and epidermal phenotypes expressed by flies of the different genotypes. Are epistatic relationships evident?

Expected results. Suppression of *fred* results in an increase of sensory organs in the dorso–central region of the thorax. In addition, holes that are due to the loss of epidermal cells can be observed. Ectopic expression of *E(spl)m7*, mimicking an activated Notch signal, causes suppression of SOP formation and hence absence of sensory organs. Ectopic expression of *E(spl)m7* completely suppresses the *fred* RNAi phenotype (Figure 22.5).

DISCUSSION

Neurogenesis in *Drosophila* requires the function of the proneural genes such as members of the *achaete-scute* complex and *atonal*. Loss of any of the proneural genes results in the loss of neural tissue (mutant phenotype), whereas gain of proneural function leads to an increase of neural progenitor cells. All proneural genes encode members of the bHLH family of transcription factors that need to dimerize in order to function. An important target of the proneural genes is the neurogenic gene *Delta*, which initiates mutual (or, later, lateral) inhibition (Figure 22.1). Vertebrate homologs of the proneural genes (*MASH*, murine achaete scute homologs, and neurogenins) also function in a similar manner during vertebrate neurogenesis.

It is important to note that proneural proteins not only confer neural fate but also impose some fate restrictions. In addition the so-called prepattern genes help to determine progenitor specificity. Once specific progenitor cells (SOP or NB) have been determined, division patterns must be controlled to generate well-defined lineages. In *Drosophila* the lineages of all neuroblasts of the ventral nerve cord have been determined and, using genetic tools, the genes that control the first-born fate versus the second- or third-born fate are being identified.

Loss of any of the neurogenic genes generally results in neural hyperplasia. Neurogenic genes confer mutual or lateral inhibition within a group of cells expressing a given proneural gene and thus restrict the number of proneural progenitor cells. The signaling process implies that the later neural cell inhibits its neighbors from adopting a neural fate. Accordingly neurogenic genes can either act non–cell-autonomously in the

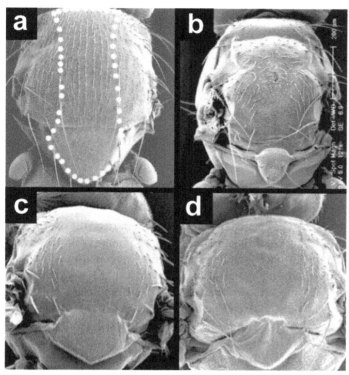

Figure 22.5. Functional interactions of *fred* and *E(spl)m7*. Thorax of (a) +/+; *pnr-Gal4*/+, (b) *UAS-fred RNAi/pnr-Gal4*, (c) *UAS-E(spl)m7*/+; *pnr-Gal4*/+ and (d) *UAS-E(spl)m7*/+; *UAS-fred RNAi/pnr-Gal4* flies. White dots in (a) mark *pnr* expression domain. Reprinted from Chandra et al. (2003). Copyright (2003), with permission from Elsevier.

signaling cell or cell-autonomously in the signal-receiving cell. The neurogenic genes encode a conserved signaling cascade (Figure 22.2) that is used in other developmental contexts and species. The best known neurogenic genes are *Notch* and *Delta*, which encode a receptor and its ligand, respectively. Activation of Notch keeps a cell in an uncommitted state and ensures that only one neural progenitor will develop from one proneural cluster. The membrane-bound Notch receptor undergoes a series of proteolytic cleavages, ultimately releasing its cytoplasmic domain. This domain travels to the nucleus where it acts as a transcriptional regulator. The recent finding that proteases involved in Notch processing are also involved in Alzheimer pathogenesis has further boosted interest in analyzing members of the Notch pathway.

TIME REQUIRED FOR THE EXPERIMENTS

Experiments on embryonic nervous system development can be performed in about 1 week. To perform an experiment addressing adult PNS development, at least 14 days are required before data collection and analyses. Three weeks is the recommended time frame. About 4–6 weeks should be allocated ahead of this time to expand the required stocks and to collect the required virgin flies.

TEACHING CONCEPTS

Genetic control of early neurogenesis. Differences between proneural and neurogenic genes should be learned. Accompanying lectures and seminars should introduce that similar genetic strategies are used in other systems.

Relation of genotype and phenotype.

Simple genetic techniques can be used to determine epistatic relationships.

ALTERNATIVE EXERCISES

PROPOSED EXPERIMENTS

➤ F1 animals can be raised at 29°C. Higher temperatures generally result in stronger activity of the Gal4/UAS system, and thus more pronounced phenotypes. This modification can also be used to shorten the time frame of this exercise.

➤ Analysis of SOP formation in the imaginal discs of 3rd instar larvae can be performed, e.g. utilizing the SOP marker *neur-lacZ A101*. To this end the Gal4 driver line *neur-lacZ A101, pnr-Gal4/TM6, Tb* can be substituted in Crosses 1 to 4. This line carries, in addition to *pnr-Gal4*, the sensory precursor marker *neur-lacZ A101*. Expression of the *neur-lacZ A101* marker can be detected either with X-gal staining or immunohistochemistry using anti-βGal (Cappel). Third instar larvae of the appropriate genotype can be identified due to the absence of *Tb*.

➤ A similar analysis can also be performed for other genes of the Notch signaling pathway such as *Su(H)*, *H*, *Dl*, etc. Appropriate UAS-constructs are available from the Bloomington Drosophila Stock Center (http://fly.bio.indiana.edu).

QUESTIONS FOR FURTHER ANALYSIS

➤ How does the pattern of SOPs change in response to the suppression of *fred* gene function?

➤ How does ectopic expression of Notch signaling pathway genes change the SOP pattern in a *fred* RNAi background?

➤ Determine the respective epistatic relations.

REFERENCES FOR FURTHER READING

Ashburner, M. (1989). *Drosophila, a Laboratory Handbook*. New York: CSHL press.

Campos-Ortega, J. A. (1993). Early neurogenesis in *Drosophila melanogaster*. In *Development of Drosophila*, eds. C. M. Bate and A. Martínez Arias, pp. 1091–1129. New York: Cold Spring Harbor Laboratory Press.

Chandra, S., Ahmed, A., and Vaessin, H. (2003). The *Drosophila* IgC2 domain protein friend-of-echinoid, a paralogue of echinoid, limits the number of sensory organ precursors in the wing disc and interacts with the *Notch* signaling pathway. *Dev. Biol.*, 256, 302–16.

de la Concha, A., Dietrich, U., Weigel, D., and Campos-Ortega, J. A. (1988). Functional interactions of neurogenic genes of *Drosophila melanogaster*. *Genetics*, 118, 499–508.

Greenspan, R. (1997). *Fly Pushing*. New York: CSHL Press.

Isshiki, T., Pearson, B., Holbrook, S., and Doe, C. Q. (2001). *Drosophila* neuroblasts sequentially express transcription factors which specify the temporal identity of their neuronal progeny. *Cell*, 106, 511–21.

Justice, N. J., and Jan, Y. N. (2002). Variations on the Notch pathway in neural development. *Curr. Opin. Neurobiol.*, 12, 64–70.

Jan, Y. N., and Jan, L. Y. (1993). The peripheral nervous system. In *The Development of Drosophila melanogaster*, eds. M. Bate and A. Martínez Arias, pp. 1207–44. New York: Cold Spring Harbor Laboratory Press.

Kennerdell, J. R., and Carthew, R. W. (2000). Heritable gene silencing in *Drosophila* using double-stranded RNA. *Nat. Biotechnology*, 18, 896–8.

Novotny, T., Eiselt, R., and Urban, J. (2002). Hunchback is required for the specification of the early sublineage of neuroblast 7–3 in the *Drosophila* central nervous system. *Development*, 129, 1027–36.

Weigmann, K., Klapper, R., Strasser, T., Rickert, C., Technau, G. M., Jäckle, H., Janning, W., and Klämbt, C. (2003). FlyMove – a new way to look at development of *Drosophila*. *Trends in Genetics*, 19, 310–1. URL: http://flymove.uni-muenster.de/.

23 Role of the *achaete-scute* complex genes in the development of the adult peripheral nervous system of *Drosophila melanogaster*

S. Sotillos and S. Campuzano*

OBJECTIVE OF THE EXPERIMENT Formation of the pattern of external sensory organs (SOs) that cover the adult cuticle of *Drosophila melanogaster* is controlled by the proneural *achaete* (*ac*) and *scute* (*sc*) genes of the *achaete-scute complex* (*AS-C*) (reviewed in Campuzano and Modolell, 1992; Modolell and Campuzano, 1998). *ac* and *sc* encode transcription factors of the basic helix-loop-helix (bHLH) family. Both genes are co-expressed in the imaginal discs (the precursors of the adult epidermis) in a reproducible pattern of clusters of cells (the proneural clusters) from which one cell is determined to become the precursor cell of the SO, the sensillum mother cell (SMC). The objectives of the experiments outlined in this chapter are to demonstrate the requirement of the spatially restricted expression of *ac/sc* for the generation of the wild-type pattern of SOs and to analyze the control of *ac/sc* expression.

DEGREE OF DIFFICULTY The experiments are moderately difficult.

INTRODUCTION

The peripheral nervous system of *Drosophila* is a very suitable model in which to analyze the genetic control of pattern formation. The cuticle of *Drosophila melanogaster* contains thousands of external mechano- or chemoreceptor SOs – the bristles or chaetae and other sensilla – which are arranged according to a stereotyped pattern (Figure 23.1a). Each external SO is derived from a single cell, the sensillum mother cell (SMC), which, in general, undergoes two differential divisions. Each of the four resulting cells differentiates into one component of the SO: the bristle or another type of external structure, the socket, a neuron and a support cell. SMCs for the wing and thorax SOs are specified within the population of epithelial cells of the imaginal wing discs (Figure 23.1b). SMCs can be visualized by their β-galactosidase expression in the imaginal discs

* Corresponding author

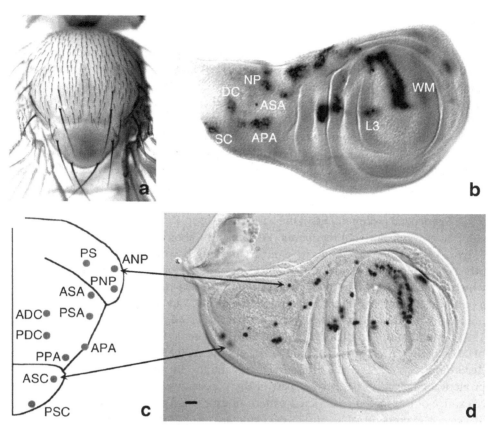

Figure 23.1. (a) Mesothorax (notum) of a wild-type *Drosophila melanogaster* adult fly. (b) Pattern of expression of *scute* in the imaginal wing disc detected by whole-mount staining with anti-SC antibody. (c) Scheme of an heminotum showing the position of the macrochaetae. Microchaetae are not indicated. Nomenclature of the macrochaetae and of their corresponding proneural clusters: ADC, anterior dorsocentral; ANP, anterior notopleural; ASA, anterior supralar; APA, anterior postalar, ASC, anterior scutellar, PDC, posterior dorsocentral; PNP, posterior notopleural; PPA, posterior postalar; PSA, posterior supralar; PS, presutural; PSC, posterior scutellar. L3, sensilla of the longitudinal vein L3. AWM, anterior wing margin. (d) SMCs pattern in the imaginal wing disc from *neuralized-lacZ* third instar larvae detected by X-gal staining. Arrows indicate the correspondence between certain macrochaetae and their SMCs. Bar = 10μm.

of *neuralized-lacZ* (*neur-lacZ*) larvae (Figure 23.1d). These larvae carry a transposon containing the *E. coli lacZ* gene inserted at the *neuralized* locus, which causes it to be expressed in the SMCs, in the same pattern as the *neuralized* gene. Staining with the chromogenic β-galactosidase substrate X-gal allows visualization of the SMCs. SMCs appear at highly reproducible positions in the imaginal discs. Thus, the problem of SO positioning resides in understanding the molecular basis for the site-specific determination of the SMCs in the imaginal discs.

Pattern formation during development relies on the spatially and temporally restricted expression of genes. Formation of the pattern of external SOs is controlled by the proneural *ac* and *sc* genes of the *ac-sc* complex (reviewed in Campuzano and Modolell, 1992; Modolell and Campuzano, 1998). *ac/sc* mutations were initially isolated by their loss-of-function phenotype, the loss of bristles. This showed that *ac/sc* are required for bristle development. Other mutations, *Hairy-wing*, cause development of extra SOs at ectopic positions. The *Hairy-wing* mutations have been identified as gain-of-function *ac/sc* mutations, providing further evidence that *ac/sc* play a role in the determination of the SMCs (García-Bellido, 1979; Campuzano et al., 1986; García-Alonso and García-Bellido, 1986; Balcells, Modolell, and Ruiz-Gómez, 1988). In the following set of experiments, we will first analyze the wild-type pattern of SOs in the mesothorax and that of the SMCs in the imaginal wing disc and will establish the correspondence between the adult and imaginal disc phenotypes by comparing the bristles and the SMCs that develop in several *ac/sc* loss-of-function mutations.

   ac/sc encoded proteins contain the basic Helix-Loop-Helix (bHLH) motif that defines a family of transcriptional regulators (Villares and Cabrera, 1987). This motif consists of two α-helices, separated by a loop of variable length, which allow dimerization with other members of the family and an adjacent basic region required for DNA binding, which confers binding specificity. Dimerization is independent of DNA binding but is a prerequisite for DNA binding. *ac/sc* are co-expressed in small clusters of cells in the imaginal wing disc from which SMCs are later selected (Figure 23.1b) (Cubas et al., 1991; Skeath and Carroll, 1991). AC/SC bHLH proteins form heterodimers with the bHLH protein Daughterless, and confer on cells the ability to become neural precursors. Accordingly, *ac/sc* are named proneural genes and the clusters of cells expressing *ac/sc*, proneural clusters. In the second part of the practice we will confirm the functional relevance of the spatially restricted pattern of expression of *sc* by looking at the effect of ectopic expression of *sc* on SO development. As we will see, the experimentally induced, generalized *sc* expression leads to the development of extra SOs at ectopic positions.

   One question to pursue about *ac/sc* is: how is their expression controlled? Genetic and molecular analyses have shown that expression of *ac/sc* is controlled by *cis*-acting enhancer elements, interspersed along the more than 100 kb that encompass the AS-C. (Ruiz-Gómez and Modolell, 1987; Gómez-Skarmeta et al., 1995; Culí and Modolell, 1998; García-García et al., 1999; Figure 23.2). Enhancers provide the binding sites for transcriptional regulators, which are thought to be expressed in the imaginal disc in broad regions. Thus, different enhancers will respond to a combination of these transcriptional regulators. The genes coding for the regulators of *ac/sc* expression belong to the class of the prepattern genes. Some of them, such as the genes of the *Iroquois* complex, *pannier* (*pnr*) and *u-shaped*, have already been identified (Gómez-Skarmeta et al., 1996; Cubadda et al., 1997; García-García et al., 1999; Figure 23.2). In the third part of this exercise, we will visualize the ability of a fragment of the genomic AS-C DNA to drive expression of a reporter gene in one of the proneural clusters, the dorsocentral (DC) cluster. This fragment of DNA contains the "enhancer element" that drives *ac/sc* expression in the DC proneural cluster (García-García et al., 1999). Finally, we will demonstrate the role of the *ac/sc* regulator Pannier in the activation of *ac/sc* expression through the DC enhancer.

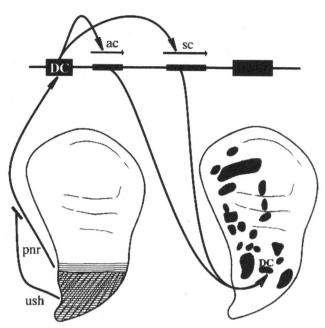

Figure 23.2. Control of *ac/sc* expression. Co-expression of *ac* and *sc* in the different proneural clusters (shown in the imaginal disc on the right) responds to a combination of transcriptional factors, the products of the prepattern genes, which act through the *cis*-control enhancer elements (black boxes). The imaginal disc on the left shows the pattern of expression of two prepattern genes: *pannier* (horizontal hatching) and *u-shaped* (transversal hatching). The DC- enhancer will be only active in cells expressing *pannier* and devoid of U-shaped.

MATERIALS AND METHODS

EQUIPMENT AND MATERIALS

The equipment necessary for handling the flies and the fly food is described in Chapter 14.

Per student

> Dissecting microscope
> Cold-light lamp
> Compound microscope
> Timer
> Slides with a shallow depression
> Slides, 24 × 24 mm coverslips
> Glass wells
> A pair of dissection forceps Dumont #5 (Fine Science Tools)
> Fine tungsten wire needles attached to any kind of holder (we currently fix them
>   to a glass rod)
> Tissue culture 24-well plates
> Pasteur pipettes

Eppendorff tubes

5-ml tubes

Gilson pipettes (p20, p200, and p1,000) and tips

Baskets to transfer the imaginal disc complexes from one solution to another. (To make them, cut in half a 1.5-ml Eppendorff tube with a prewarmed razor blade. Discard the lower conical half. Remove the lid. Cut a small piece of metal mesh and fit it onto one side of the tube using a hot plate to melt the plastic onto the metal mesh.)

Per practical group

Water bath

Eppendorff microfuge

Hot plate

Oven

Rocker platform

Microscope with photographic equipment

Biological material

FLY STRAINS. The *Drosophila melanogaster* strains used are:

1. For the phenotypic analysis of *ac/sc*
    ac^3
    sc^7
    sc^6
    sc^4
2. For SMC visualization
    neuralized-lacZ/ TM6B
3. For ectopic *sc* expression
    Heat-shock promoter-sc (HS-sc)
4. For the analysis of the *cis*-control of *sc* expression
    AS 1.4 DC-lacZ
5. For the analysis of the function of *pnr* on *sc* expression
    pannier-Gal4
    UAS-lacZ
    UAS-pnr/CyO
    C-734 Gal4/SM6a-TM6b/AS 1.4 DC-lacZ
    AS 1.4 DC-LacZ/CyO; pnr^{vx6} /TM6B

STOCK REQUEST. Flies can be obtained from the authors or from international stock centers (see Chapter 14).

REAGENTS AND KITS

*N,N,*Dimethyl formamide (Merck)   Potassium ferrocyanide (Merck)

X-Gal (Promega)   Potassium ferricyanide (Merck)

Glycerol (Merck)   Triton X100 (Roche)

Glutaraldehyde (Fluka)

PREVIOUS TASKS FOR STAFF

Maintenance of fly strains and preparation of fly food are described in Chapter 14.

Solutions. All solutions are prepared by dissolving the reagents in distilled water except when indicated.

10 × PBS (phosphate buffered saline 10 times concentrated): 1.3 M NaCl, 70 mM Na_2HPO_4, 30 mM NaH_2PO_4

1 × PBS

100 mM Potassium ferrocyanide

100 mM Potassium ferricyanide

10% Triton × 100.

8% X-gal. Prepare a solution of 80 mg/ml of X-gal in dimethyl formamide in an Eppendorff tube. Wrap it in aluminum foil and store it at −20°C. HANDLE THE TUBE WITH GLOVES.

PROCEDURES

X-gal staining to reveal β-galactosidase activity

1. Place a 24-well tissue culture plate in a container with ice. Fill one of the wells with 1 ml PBS and place one of the baskets in the PBS solution. Also fill a glass well with PBS.
2. To begin the dissection, take a third instar larvae (crawling larvae) from the wall of the food vial and wash it briefly in PBS in the glass well.
3. Transfer the larvae to a slide containing a drop of PBS.
4. Under the stereomicroscope, use the dissecting forceps to dissect the imaginal discs (a black background is essential to see the discs). Hold the head of the larva below the mouth hooks with one of the forceps and pull the hooks away from the rest of the body using the other pair of forceps. The imaginal discs will remain in the "cephalic" one-third of the larva along with the brain and the salivary glands. Do not try to clean the discs because they will be lost during the subsequent procedures if they are not attached to the larval body wall and to the trachea.
5. Transfer the larval head to the basket immersed in cold PBS and repeat the procedure with the next larva.
6. Dissect larvae for no more than 20 min. The following incubations will be preformed in the multi-well plate, with gentle shaking, filling successive wells with the indicated solutions (1 ml save when indicated), and transferring the basket from one well to the next one.
7. Fixation. Place larval heads in 0.5% glutaraldehyde in PBS 2 min at 4°C. (Place the plate over ice and in the hood.)
8. Wash larval heads two times 5 min with 1 × PBS.
9. While washing, prepare the staining solution:

Mix in an Eppendorff tube

100 mM Potassium ferrocyanide	25 μl
100 mM Potassium ferricyanide	25 μl
1 × PBS	422 μl
10% Triton X100	15 μl

Incubate 2 min at 37°C in a water bath.

Add 12.5 μl of 8% X-gal.

Incubate 2 min at 37°C.

Centrifuge 1 min.

Transfer the supernatant to a well of the plate.

10. Place the basket containing the heads in this solution, seal the plate with Parafilm and incubate at 37°C in the oven.

11. Check appearance of the characteristic blue staining under the stereomicroscope after 15 min, 1 h, 2 h, and, if undetectable, leave it overnight.

12. To stop the reaction, wash three times for 5 min each with 1 × PBS.

Mounting of the discs

1. Transfer the larval heads to 50% glycerol.
2. Place a drop of 90% glycerol on a slide and transfer one of the heads to the drop.
3. Under the stereomicroscope, look for the pair of imaginal wing discs.
4. Clean them of other larval material with the help of the fine tungsten needle and transfer them to another slide with 90% glycerol.
5. Proceed with another head. When there are several discs on the second slide, cover the glycerol drop with a coverslip and flatten the discs by placing over the coverslip a 5-ml tube filled with water. Let it stand for 2 hours.

Heat shock treatment to staged larvae and pupae. Flies are reared at 25°C. *HS-sc* flies are allowed to lay eggs for a 24 h period (for example, from Monday morning to Tuesday morning) (vial 1) and then they are transferred to a new vial every 24 h, that is on Wednesday to vial 2, and on Thursday to vial 3. Using this setup, by Friday morning there will be three vials with larvae of the following developmental ages:

vial 3   24–48 h after egg laying (AEL)

vial 2   48–72 h AEL

vial 1   72–96 h AEL

On Friday morning, subject all of these larvae to the heat shock regime (1 h at 37°C, 2 h at 25°C, 1 h at 37°C, 2 h at 25°C, 1 h at 37°C) by immersing the vials, tied with elastic bands to a grid for 50-ml Falcon tubes, in a water bath. Be sure that the vials are covered in water up to the cotton plug because the larvae tend to avoid the heat shock by climbing to the upper part of the tube.

To obtain older staged larvae/pupae, *HS-sc* flies are allowed to lay eggs during 24 h (from Tuesday to Wednesday, vial 4). On Wednesday morning adults are transferred to another vial (vial 5). Once again adults are removed after 24 h. On the following Monday, vial 5 will contain 96–120 h AEL larvae and vial 4 will contain 0–24 h old pupae. At this point, the vials can be subjected to the heat shock regime above indicated.

OUTLINE OF THE EXPERIMENTS

EXPERIMENT 1. ADULT PHENOTYPE OF *achaete-scute* MUTANTS

Method: Direct observation of the flies under the stereomicroscope. Observe the adult pattern of SOs in the notum of wild-type flies and compare it with Figures 23.1a and

23.1c to become familiar with the arrangement of the bristles (macrochaetae). Note the regular arrangement in rows and columns of the small bristles (microchaetae). Next, look at the bristle phenotype of several *ac/sc* mutants (*ac³*, *sc⁷*, *sc⁶*, *sc⁴*).

Data recording. At least 10 flies (20 heminota) of each genotype will be examined. Use copies of the scheme of Figure 23.1c to record the bristles present in every heminotum. Record the frequency of appearance of the different macrochaetae (this enables you to establish the penetrance of the phenotype). Note that different *ac/sc* mutations affect the development of different SOs.

Conclusion. The *achaete-scute* genes are required for the development of the adult SOs.

EXPERIMENT 2. DEVELOPMENT OF SMCs IN *achaete-scute* MUTANTS

Method. The different *ac/sc* mutants indicated above will be mated with the *neur-lacZ* stock. Development of SMCs will then be followed by X-gal staining of the corresponding imaginal wing discs.

The *ac/sc* complex is located on the X chromosome and the *neur-lacZ* insertion is located in the third chromosome. *neuralized-lacZ* is homozygous lethal and maintained in a heterozygous condition over the balancer chromosome *TM6B*. This chromosome carries the dominant mutation *Tubby* (*Tb*) that results in larvae that are shorter than the wild type larvae.

The general scheme of the crosses will be as follows:

5 virgin females of every *ac/sc* mutant (for instance *sc⁶*) will be mated with 5 *neur-lacZ/TM6B* males in a small fly-food vial.

Adults will be changed to a new vial after two days at 25°C.

On the fifth day after egg laying (approximately 100–120 hours AEL) larvae will start crawling up the walls of the vials. These larvae can then be analyzed for the *lacZ* pattern in the imaginal discs.

The genotype of the progeny of the cross will be:

sc⁶/sc⁶ females X *sc⁺/Y; neur-lacZ/TM6B* males
sc⁶/sc⁺; neur-lacZ/+ females (normal size)
sc⁶/sc⁺; TM6B/+ females (shorter than normal)
sc⁶/Y; neur-lacZ/+ males (normal size)
sc⁶/Y; TM6B/+ males (shorter than normal)

Since *ac* and *sc* mutations are recessive, *sc⁶/sc⁺; neur-lacZ/+* female larvae can be considered as the control. The experimental class is the *sc⁶/Y; neur-lacZ/+* male larvae.

Remove third instar larvae from the wall of the vials. Separate male and female larvae by looking at the gonads, a pair of bilateral rounded structures located in the abdominal region and which are bigger in males than in females, and afterwards select the Non-*Tb* larvae for X-gal staining (see described above protocol).

Data recording. Once the imaginal discs are mounted, compare the pattern of SMCs in wing discs from *sc⁶/Y; neur-lacZ/+* male larvae with that shown in Figure 23.1d. Record the positions of the SMCs on a schematic of the imaginal wing disc (like that shown in Figure 23.1d) for the control larvae.

Next, compare the pattern of SMCs between sc^6/Y; *neur-lacZ*/ + and sc^6/sc^+; *neur-lacZ*/ + larvae. Record the positions of the SMCs on a schematic of the imaginal wing disc for the *ac/sc* mutants.

Expected results. Most of the control wing discs will display the pattern of β-*galactosidase* expressing cells shown in Figure 23.1d. However, some imaginal discs may have fewer or more SMCs than the one shown in Figure 23.1d. This is due to the sequential appearance of the SMCs. Thus, progressively older imaginal discs show more new SMCs, while the earliest appearing ones, and their progeny, continue expressing *lacZ*. The earliest appearing SMCs are PDC, PSC, and APA. The latest appearing, not shown in Figure 23.1d because they develop around pupariation, are the PS, ASA, and PPA (see Cubas *et al.*, (1991) for more details).

In accordance with the adult phenotype, note the absence of different SMCs in the different *ac/sc* mutations.

Conclusion. *ac/sc* genes are required for the determination and/or differentiation of the SMCs.

EXPERIMENT 3. PRONEURAL ROLE OF *sc*

ac/sc genes are expressed in the wing imaginal discs in defined clusters of cells, the proneural clusters (Figure 23.1b). In this experiment, the effect on SO development of a generalized over-expression of *sc* can be observed.

Method. Transgenic flies harbouring the *sc* gene under the control of the heat shock promoter (*HS-sc* flies, Rodríguez et al., 1990) will be subjected to a heat shock regime, as indicated above, at different stages of larval and pupal development as follows:

    24–48 h after egg laying (AEL)
    48–72 h AEL
    72–96 h AEL
    96–120 h AEL
    0–24 h old pupae

Data recording. Adult flies will be examined for the presence of SOs.

Expected results. The generalized and ectopic expression of *sc* causes the development of extra SOs in the notum only when it is expressed in late third instar larvae (96–120 h AEL) or one-day-old pupae. Note the different time windows (competence periods) for the induction of extra macrochaetae or extra microchaetae. Note the different sensitivity of the different regions of the body to the presence of *sc* (see Rodríguez et al., 1990 for more details).

Conclusion. The spatially restricted expression of *sc* is required for the development of the wild-type pattern of SOs.

EXPERIMENT 4. *CIS*-REGULATION OF *ac/sc* EXPRESSION: THE DC ENHANCER

The expression of *ac/sc* is controlled by *cis*-acting enhancer-like elements that drive *ac/sc* expression in the proneural clusters (Ruiz-Gómez and Modolell, 1987). Transgenic flies that are carrying fragments of the *AS-C* genomic DNA (including enhancer elements) cloned upstream of the bacterial *lacZ* gene express *lacZ* under the control of these enhancer elements. Thus, the *lacZ* expression is seen at the position of individual proneural clusters (Gómez-Skarmeta et al., 1995; Culí and Modolell, 1998; García-García et al., 1999). As an example, we will examine the pattern of expression of *lacZ* in the imaginal wing disc driven by a 1.4 kb fragment of the *AS-C* DNA that contains the dorsocentral (DC) enhancer.

Method. X-gal staining of the imaginal wing discs of *AS 1.4 DC-lacZ* third instar larvae.

Data recording. Take pictures of the stained imaginal discs.

Expected results. Wing discs will display only a cluster of stained cells located at the position of the DC proneural cluster (DC in Figure 23.1b).

EXPERIMENT 5. *TRANS*-REGULATION OF *ac/sc* EXPRESSION: ACTIVATION OF THE DC ENHANCER BY PANNIER

Pannier(Pnr), a transcription factor of the GATA-1 family, is one of the prepattern factors that are expressed in broad regions of the imaginal discs and control the more restricted expression of *ac/sc* (García-García et al., 1999). The expression domain of *pnr* encompasses the most medial part of the presumptive notum and includes the DC proneural cluster (Figure 23.2). We will analyze the dependence of the development of the DC macrochaetae on *pnr* activity.

EXPERIMENT 5A. EFFECT OF *pnr* LOSS-OF-FUNCTION ON BOTH THE ADULT PATTERN OF SOS AND THE DC ENHANCER

Method. *pnr-Gal4/pnr^{vx6}* is a viable hypomorphic *pnr* combination.

1. Set up the cross of *AS 1.4 DC-LacZ/SM6a-TM6b/pnr^{vx6}* females with *pnr-Gal4* males. (The *SM6a-TM6b* balancer chromosome carries the dominant markers *CyO* and *Tb*).
2. Stain the imaginal discs from the resulting non-*Tb* third instar larvae with X-gal. Do not dissect all of the larvae. Let some of them develop into adulthood.
3. Observe the *AS 1.4 DC-LacZ; pnr-Gal4/pnr^{vx6}* adult flies (*CyO$^+$ Tb$^+$*).

Data recording. Take pictures of the stained imaginal discs.

Expected results. Adult flies will lack DC bristles. *DC-lacZ* expression will be undetectable in the imaginal wing disc.

EXPERIMENT 5B. EFFECT OF THE OVER-EXPRESSION OF *pnr* ON BOTH THE ADULT PATTERN OF SO AND THE DC ENHANCER

Method: Gal4/UAS ectopic expression of *pnr*. In the Gal4 technique (Brand and Perrimon, 1993), Gal4, a yeast transcription factor, activates the transcription of any gene containing in its promoter region the Gal4 binding sequences, denoted as UAS. Two transgenic lines of flies are established; one containing Gal4, driven by a tissue-specific promoter; the other containing UAS linked to the gene of interest. To visualize where the Gal4 promoter is driving ectopic expression, another line of flies is used: *UAS-lacZ*. In this line, *lacZ* will be expressed wherever Gal 4 is translated and binds to the UAS. In this experiment, we will use the *C-734 Gal4* line. To visualize the domain of expression driven by *C-734 Gal4*:

> Set up the cross of *C-734 Gal4* females with *UAS-lacZ* males.
> Stain the imaginal discs from the resulting third instar larvae with X-gal.

> Simultaneously set up the cross of *C-734 Gal4/SM6a-TM6b/AS 1.4 DC-lacZ* females with *UAS-pnr* males and proceed as in Experiment 5a, steps 2 and 3. In these larvae, *pnr* will be expressed in the same pattern as *lacZ* in *C-734 Gal4/UAS-lacZ* larvae.

Expected results. Adult flies will present extra DC bristles. *DC-lacZ* expression will be expanded in the imaginal wing disc.

Conclusion. Pnr is required for DC development because it regulates *ac/sc* expression through the DC enhancer. Note, however, that Pnr does not activate the DC-enhancer in all of the cells where it is expressed (Figure 23.2). The activity of Pnr is repressed in the most proximal part of the wing disc by heterodimerization with the transcription factor U-shaped (Figure 23.2, Haenlin et al., 1997, and García-García et al., 1999.

GENERAL EXPECTED RESULTS

> *ac/sc* mutations cause the loss of adult external SOs and of their corresponding SMCs.
> Generalized expression of *sc* leads to the development of extra ectopic SOs.
> Loss-of-function of *pnr* eliminates the DC macrochaetae and the expression of the reporter gene driven by the DC enhancer.
> Over-expression of *pnr* produces development of extra DC macrochaetae and expands the domain of expression of the reporter gene under the control of the DC enhancer.

DISCUSSION

The results described above indicate that *ac/sc* genes are required for the selection of the SMCs from the population of undifferentiated cells of the imaginal wing discs. A restricted expression of *ac/sc* in the proneural clusters is required for the differentiation of the SMCs at their specific positions, and hence for the development of the wild-type pattern of external SOs. Such expression is controlled by *cis*-acting regulatory enhancer elements, activated by a combination of transcription factors present in the imaginal disc that constitute the prepattern. Pnr is one of the components of the prepattern and regulates *ac/sc* expression through the DC enhancer element.

TIME REQUIRED FOR THE EXPERIMENTS

Three weeks. Staff may amplify stock previous to the practice. During the first week, students will set up the crosses and let the HS-sc flies lay eggs as indicated. On Friday, they will subject the staged larvae to the heat shock. In the second week, they will continue with the heat shock experiment and carry out the X-gal staining of the imaginal discs. During the end of the second week and the third week, they will look at the adult phenotypes and analyze and discuss the results obtained.

Potential sources of failure. Care should be taken to differentiate male from female larvae in experiment 2, to select Tb from Tb^+ larvae, and to properly dissect and mount the imaginal discs.

TEACHING CONCEPTS

Proneural genes. Genes whose expression confers on cells the ability to become neural precursor cells.

Proneural clusters. Groups of cells expressing the proneural genes.

Neural commitment. Selection of a cell to become a neural precursor from a population of undifferentiated cells.

***Cis*-regulation of gene expression.** Regulation of gene expression by adjacent genomic sequences.

***Trans*-regulation of gene expression.** Regulation of gene expression by transcription factors.

Prepattern genes. Genes expressed in domains broader than that of the proneural genes. These genes regulate proneural gene expression.

Regulation of the activity of transcription factors by heterodimerization. Heterodimerization of a transcription factor with another one may change the former from a transcriptional activator to a repressor.

ALTERNATIVE EXERCISES

QUESTIONS FOR FURTHER ANALYSIS

➤ Are all cuticular cells always competent for chaetae formation? How could you refute this?

➤ Does Pannier regulate the development of bristles other than the dorsocentral bristles?

ADDITIONAL INFORMATION

bHLH proteins control neurogenesis as well as myogenesis and are evolutionarily conserved (reviewed in Jan and Jan, 1993 and Bertrand, Castro, and Guillemot, 2002).

REFERENCES

Balcells, L., Modolell, J., and Ruiz-Gómez, M. (1988). A unitary basis for different *Hairy-wing* mutations of *Drosophila melanogaster. EMBO J.,* 7, 3899–906.

Bertrand, N., Castro, D. S., and Guillemot, F. (2002). Proneural genes and the specification of neural cell types. *Nat. Rev. Neurosci.,* 3, 517–30.

Brand, A. H., and Perrimon, N. (1993). Targeted gene expression as a means of altering cell fates and generating dominant phenotypes. *Development,* 118, 401–15.

Campuzano, S., Balcells, L., Villares, R., Carramolino, L., García-Alonso, L., and Modolell, J. (1986). Excess function *Hairy-wing* mutations caused by *gypsy* and *copia* insertions within structural genes of the *achaete-scute* locus of *Drosophila. Cell,* 44, 303–12.

Campuzano, S., and Modolell, J. (1992). Patterning of the *Drosophila* nervous system: The *achaete-scute* gene complex. *Trends Genet.,* 8, 202–7.

Cubas, P., de Celis, J. F., Campuzano, S., and Modolell, J. (1991). Proneural clusters of *achaete-scute* expression and the generation of sensory organs in the *Drosophila* imaginal wing disc. *Genes Dev.,* 5, 996–1008.

Cubadda, Y., Heitzler, P., Ray, R. P., Bourouis, M., Ramain, P., Gelbart, W., Simpson, P., and Haenlin, M. (1997). *u-shaped* encodes a zinc finger protein that regulates the proneural genes *achaete* and *scute* during formation of bristles in *Drosophila. Genes Dev.,* 11, 3083–95.

Culí, J., and Modolell, J. (1998). Proneural gene self-stimulation in neural precursors: An essential mechanism for sense organ development that is regulated by *Notch* signaling. *Genes Dev.,* 12, 2036–47.

García-Bellido, A. (1979). Genetic analysis of the *achaete-scute* system of *Drosophila melanogaster. Genetics,* 91, 491–520.

García-Alonso, L., and García-Bellido, A. (1986). Genetic analysis of the *hairy-wing* mutations. *Roux's Arch. Dev. Biol.,* 195, 259–64.

García-García, M. J., Ramain, P., Simpson, P., and Modolell, J. (1999). Different contributions of *pannier* and *wingless* to the patterning of the dorsal mesothorax of *Drosophila. Development,* 126, 3523–32.

Gómez-Skarmeta, J. L., Rodríguez, I., Martínez, C., Culí, J., Ferrés-Marcó, M. D., Beamonte, D., and Modolell, J. (1995). Cis-regulation of *achaete* and *scute*: Shared enhancer-like elements drive their coexpression in proneural clusters of the imaginal discs. *Genes Dev.,* 9, 1869–82.

Gómez-Skarmeta, J. L., Diez del Corral, R., de la Calle-Mustienes, E., Ferrés-Marcó, D., and Modolell, J. (1996). *araucan* and *caupolican*, two members of the novel Iroquois complex, encode homeoproteins that control proneural and vein forming genes. *Cell,* 85, 95–105.

Haenlin, M., Cubadda, Y., Blondeau, F., Heitzler, P., Lutz, Y., Simpson, P., and Ramain, P. (1997). Transcriptional activity of Pannier is regulated negatively by heterodimerization of the GATA DNA-binding domain with a cofactor encoded by the *u-shaped* gene of *Drosophila. Genes Dev.,* 11, 3096–108.

Jan, Y. N., and Jan, L. Y. (1993). HLH proteins, fly neurogenesis and vertebrate myogenesis. *Cell,* 75, 827–30.

Modolell, J., and Campuzano, S. (1998). The *achaete-scute* complex as an integrating device. *Int. J. Dev. Biol.,* 42, 275–82.

Rodríguez, I., Hernández, R., Modolell, J., and Ruiz-Gómez, M. (1990). Competence to develop sensory organs is temporally and spatially regulated in *Drosophila* epidermal primordia. *EMBO J.,* 9, 3583–92.

Ruiz-Gómez, M., and Modolell, J. (1987). Deletion analysis of the *achaete-scute* locus of *D. melanogaster. Genes Dev.*, 1, 1238–46.

Skeath, J. B., and Carroll, S. B. (1991). Regulation of *achaete-scute* gene expression and sensory organ pattern formation in the *Drosophila* wing. *Genes Dev.*, 5, 984–95.

Villares, R., and Cabrera, C. V. (1987). The *achaete-scute* gene complex of *D. melanogaster.* Conserved domains in a subset of genes required for neurogenesis and their homology to myc. *Cell,* 50, 415–24.

24 The conservation of the genome and nuclear reprogramming in *Xenopus*

J. B. Gurdon

OBJECTIVE OF THE EXPERIMENT A fundamental question in developmental biology is whether the processes of development and cell differentiation involve a stable change in the genome of cells, as was thought to be the case many years ago. An alternative idea is that the genome remains constant in cells of all different types and that the readout of the genome, namely transcription and translation, is modified according to cell type.

The first experiments to address this question and to provide most of an answer to it were carried out with amphibian eggs and embryos, using the technique of nuclear transplantation. Amphibian eggs have provided the most favoured material for embryological research since the late 1800s until the last two or three decades. The classical embryological experiments from Spemann onwards were conducted with this material. Nuclear transplantation exemplifies the value of amphibian eggs for embryological experiments involving manipulation of embryos and cells.

The aim of the experiment described in this chapter is to test the proposition that cells can embark on a particular pathway of differentiation while retaining a complete, totipotent, or at least multipotent, genome in their cells.

Further technical details required for this experiment may be found in Gurdon, 1991.

DEGREE OF DIFFICULTY From an intellectual point of view the experiment is extremely simple. It aims to replace the zygote nucleus of the fertilized egg, i.e. the egg and sperm pronuclei, with the nucleus of a cell that has clearly embarked on one direction of differentiation. From a practical point of view it is quite a lot more difficult, since it requires careful control of a microinjection pipette.

INTRODUCTION

The first real success in transplanting nuclei from somatic cells to enucleated eggs was achieved by Briggs and King in 1952. They succeeded in obtaining normal tadpoles by transplanting a nucleus from a blastula cell to an enucleated egg of *Rana pipiens*. However, when Briggs and King transplanted nuclei from differentiating endoderm cells of neurula stage embryos they were unable to obtain any normal tadpole development. This suggested that, as cells differentiate, their nuclei may indeed undergo some loss or stable inactivation of genes.

Soon after these early experiments, nuclear transplantation succeeded with eggs and embryos of *Xenopus laevis*. In this case, it was found that even when cells had become fully differentiated as intestinal epithelium cells, some of them contained nuclei which could support entirely normal development yielding fertile adult frogs. Since the *Xenopus* experiments carried a genetic marker to eliminate any possibility that the enucleation of recipient eggs had been unreliable, these experiments established the principle that, as cells differentiate, their nuclei can remain totipotent (Gurdon, 1962; Gurdon and Uehlinger, 1966).

MATERIALS AND METHODS

EQUIPMENT AND MATERIALS

Per student

> Pairs of fine no. 5 forceps (BDH; Jencons)
> A stereomicroscope with magnification up to × 25 (BDH; Jencons)
> A good incident light source (BDH; Jencons)
> A syringe connected by plastic tubing to a micromanipulator reducing hand movements by about 5 times (BDH; Jencons)
> Liquid paraffin oil to fill the syringe and plastic tubing (Sigma)
> An ultraviolet source of bacteriocidal wavelength (BDH; Jencons)
> A short wave length ultraviolet light source (BDH; Jencons)
> Equipment for making a microinjection pipette (BDH; Jencons)
> Plastic or glass petri dishes for culturing embryos (BDH; Jencons)

Biological material. *Xenopus laevis* frogs to provide donor embryos and unfertilized recipient eggs (Nasco).

PREVIOUS TASKS FOR STAFF

Preparation of full-strength Modified Barth Saline solution (MBS) (see Table 24.1 for details).

Preparation of Modified Barth Saline solution lacking calcium and magnesium ions, and containing 0.5 millimolar EDTA pH 8.2 (dissociation medium).

Solutions. Composition of Modified Barth Saline (see Table 24.1). Composition of high-salt medium, for egg laying (see Table 24.1).

Preparation of microinjection pipettes. Any micropipette pulling machine should be able to produce pipettes from approximately 1 mm glass tubing. The pipettes need to have an opening of about 20 microns in diameter and to be parallel sided for about 1 cm

Table 24.1. Composition of salt solutions used in nuclear transplantation experiments

	Modified Barth's Saline (MBS) (mM)	High salt MBS (mM)	Normal amphibian ringer's (NAM) (mM)
NaCl	88	110	110
KCl	1	2	2
CaCl$_2$	0.41	–	1
Ca(NO$_3$)$_2$	0.33	–	–
MgSO$_4$	0.82	1	1
NaHCO$_3$	2.4	2	0.5
HEPES (NaOH), pH 7.4	10	–	–
Tris base, pH 7.6, acetic acid	–	15	–
EDTA	–	–	0.1
Sodium phosphate, pH 7.4	–	0.5	1
Penicillin, benzyl	10 mg/litre	–	100 units/ml
Streptomycin sulfate	10 mg/litre	–	60 units/ml
Nystatin	–	–	2 mg/litre

Notes: For MBS, see Gurdon (1977). For NAM, see Slack and Forman (1980).

from the tip. To produce the best results, the end of such a micropipette needs to be shaped like a hypodermic syringe needle using a microforge (Gurdon, 1991). However pipettes that are broken and happen to have a sharp point are sufficient.

OUTLINE OF THE EXPERIMENTS

PREPARATION OF DONOR CELLS

Embryos obtained by normal or artificial fertilization should be grown to an appropriate stage, for example the tailbud stage (stage 26 of *Xenopus*). Using two pairs of fine forceps, one can dissect such an embryo so that the endoderm portion is cut out. This is then incubated in calcium-free, magnesium-free MBS containing EDTA for about 20 minutes at room temperature. This dissociation should be done in 2-cm plastic dishes containing a thin (1 mm) layer of 1% agarose in the dissociation medium. After about 20 minutes the endoderm part of the embryo can be gently pipetted up and down so as to release large numbers of single cells. These can then be kept in a shallow plastic dish containing calcium-free, magnesium-free MBS, on an agarose base, as a source of nuclei for transplantation.

PREPARATION OF RECIPIENT EGGS

A female should lay its eggs in high-salt medium in the absence of a male. These unfertilized eggs should be collected into a 5-cm plastic dish containing the high-salt medium. Individual eggs are then placed with forceps on a small square (0.5 cm × 0.5 cm) of moist but not wet blotting paper. It is essential that the eggs are so placed that the egg nucleus, visible as a small white patch in the dark pigmented animal pole of the egg, is placed uppermost. Usually 3 or 4 such eggs are placed separately on the small piece of blotting paper, which is then placed on a standard microscope slide. The eggs are not covered by medium but remain moist because they are surrounded by a jelly coat. The microscope slide carrying the recipient eggs is then placed under the bacteriocidal ultraviolet lamp for 30 s to give a UV dose of $16 \text{ W} \neq \text{m}^2$, with the UV light, shining directly on the animal pole of the egg. The same slide is then exposed for between 5 and 10 s to a short-wave-length ultraviolet light that dissolves the jelly and makes the vitelline membrane penetrable. The pronuclei of these recipient eggs have now been destroyed by the ultraviolet light and can be placed, still on the blotting paper, on the microscope stage, next to the preparation of donor cells, for nuclear transplantation.

NUCLEAR TRANSPLANTATION

A single dissociated endoderm cell is selected such that it is just too large to fit into the tip of the micropipette. The pipette tip is placed next to the selected endoderm cell and, by use of the syringe, suction is applied. The donor cell is sucked into the tip of the injection pipette so that it is squeezed into a sausage shape, its diameter being about half or one-third its length. As a result, the donor cell membrane is ruptured but the nucleus is still protected by surrounding cytoplasm. Because of the large amount of yolk in these endoderm cells, the nucleus cannot be seen. In order to follow the injection of a donor nucleus into the opaque egg, it is necessary to have a bubble of air separating the paraffin oil from the saline medium in the shaft of the injection pipette. The pipette carrying the ruptured donor cell in its tip is then lifted from the dish containing the donor cells and inserted into a recipient egg. The injection pipette is inserted into the egg so that its tip is approximately one-third of the distance from the animal pole to the vegetal pole. The movement of the air bubble can then be used to judge the expressing of the donor nucleus from the tip of the pipette into the recipient egg. The syringe is then used to slowly move the air bubble marker in the shaft of the pipette toward its tip, thereby depositing the donor cell and surrounding medium in the middle of the recipient egg. The position in which the donor cell is deposited is not critical, since eggs have the ability to move an injected nucleus (just as they do a sperm nucleus) to the appropriate part of the egg. As soon as the pipette has delivered a donor cell it is gently withdrawn. The next egg on the slide can then be treated similarly. As soon as the recipient eggs have each received a nucleus, the small piece of blotting paper carrying the recipient eggs is now inserted into a dish containing $1 \times$ MBS saline medium.

The number of recipient eggs that can be placed on a piece of blotting paper depends on the speed of the operator. At first it is best to work with only 1 recipient egg at a time. As the operator becomes more skilful, up to 6 or even 8 eggs can be irradiated and injected with nuclei within a few minutes. It is critical that the recipient eggs do not become dry since they will certainly not survive this.

EXPECTED RESULTS

The nuclear transplant embryos should be cultured in $1 \times$ MBS until the blastula stage and then transferred to one-tenth MBS from then onward. The temperature of culture can be anywhere between 15 and 23°C.

The results should be recorded as a table of survival showing what proportion of total transfers reach various stages of development. If the recipient eggs are of good quality and if the nuclear transplantation procedure is carried out expertly, nuclei from endoderm cells of tailbud embryos should yield completely cleaved blastulae in 35% of the cases, muscular response stage embryos (ones that react when touched with a needle) in 20% of cases and swimming tadpoles in 15% of total transfers (Gurdon, 1960).

The progressive mortality of somatic cell nuclear transplant embryos is thought to be a result of genetic damage to nuclei after transfer. The slow dividing cells of most somatic tissues contain nuclei that must initiate and complete DNA replication within 1.5 hours after transplantation to an egg. This is because an egg, once activated by penetration, will divide at its normal time whether the introduced nucleus has or has not replicated its DNA. It is known that somatic nuclei often fail to complete their replication at the time of the first egg division into two cells.

DISCUSSION

Endoderm cells become committed soon after development starts to the endodermal pathway of differentiation. When such cells are transplanted to other parts of an embryo, they can form only endodermal derivatives. Therefore, these cells have embarked irreversibly on their pathway of differentiation. By the feeding tadpole stage some endoderm cells have formed the intestinal epithelium, and these cells are fully differentiated since they have the so-called striated border specialized for absorption.

The experiment described here shows that, even though the endoderm cells of a tailbud tadpole are committed to their specialized fate, the nuclei of such cells can often participate in the formation of functional muscle and nerve cells after nuclear transplantation to enucleated eggs (Gurdon, 1962). Some of these tailbud endoderm nuclei form entirely normal embryos and grow into adults. It has even been found that some of the nuclei of fully differentiated intestinal epithelium cells can, after nuclear transplantation, form fertile male and female adult frogs (Gurdon and Uehlinger, 1966).

These results illustrate the general principle that the nuclei of some determined or specialized cells can remain totipotent. This means that they must carry the complete range of genes in the genome, which has been unchanged in the course of development and cell differentiation. Furthermore, the same experiment shows that the nuclei of specialized cells can be reprogrammed to embryo patterns of gene expression; they are, in effect, rejuvenated to an embryonic condition or set back to the beginning of life. Therefore, this experiment demonstrates the principles of the conservation of the genome during cell differentiation and the ability of egg cytoplasm to reprogram nuclei from differentiated cells.

TEACHING CONCEPTS

Nuclear genes control development. Enucleated eggs cannot develop without a nucleus. The ability of nuclear transplant embryos to develop normally requires the transplanted nucleus to be reprogrammed so that it expresses normal embryo genes at the right time and in the right place. Experiments of the kind described can be analyzed biochemically to show gene expression in transplanted nuclei.

The cytoplasmic control of nuclear activity. The stages in development at which a transplanted nucleus replicates its DNA and at which it expresses its various genes are controlled entirely by the cytoplasm.

ALTERNATIVE EXERCISES

If biochemical facilities are available, it is instructive to measure the kind and amount of gene activity in various stages of nuclear transplant embryo development. This can be done using RT-PCR analysis of transcripts. It will be found that nuclear transplant embryos analyzed at the gatsrula stage express normal gastrula genes and do not continue to express tail-bud endoderm genes.

REFERENCES

Briggs, R., and King, T. J. (1952). Transplantation of living nuclei from blastula cells into enucleated frogs' eggs. *Proc. Natl. Acad. Sci. U.S.A.* 38, 455–63.

Gurdon, J. B. (1960). The developmental capacity of nuclei taken from differentiating endoderm cells of *Xenopus laevis. J. Embryol. Exp. Morphol.* 8, 505–26.

Gurdon, J. B. (1962). The developmental capacity of nuclei taken from intestinal epithelium cells of feeding tadpoles. *J. Embryol. Exp. Morphol.* 10, 622–40.

Gurdon, J. B. (1977). Methods for nuclear transplantation in Amphibia. In *Methods in Cell Biology* 16, eds. G. Stein, J. Stein, and K. J. Kleinsmith, pp. 125–39. New York: Academic Press.

Gurdon, J. B. (1991). Nuclear transplantation in *Xenopus.* In *Methods in Cell Biology* 36, eds. B. K. Kay and H. B. Peng, pp. 299–309. San Diego: Academic Press.

Gurdon, J. B., and Uehlinger, V. (1966). "Fertile" intestine nuclei. *Nature* 210, 1240–1.

Slack, J. M. W., and Forman, D. (1980). An interaction between dorsal and ventral regions of the marginal zone in early amphibian embryos. *J. Embryol. Exp. Morphol.* 56, 283–99.

SECTION X. CELL CULTURE

25 *In vitro* culture and differentiation of mouse embryonic stem cells

A. Rolletschek, C. Wiese and A. M. Wobus[*]

OBJECTIVE OF THE EXPERIMENT The aim of the experiments is (1) to establish mouse embryonic stem (ES) cell lines and (2) to study the differentiation of ES cells into cardiac and neuronal cells.

DEGREE OF DIFFICULTY Determination of alkaline phosphatase activity: easy. Immunofluorescence staining: moderate. All experiments concerning the establishment of ES cell lines and in vitro differentiation: difficult.

INTRODUCTION

Embryonic stem (ES) cells have been established as undifferentiated cell lines from the inner cell mass (ICM) of mouse and human blastocysts (Evans and Kaufman, 1981; Thomson et al., 1998). These pluripotent ES cells are characterized by a high proliferative capacity and the ability to develop into terminally differentiated cells of all three primary germ layers: the endodermal, ectodermal and mesodermal lineage (Figure 25.1). Mouse and human ES cells can differentiate in vitro to give rise to various somatic cell types (reviewed in Schuldiner et al., 2000; Wobus, 2001). In vivo, after transfer into blastocysts or by aggregation with blastomeres, mouse ES cells participate in the development of the embryo, including the germ line (Bradley et al., 1984), whereas human ES cells after transplantation into nude mice give rise to teratomas (Thomson et al., 1998).

Besides ES cells, two other types of pluripotent embryonic cell types have been established as permanent lines: embryonic carcinoma (EC) cells derived from teratocarcinomas and embryonic germ (EG) cells isolated from primordial germ cells of early embryos (Stewart, Ghadi, and Bhatt, 1994; Shamblott et al., 1998; see Figure 25.1).

---

[*] Corresponding author

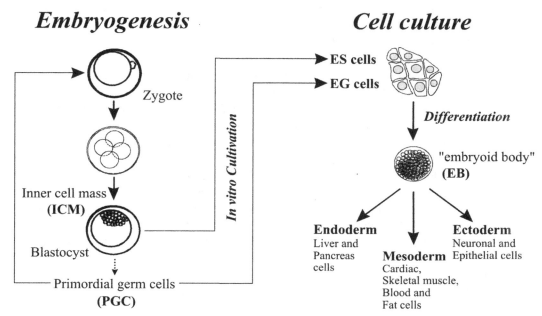

Figure 25.1. Cell types involved in early embryogenesis and after in vitro cultivation and differentiation of embryonic stem and embryonic germ cells. Comparison of early embryonal development in vivo (left) and cultivation of pluripotent embryonic stem (ES) and embryonic germ (EG) cells in vitro (right) derived from the inner cell mass (ICM) of mouse blastocysts and primordial germ cells, respectively (modified according to Wobus, 2001).

Undifferentiated ES cells are characterized by specific properties such as expression of the cell surface antigen SSEA-1 (Solter and Knowles, 1978), the transcription factor Oct-4 (Schöler et al., 1989), as well as alkaline phosphatase (ALP, Wobus et al., 1984) and telomerase activity.

During the last few years, protocols for the differentiation of mouse ES cells into various cell types, including cardiomyocytes and neuronal cells, have been established. Usually, the first step to differentiate ES cells in vitro is the formation of embryo-like aggregates called embryoid bodies (EBs). Within these EBs, ES cells differentiate into endodermal, ectodermal, and mesodermal cell types.

The differentiation potential is influenced by the following parameters: (1) the number of cells differentiating in the EBs, (2) the quality of fetal calf serum, (3) the addition of growth and extracellular matrix factors, (4) the ES cell lines used, and (5) the time of EB plating (Wobus et al., 2002). In principle, EBs spontaneously develop into various differentiated cell types; however, to obtain a maximal differentiation of a specific cell type, selective culture conditions, substitution by differentiation factors and/or introduction of tissue-specific genes have to be applied.

In this chapter, we describe methods to establish mouse ES cell lines, to differentiate pluripotent ES cells into cardiomyocytes and neuronal cells, and to characterize these phenotypes by immunofluorescence analysis.

MATERIALS AND METHODS

EQUIPMENT AND MATERIAL

Per student

Tissue culture plates: 35 mm, 60 mm, 100 mm (Nunc).

Pasteur pipettes: 2, 5, 10, 20 ml pipettes.

Bacteriological petri dishes (Greiner): 60 mm for suspension culture of EBs, 100 mm for hanging drop culture of EBs.

Tissue culture plates (60 mm) containing sterilized coverslips (n = 4, 16 mm × 16 mm) for immunofluorescence.

2 ml glass pipettes for preparing single-cell suspensions.

For feeder layer culture: sterile dissecting instruments, screen or sieve (about 0.3 mm diameter for filtration of cell suspensions and removing tissue aggregates), Erlenmeyer flasks with stir bars, centrifuge tubes.

0.5 and 1.5 ml microtubes (Eppendorf) and 20, 100, 1,000 μl filtertips (Biozym).

Per practical group

Tissue culture incubator with 37°C temperature and 5% CO_2 atmosphere.

Fluorescence microscope (Nikon).

Biological material

Mice of inbred strain 129/ter Sv for the establishment of ES cells, NMRI or CD-1 outbred mice for the preparation of feeder cells (Wobus et al., 2002).

ES cell line R1 (Nagy et al., 1993).

PREVIOUS TASKS FOR STAFF

Media and stock solutions (to be prepared sterile). As flushing buffer for the isolation of mouse embryos, a modification of PB-1 buffer is prepared (see Spielmann and Eibs, 1977):

Stock solution I: Dissolve 3.905 g NaCl, 0.1 g KCl, 0.95 g KH_2PO_4 in 500 ml Aqua tridest.

Stock solution II: Dissolve 9.8 mg $NaHCO_3$, 3.6 mg Na-pyruvate, 0.1 g glucose, 10 mg $CaCl_2$, 0.143 g Na_2HPO_4, 10 mg $MgCl_2 \times 6H_2O$, 8 mg penicillin, 5 mg streptomycine, 1 mg Phenol Red (Sigma) to make up a volume of 100 ml with stock solution I, sterilize through a 0.22 μm filter, and store at 4°C.

Before use, add 2 ml fetal calf serum (FCS) to 8 ml stock solution II.

Solutions and media for cell culture

Phosphate-buffered saline (PBS): 10 g NaCl, 0.25 g KCl, 1.44 g Na_2HPO_4, 0.25 g $KH_2PO_4 \times 2H_2O/l$, filter-sterilized!

Trypsin solution: 0.2% trypsin (Gibco) in PBS, filter-sterilized!

EDTA solution: 0.02% EDTA (Sigma) in PBS, filter-sterilized!

Trypsin-EDTA solution: Mix trypsin solution and EDTA solution = 1:1.

Gelatin solution: 1% gelatin (Fluka) in double-distilled water, autoclaved, and diluted 1:10 with PBS. Coat tissue culture dishes with 0.1% gelatin solution for 1–24 hours at 4°C before use.

Poly-L-ornithine solution: Add 0.1 g poly-L-ornithine hydrochloride (Sigma) to 100 ml boric acid buffer (pH 8.4; 10 mM, Sigma) to form a stock solution of 1.0 mg/ml, filter sterile through a 0.22 μm filter. For use, dilute 1 ml stock solution to a final volume of 10 ml with sterile Aqua tridest. to prepare a working solution of 0.1 mg/ml. Cover 60 mm tissue culture dishes containing 4 coverslips with 5 ml poly-L-ornithine solution and incubate at 37°C for 3h. Rinse 3 × with Aqua tridest. and incubate at room temperature for 12 h. Rinse 3 × with Aqua tridest. and dry at 40°C.

Laminin solution: Dilute 0.001 mg/ml laminin (Sigma) in PBS, filter sterile through a 0.22 μm filter. Cover 60 mm tissue culture dishes containing 4 poly-L-ornithine–coated coverslips with 5 ml laminin solution and incubate at 37°C for 3 h.

Mitomycin (MC) solution: Dissolve 2 mg Mitomycin C (Serva) in 10 ml PBS, filter sterilize through a 0.22 μm filter. From this stock solution dilute 300 μl into 6 ml of PBS (final concentration is 0.01 mg/ml). MC stock solution should be freshly prepared at weekly intervals and stored at 4°C. CAUTION! MC IS CARCINOGENIC!

β-mercaptoethanol (β-ME, Serva): Prepare a stock solution from 7 μl of β-ME in 10 ml of PBS (stock concentration = 10 mM). Make fresh at weekly intervals and store at 4°C.

Cultivation medium I: Dulbecco's modification of Eagle's medium (DMEM, 4.5 g/l glucose, Gibco) supplemented with 15% FCS for cultivation of feeder layer cells.

Additives I: To 100 ml media add 1 ml of 200 mM L-glutamine stock (100 ×, Gibco), 1 ml of β-ME stock, 1 ml of nonessential amino acids (NEAA) stock (100 ×, Gibco).

Cultivation medium II for ES cells: Cultivation medium I, additives I and 10 ng/ml leukemia inhibitory factor (LIF, Chemicon International, Inc.). For the establishment of mouse ES cells and the cultivation of R1 ES cells (Nagy et al., 1993), FL cells and LIF are used to maintain ES cells in the undifferentiated state.

Differentiation media

1. EB DIFFERENTIATION

For EB differentiation, Differentiation medium I is used: Iscove's modification of DMEM (IMDM, Gibco) supplemented with 20% FCS and additives I. In additives I, monothioglycerol (MTG, 3-mercapto-1,2-propanediol, Sigma) instead of β-ME is used. Prepare a stock solution from 13 μl of MTG into 1 ml of IMDM and filter sterile through a 0.22 μm filter. Make fresh before use. To 100 ml media, add 300 μl of stock solution (final concentration is 450 μM).

2. CARDIAC DIFFERENTIATION

Differentiation medium I (see EB differentiation)

3. NEURONAL DIFFERENTIATION

Differentiation medium II for selection of nestin-positive cells: DMEM/F12 medium supplemented with 5 μg/ml insulin, 30 nM sodium selenite (both from Sigma), 50 μg/ml transferrin, and 5 μg/ml fibronectin (both from Gibco).

Differentiation medium III for expansion of nestin-positive precursor cells: DMEM/F12 containing 20 nM progesteron, 100 μM putrescin, 1 μg/ml laminin (all from Sigma), 25 μg/ml insulin, 50 μg/ml transferrin and 30 nM sodium selenite, 10 ng/ml basic fibroblast growth factor (bFGF), and 20 ng/ml epidermal growth factor (EGF; all from Strathmann Biotech).

Differentiation medium IV for induction of neuronal differentiation: Neurobasal medium plus 2% B27 media supplement (all from Gibco) and 10% FCS.

Solutions for alkaline phosphatase (ALP) staining

3.7% Paraformaldehyde (PFA, Serva): Dissolve 3.7 g PFA in PBS and adjust to 100 ml with PBS, heat the mixture to 95°C, stir until the solution becomes clear, and cool to room temperature. (PFA IS TOXIC! WORK UNDER THE HOOD AND USE GLOVES!) Make fresh before use!

Tris-maleate (TM) buffer: 1 M maleic acid, add 3.6 g Tris (all from Roth), add 1 l H_2O, and adjust to pH 9.0 with 1 M maleic acid.

PBS (see Solutions and Media for cell culture)

Staining solution: Dissolve 25 mg Fast Red TR salt and 10 mg naphthol AS-MX phosphate (all from Sigma) in 25 ml TM-buffer for 15–20 min. Add finally 200 μl of 10% $MgCl_2$ solution. Make fresh before use!

Solutions and reagents for immunofluorescence analysis. The preparation of solutions and reagents for immunofluorescence analysis is listed in Wobus et al. (2002) for cardiac and in Rolletschek et al. (2001) for neuron-specific proteins.

OUTLINE OF THE EXPERIMENTS

EXPERIMENT 1: FEEDER LAYER (FL) CULTURE

To maintain ES cells in the undifferentiated state, cells are cultured on a FL of mouse embryonic fibroblasts. To be suitable for ES cell culture, feeder cells have to be inactivated by mutagenic treatment using mitomycin C (MC) or by X-irradiation. FL cells retain metabolic activity and provide extracellular matrix and growth factors as well as soluble differentiation inhibitory activity.

1. Remove the fetuses from a pregnant mouse between embryonic day 15 to 17, rinse in PBS, and remove placenta and fetal membranes, head, liver, and heart. Rinse the carcasses in trypsin solution.
2. Mince the embryonic tissue in 5 ml of freshly prepared trypsin solution with scissors and scalpels, and transfer the minced tissue to an Erlenmeyer flask containing a stir bar.
3. Stir on magnetic stirrer for 25 to 45 min (use longer incubation time if the embryos are older), filter the suspension through a sieve or a screen in order to prepare a

cell suspension without tissue particles, add 10 ml of culture medium I, and spin down.

4. Resuspend the pellet in about 3 ml of culture medium I and transfer the cell suspension on 100 mm tissue culture plates (about 2×10^6 cells per 100 mm dish) containing 10 ml culture medium I, incubate at 37°C and 5% CO_2 for 24 hours.

5. After 1 day in culture, change the medium to remove debris, erythrocytes, and unattached tissue aggregates and cultivate for an additional 1 to 2 days.

6. Passage the primary culture of mouse embryonic fibroblasts by splitting 1:2 to 1:3 on 100-mm tissue culture plates, and grow in culture medium I for 1 to 3 days. The cells in passages 2 to 4 are most suitable as feeder layer for the culture of undifferentiated ES cells.

7. Before use, incubate feeder layer cells with MC buffer for 2.5 to 3 hours, aspirate the MC solution, wash 3 × with PBS, trypsinize feeder cells with trypsin-EDTA solution (1:1), and replate to new gelatin (0.1%)–treated microwell plates or to petri dishes. Feeder layer cells prepared one day before ES cell subculture are optimal.

EXPERIMENT 2: ESTABLISHMENT OF ES CELL LINES

Isolation of blastocysts and establishment of ES cell lines from the inner cell mass of mouse embryos. The original method described by Evans and Kaufman (1981) is recommended (see Wobus et al., 1984; Doetschmann et al., 1985; Robertson, 1987). The most efficient method to establish mouse ES cell lines is the culture of blastocysts on mouse embryonic fibroblast FL cells supplemented by LIF.

1. Isolate early blastocysts by flushing the uterus horns of mice 3 days after detection of vaginal plugs (day of v.p. = 0.5 d), with 1 ml of PB-1 medium, collect the flushing medium in a "watchglass." The blastocysts will accumulate in the center.

2. Place blastocysts (n = 7 to 10) with a mouth-controlled transfer pipette to gelatine-coated wells filled with FL cells the day before (see above), and replaced with 0.3 to 0.5 ml of fresh cultivation medium II (see Solutions and Media for cell culture).

3. Cultivate the blastocysts for 3 to 4 days in the incubator with 5% CO_2 at 37°C. The blastocysts hatch from the Zona pellucida and will attach and grow out on the FL cells. Observe the cultures only after 2 to 3 days. Do not disturb early blastocyst outgrowths!

4. For subculture and disaggregation, use only blastocyst outgrowths showing the typical ICM morphology (small cell size, high nucleo-cytoplasmic ratio, impossible to identify individual cells, compacted undifferentiated stem cells in contrast to distinct endoderm-like and giant trophoblast cells). Disaggregate ICM clones by a short treatment with trypsin-EDTA solution (about 30 s), add about 0.3 ml of cultivation medium II, and resuspend the ICM cells with a finely drawn pipette. Transfer the cell clumps to a new well with freshly prepared feeder cells. At the first passage, dissociate cell clumps gently.

5. Every second or third day, transfer the undifferentiated stem cells onto a new well with freshly prepared FL cells. During establishment of a culture, the quality of FL cells is important (use 2nd or 3rd passage FL cells only). If the ES cells proliferate and

have adapted to growth in culture, clone ES cells by the "limited dilution" technique or by single-cell cloning on 96- or 24-well microwell plates.

6. Propagate a cloned colony (showing the typical undifferentiated "stem cell-like" morphology) to a sufficient cell number (a sub-confluent 60 mm tissue culture plate corresponds to about 1×10^6 cells), replate 1:3, freeze one culture, and propagate the others. During the next passages, freeze a stock of the cell line.

EXPERIMENT 3: CULTURE AND CHARACTERIZATION OF UNDIFFERENTIATED ES CELLS

Routine subculture of ES cells. The R1 ES cell line (Nagy et al., 1993) will be used for in vitro differentiation into cardiac (Wobus et al., 2002) and neuronal cells (Rolletschek et al., 2001).

It is important to passage ES cells every 24 or 48 hours. ES cells grow as compacted colonies on FL cells. Do not cultivate longer than 48 hours without passaging, or the cells may differentiate and be unsuitable for differentiation studies. Selected batches of FCS have to be used for ES cell culture (see below).

1. Change the medium 1 to 2 hours before passaging.
2. Aspirate the medium, add 2 ml of trypsin-EDTA solution (1:1), incubate at room temperature for 30 to 60 seconds.
3. Remove carefully the trypsin-EDTA solution, add 2 ml of fresh cultivation medium II.
4. Resuspend the cell population with a 2 ml glass pipette into a single-cell suspension and split 1:3 to 1:5 to freshly prepared (60 mm) FL plates.

Good quality FCS is critical for long-term culture of ES cells, and failure to acquire good quality serum may be one reason why ES cells fail to differentiate appropriately. Therefore, extensive serum testing is necessary to achieve good results. The most sensitive tests for FCS include (1) comparative plating efficiency tests with different serum concentrations, (2) alkaline phosphatase activity in undifferentiated ES cells (see *Determination of alkaline phosphatase activity*), and (3) test of in vitro differentiation capacity of ES cells (see Exp. 4) after 3 to 5 passages in selected serum batches (Robertson, 1987).

Determination of alkaline phosphatase (ALP) activity

1. Plate R1 ES cells in cultivation medium II on FL into 35 mm gelatine-coated tissue culture plates for 2 days, or plate ES cells at low density (1,000 cells/60 mm petri dish), and incubate (without moving the dishes) for one week for clonal analysis.
2. Aspirate the medium.
3. Fix the cultures with 2 ml 3.7% PFA at room temperature for 20 min.
4. Rinse the cells 3 × in TM-buffer for 10 min.
5. Aspirate the buffer and stain the cultures with freshly prepared staining solution for 20 min.
6. Aspirate the staining solution and rinse with PBS.

EXPERIMENT 4: DIFFERENTIATION OF ES CELLS AND CHARACTERIZATION OF DIFFERENTIATED CELL TYPES

For the development of ES cells into differentiated phenotypes, cells must be cultivated in three-dimensional aggregates or "embryoid bodies" (EB) by the "hanging drop" method (Robertson, 1987; Wobus, Wallanat, and Heschler, 1991) or by "mass culture" (Doetschman et al., 1985). The in vitro differentiation of ES cells into cardiac and neuronal cells requires different conditions as demonstrated in Figure 25.2. The "hanging drop" method generates EBs of a defined cell number and size. Therefore this technique is recommended for developmental studies, because the differentiation pattern is dependent on the number of ES cells that differentiate within the EBs.

Preparation of EBs

1. Prepare a cell suspension containing a defined ES cell number of 400 (cardiac differentiation) or 200 (neuronal differentiation) cells in 20 μl of Differentiation medium I.
2. Place 20 μl drops (n = 50 to 60) containing ES cells onto the lids of 100 mm bacteriological petri dishes and put them upside down on the plates containing 10 ml PBS.
3. Cultivate the ES cells in hanging drops for 2–3 days. The cells will aggregate and form one EB per drop (Figure 25.3c).
4. Rinse the aggregates carefully from the lids with 2 ml of medium and transfer them into a 60 mm bacteriological dish with 5 ml of Differentiation medium I. Continue cultivation in suspension for 2 (neuronal differentiation) and 3 (cardiac differentiation) days until the time of plating.
5. Transfer 20 to 40 EBs/dish onto 60 mm tissue culture dishes, each containing 4 (sterile!) coverslips (10 × 10 mm) for immunofluorescence analyses.

Cardiac muscle cell differentiation

1. Aspirate gelatin solution from 60 mm tissue culture dishes containing 4 coverslips and add 4 ml Differentiation medium I.
2. Transfer 20 to 30 EBs (5 d old, see Preparation of EBs) to the dishes.
3. Replenish the medium every 2 to 3 days.
4. The first beating cardiomyocytes appear two to three days after plating of 5-d-old EBs, and maximal cardiac differentiation is achieved between 7 to 12 days after EB plating (Maltsev et al., 1993).

Neuronal differentiation. There are several possibilities for inducing neuronal differentiation, i.e. by retinoic acid induction (Guan et al., 2001) or by lineage-restricted differentiation (Rolletschek et al., 2001). Here, we present a protocol for lineage-restricted neuronal differentiation that includes the following four steps: (1) formation of cells of all three primary germ layers in EBs, (2) selective differentiation of neuroectodermal and inhibition of mesodermal cells by growth factor removal (serum depletion), (3) induction of proliferation of neural precursor cells by bFGF and EGF, and (4) induction of differentiation and maintenance of functional neurons by withdrawal of bFGF/EGF and cultivation in specific neuronal differentiation media (see Figure 25.2).

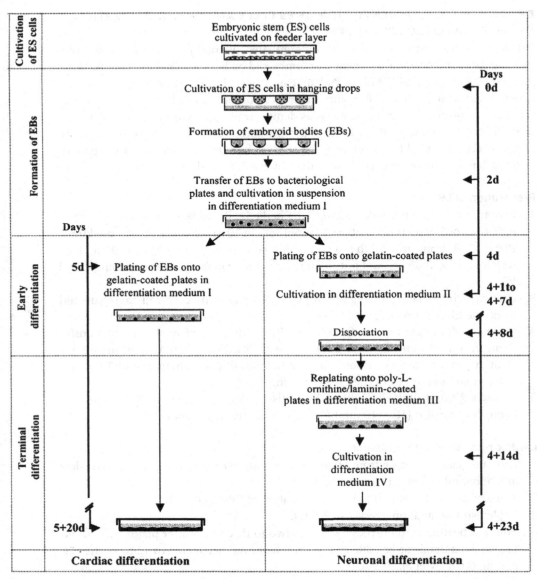

Figure 25.2. General scheme of ES cell differentiation into cardiac and neuronal cells in vitro. Undifferentiated ES cells are cultivated on feeder layer and differentiated as embryoid bodies (EBs) in hanging drops and in suspension. Early differentiation processes are induced within the EBs and after plating of EBs on tissue culture plates. Cardiac differentiation is maintained by medium change until the terminal differentiation stage. Terminal differentiation of neuronal cells is achieved after dissociation of ES cell–derived aggregates, replating onto poly-L-ornithine/laminin–coated plates and by further cultivation in the presence of differentiation factors.

1. Plate 20 to 30 EBs (4 d old, see Preparation of EBs) onto 60 mm gelatin-coated tissue-culture plates in Differentiation medium I. Allow attachment of EBs for 1 d.
2. At day 4 + 1, exchange the medium to Differentiation medium II for selection of nestin-positive precursor cells.
3. Cultivate the cells for 7 days, replenishing the medium every two days.
4. At day 4 + 8, rinse 60 mm tissue culture dishes containing poly-L-ornithine/laminin–coated coverslips (n = 4) with PBS (2 ×), dissociate EBs by trypsin-EDTA solution (1:1) for 1 min, collect by centrifugation, and replate into Differentiation medium III for proliferation of nestin-positive precursor cells for six days until day 4 + 14.
5. Replenish the medium every 2 days, but maintain the correct concentration of bFGF and EGF daily.
6. At day 4 + 14, induce the differentiation and maturation of neurons by adding Differentiation medium IV for 1 to 2 weeks until day 4 + 23 (see Rolletschek et al., 2001).
7. Exchange half of the medium with fresh Differentiation medium IV every 2 to 3 days.

Immunofluorescence protocols. For characterization of ES cell-differentiated pheno-types, antibodies against tissue-specific intermediate filament proteins, sarcomeric proteins or neuro-transmitters are applied. Protocols for immunofluorescence analysis are described in Wobus et al. (2002) for cardiac, and in Rolletschek et al. (2001) for neuron-specific antibodies.

EXPECTED RESULTS

CHARACTERIZATION OF UNDIFFERENTIATED ES CELLS

Undifferentiated ES cells grow on FL as compacted cell clusters, as shown in Figure 25.3a. These clusters are characterized by alkaline phosphatase activity resulting in intensive red staining (Figure 25.3b). To estimate the percentage of ALP-positive clones after colony formation, the number of ALP-positive colonies per culture dish is determined in comparison to unlabelled colonies. A value of 98 to 100% ALP-positive colonies characterizes an undifferentiated ES cell population that is suitable for differentiation experiments. FL cells sometimes show a very faint, but nonspecific staining.

CARDIAC DIFFERENTIATION

Differentiation of ES cells into cardiomyocytes will be demonstrated by morphological characteristics and immunofluorescence analysis. Specific cardiac cell types have been found to develop from ES cells in vitro, such as pacemaker-, atrium-, ventricle-, sinusnodal-, and Purkinje-like cells (Hescheler et al., 1997). Early pacemaker-like cells represent the earliest ES-derived cardiac cell type forming in vitro. Later during ES cell differentiation, more terminally differentiated cell types, such as atrium-, ventricle-, and

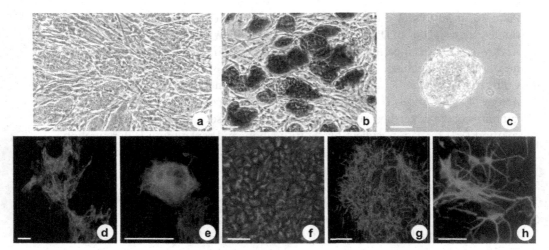

Figure 25.3. Morphological, biochemical, and immunocytochemical characterization of ES cells and differentiated derivatives. (a) Colony morphology of R1 ES cells (phase contrast) and (b) alkaline phosphatase activity (red colonies) of R1 cells cultured on mouse embryonic fibroblasts. (c) Morphology of an embryoid body (EB) after 2 days in suspension culture. (d) Desmin-positive cardiomyocytes shown by immunofluorescence after cardiac differentiation of ES cells at day 5 + 9d. (e) A desmin-positive cardiomyocyte at higher magnification. (f) ES cell-derived nestin-positive neural precursor cells at day 4 + 10. (g) β III-tubulin–expressing neurons after neuronal differentiation at day 4 + 23. (h) A network of β III-tubulin–positive cells at higher magnification. Bar = 50 μm (a–e), 30 μm (f), 40 μm (h), 100 μm (g). (See also color plate 7).

sinusnodal-like types appear (Maltsev et al., 1993). The expected results are as follows:

➤ Detection by microscopic observation of the first spontaneously beating clusters beginning at day 5 + 2 to 5 + 3.
➤ Maximal degree (= 100%) of cardiac differentiation in EB outgrowths, determined as percentage of EBs containing beating cardiac clusters at differentiation stage 5 + 9 d.
➤ Immunofluorescence analysis at stage 5 + 9 d will show about 20–40% of desmin-positive clusters in individual EB outgrowths (Figures. 25.3d and 25.3e).

NEURONAL DIFFERENTIATION
Early steps of ES cell-derived neuronal differentiation in EB outgrowths are characterized by the development and proliferation of neural precursor cells. These cells are immuno-positive for the intermediate filament protein nestin.

➤ Immunofluorescence analysis at early stages (4 + 10 d) will show up to 70% nestin-positive cells (Figure 25.3f; see Rolletschek et al., 2001).

With ongoing in vitro differentiation, nestin expression is down-regulated, and neuronal cells expressing neuron-specific proteins appear:

➤ Immunofluorescence analysis at more advanced differentiation stages (4 + 16 to 4 + 23 d) shows 40–50% β III-tubulin-positive neurons (Figures 25.3g and 25.3h).

DISCUSSION

APPLICATIONS OF ES CELL TECHNOLOGY

The in vitro studies have shown that ES cells recapitulate processes of early embryonic development during differentiation in culture. This opens the possibility of using mouse ES cells for (a) the analysis of effects of teratogenic drugs and environmental factors in embryotoxicity studies, (b) drug screening in pharmacology and toxicology, (c) genetic studies using "gain-of-function" and "loss-of-function" assays in vitro, and (d) genomics analyses to identify expression profiles of ES-derived specialized cell populations (see Wobus, 2001).

TIME REQUIRED FOR THE EXPERIMENTS

In most cases, the duration of specific experimental steps is mentioned in the protocols. The following specification gives additional information:

- ➤ Preparation of FL:
    Primary preparation: 2–3 h
    Subcultivation: 3–5 days
    MC treatment: 2.5–3.5 h
- ➤ Preparation of blastocysts:
    Primary preparation: 2 h
    Subcultivation of inner-cell mass cells: 3–4 days
    Disaggregation: 1 h
- ➤ Establishment of ES cell lines: 4–6 weeks
- ➤ Routine subculture of ES cells: 1–3 min/plate
- ➤ ALP-staining: 2.5 h/10 plates
- ➤ Preparation of EBs by the hanging drop technique: 1 to 1.5 h/20 plates
- ➤ Immunofluorescence staining: 3.5 h/20 coverslips.

Most experiments require additional work with respect to the preparation of solutions and media.

TEACHING CONCEPTS

Basic knowledge of common techniques in cell and tissue culture. Sterility, culture conditions, media, and supplements.

Basic techniques of mouse ES cell culture. Cultivation and passaging of established ES cell lines, feeder layer culture, ALP-assay, establishment of primary ES cell lines.

Developmental processes during the differentiation of stem cells into specialized cell types. Commitment of undifferentiated cells into specific lineages, use of ES cells as in vitro model system for early embryogenesis and development, differentiation induction of ES cells, expression of tissue-specific genes and proteins, immunofluorescence analysis.

Stem cell biology. Proliferation and differentiation properties of pluripotent ES cells, tissue regeneration.

REFERENCES

Bradley, A., Evans, M., Kaufman, M. H., and Robertson, E. (1984). Formation of germ-line chimaeras from embryo-derived teratocarcinoma cell lines. *Nature*, 309, 255–6.

Doetschman, T. C., Eistetter, H. R., Katz, M., Schmidt, W., and Kemler, R. (1985). The in vitro development of blastocyst-derived embryonic stem cell lines: Formation of visceral yolk sac, blood islands and myocardium. *J. Embryol. Exp. Morphol.*, 87, 27–45.

Evans, M. J., and Kaufman, M. H. (1981). Establishment in culture of pluripotential stem cells from mouse embryos. *Nature*, 291, 154–6.

Guan, K., Chang, H., Rolletschek, A., and Wobus, A. M. (2001). Embryonic stem cell-derived neurogenesis. *Cell Tissue Res.*, 305, 171–6.

Hescheler, J., Fleischmann, B. K., Lentini, S., Maltsev, V. A., Rohwedel, J., Wobus, A. M., and Addicks, K. (1997). Embryonic stem cells: A model to study structural and functional properties in cardiomyogenesis. *Cardiovasc. Res.*, 36, 149–62.

Maltsev, V. A., Rohwedel, J., Hescheler, J., and Wobus, A. M. (1993). Embryonic stem cells differentiate in vitro into cardiomyocytes representing sinusnodal, atrial and ventricular cell types. *Mech. Dev.*, 44, 41–50.

Nagy, A., Rossant, J., Nagy, R., Abramow-Newerly, W., and Roder, J. C. (1993). Derivation of completely cell culture-derived mice from early-passage embryonic stem cells. *Proc. Natl. Acad. Sci. USA*, 90, 8424–8.

Robertson, E. J. (1987). Embryo-derived stem cell lines. In *Teratocarcinomas and Embryonic Stem Cells – A Practical Approach*, ed. E. J. Robertson, pp. 71–112. Oxford, Washington DC: IRL Press.

Rolletschek, A., Chang, H., Guan, K., Czyz, J., Meyer, M., and Wobus, A. M. (2001). Differentiation of embryonic stem cell-derived dopaminergic neurons is enhanced by survival promoting factors. *Mech. Dev.*, 105, 93–104.

Schöler, H. R., Balling, R., Hatzopoulos, A. K., Suzuki, N., and Gruss, P. (1989). Octamer binding proteins confer transcriptional activity in early mouse embryogenesis. *EMBO J.*, 8, 2551–7.

Schuldiner, M., Yanuka, O., Itskovitz-Eldor, J., Melton, D. A., and Benvenisty, N. (2000). From the cover: Effects of eight growth factors on the differentiation of cells derived from human embryonic stem cells. *Proc. Natl. Acad. Sci. USA*, 97, 11307–12.

Shamblott, M. J., Axelman, J., Wang, S., Bugg, E. M., Littlefield, J. W., Donovan, P. J., Blumenthal, P. D., Huggins, G. R., and Gearhart, J. D. (1998). Derivation of pluripotent stem cells from cultured human primordial germ cells. *Proc. Natl. Acad. Sci. USA*, 95, 13726–31.

Solter, D., and Knowles, B. B. (1978). Monoclonal antibody defining a stage-specific mouse embryonic antigen (SSEA-1). *Proc. Natl. Acad. Sci. USA*, 75, 5565–9.

Spielmann, H., and Eibs, H. G. (1977). Preimplantation embryos. Part 1: Laboratory equipment, preparation of media, sampling and handling of the embryos. In *Methods in Prenatal Toxicology*, eds. D. Neubert, H. J. Merker, and T. Kwasigroch, pp. 210–230. Stuttgart: George Thiem Verlag.

Stewart, C. L., Gadi, I., and Bhatt, H. (1994). Stem cells from primordial germ cells can reenter the germ line. *Dev. Biol.*, 161, 626–8.

Thomson, J. A., Itskovitz-Eldor, J., Shapiro, S. S., Waknitz, M. A., Swiergiel, J. J., Marshall, V. S., and Jones, J. M. (1998). Embryonic stem cell lines derived from human blastocysts. *Science*, 282, 1145–7.

Wobus, A. M. (2001). Potential of embryonic stem cells. *Mol. Aspects Med.*, 22, 149–64.

Wobus, A. M., Holzhausen, H., Jäkel, P., and Schöneich, J. (1984). Characterization of a pluripotent stem cell line derived from a mouse embryo. *Exp. Cell Res.*, 152, 212–9.

Wobus, A. M., Wallukat, G., and Hescheler, J. (1991). Pluripotent mouse embryonic stem cells are able to differentiate into cardiomyocytes expressing chronotropic responses to adrenergic and cholinergic agents and Ca^{2+} channel blockers. *Differentiation*, 48, 173–82.

Wobus, A. M., Guan, K., Yang, H. T., and Boheler, K. R. (2002). Embryonic stem cells as a model to study cardiac, skeletal muscle and vascular smooth muscle cell differentiation. In *Meth. Mol. Biol.* vol. 185: *Embryonic Stem Cells: Methods and Protocols*, ed. K. Turksen, pp. 127–56. Totowa, NJ: Humana Press Inc.

SECTION XI. EVO–DEVO STUDIES

26 Microevolution between *Drosophila* species

N. Skaer and P. Simpson

OBJECTIVE OF THE INVESTIGATION Ever since Darwin proposed his theory of evolution by natural selection in *The Origin of Species* there have been questions regarding how the process of evolution occurs. Originally, such questions were studied through comparative anatomy, but more recently many of these questions have been directed at the genetic level, specifically addressing which events lead to genomic evolution. It is increasingly evident that genomes are constantly evolving systems, in which genetic changes are not always mirrored by concomitant change at the phenotypic level. One way in which this concept has been studied has been by generating hybrids between closely related species, such as those of the *Drosophila* genus.

The objective of this investigation is to analyse the reason(s) for loss of large mechanosensory bristles (macrochaetes) on the thorax of hybrid progeny generated by crossing *Drosophila melanogaster* with *Drosophila simulans*. To this end, three experiments will be performed, and the progeny analysed for bristle loss in each case. The first experiment will analyse the effect on bristle loss by introducing mutations from the *Drosophila melanogaster* parent that are known to be involved in bristle development. The second experiment will establish whether hybrid bristle loss can be rescued by over-expression of a gene responsible for bristle formation, or by introducing extra copies of this gene into hybrids. The third experiment will investigate the differences in bristle loss between male and female hybrids.

DEGREE OF DIFFICULTY All crosses are straightforward and easy to perform, although they should be carried out rigorously and the analysis of progeny must be meticulous. However, scoring bristle loss from crosses is time consuming because a large number of hybrids are required for accurate analysis. Therefore, where a class is undertaking the investigation it would be sensible to divide crosses between students and compile data for the class as a whole. However, very little equipment other than standard *Drosophila* facilities is required to perform the experiments in this chapter.

INTRODUCTION

The study of evolution through developmental biology is not a novel concept, having been pioneered principally by the classical comparative embryologists Ernst Haeckel and Karl Ernst von Baer in the nineteenth century. However, there followed a long, relatively fallow period in which the two disciplines impinged on one another to only a small extent. Evolution and Development continued as parallel sciences until the beginning of the 1980s when the advent of molecular genetics caused a renaissance of evolutionary developmental biology, and the birth of the "evo–devo synthesis," as the field has come to be known. The realisation that many developmental genes, such as those of the *Hox* cluster, were conserved across diverse morphological taxa permitted previously improbable comparisons of the development of organisms with very different body plans. Yet this unexpected degree of conservation raised something of a paradox: what gives rise to the diversity if the genes are highly conserved? If it is accepted that there is a genuine link between genotype and phenotype, it is axiomatic that the diversity seen in morphologies is at some level mirrored in the genome. Hence the attention of evolutionary developmental biologists is turning increasingly to the finer resolution questions – understanding the apparent conservation in developmental mechanisms through the detailed examination of gene function and regulation. Changes in the sequences that regulate gene expression may be responsible for much of the diversity in morphology. The systems that regulate gene expression are modular – the individual elements can act, and thus evolve, independently of one another without compromising the fundamental integrity of the system. Furthermore, there is an inherently greater "evolvability" in regulatory sequence than the sequence which codes for the proteins themselves: regulatory DNA is not required to maintain a reading frame; it can function at widely varying distances from, and in either orientation to, the transcriptional units it controls. One facet of these micro-evolutionary investigations has been studying gene regulation in closely related organisms. This has revealed numerous interesting aspects of gene evolution, including a phenomenon directly opposite to the paradox described above: there can be significant evolution at the genetic level between organisms, with no concomitant change in the morphologies those genes pattern. It is this aspect of gene evolution that we shall be experimentally investigating in this chapter.

Drosophila melanogaster and *Drosophila simulans* are closely related species, being separated by between 0.8 and 3 million years (Lemeunier et al., 1984). Both species display an identical pattern of eleven large, sensory bristles (macrochaetes) on the dorsal thorax (notum) of the adult (Figure 26.1). The positions of these bristles are likely to be important for the fly's behaviour, and the pattern has been found to be evolutionarily ancient: specimens preserved in amber from up to 40 million years ago from the family Drosophilidae have almost identical patterns. The genetic control of how this pattern of macrochaetes is established has been extensively studied in *Drosophila melanogaster* (Jan and Jan, 1995). Expression of the genes of the *achaete-scute* complex (AS-C), a 100 kilobase stretch of the X-chromosome comprised of four genes and numerous regulatory elements, provides cells with neural potential, allowing them to develop as bristle organs (García-Bellido and Santamaría, 1977; Villares and Cabrera, 1987; Ghysen and

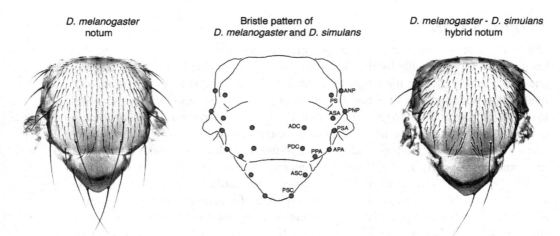

Figure 26.1. The bristle pattern on the dorsal notum of *D. melanogaster, D. simulans* and a *melanogaster/simulans* hybrid. A wild-type notum of *D. melanogaster* is shown, together with a hybrid thorax in which a number of bristles can be seen to be missing. The two parental species display an identical bristle pattern of eleven macrochaetes the names of which are abbreviated as follows: ANP, anterior notopleural; PNP, posterior notopleural; PS, presutural; ASA, anterior supraalar, PSA, posterior supraalar; APA, anterior postalar; PPA, posterior postalar; ADC, anterior dorsocentral; PDC, posterior dorsocentral; ASC, anterior scutellar; PSC, posterior scutellar. Reprinted from Skaer and Simpson (2000). Copyright (2000), with permission from Elsevier.

Dambly-Chaudière, 1988; Cubas et al., 1991; Huang, Dambly-Chaudière, and Ghysen, 1991). The AS-C genes are initially driven by autosomal transcription factors that bind various regulatory elements in the AS-C. However, to ensure acquisition of neural fate, the products of the AS-C genes must be maintained at high levels in the bristle cell progenitor (known as a Sensory Organ Precursor cell, or SOP). This involves an auto-feedback loop where the proteins of the AS-C genes drive expression of the AS-C genes by binding regulatory enhancers in the AS-C itself. If the levels of the AS-C proteins are not sufficient in the SOP, or if these high protein levels are not maintained, the SOP loses its neural potential and adopts an epidermal cell fate. This can result in loss of the bristle organ.

D. *melanogaster* and D. *simulans* will mate under normal laboratory conditions, and give rise to infertile, but otherwise viable, hybrid progeny. However, these progeny display a singular defect: elements of their notal bristle pattern are absent (Figure 26.1). If the macrochaetes are present, then they arise in the correct positions, but in each fly a variable number of the eleven are not present. Uncovering the genetic explanation of this defect is the aim of the experiments in this chapter.

MATERIALS AND METHODS

EQUIPMENT AND MATERIALS

The equipment necessary for handling the flies is described in Chapter 14.
Standard dissecting microscopes are sufficient for analysing bristle loss on hybrid progeny.

Fine forceps are required to manipulate flies when being scored.

Reliable incubators set at 25°C and 18°C are required.

A water bath capable of maintaining 37°C is required for Experiment 2.

Biological material. A wild-type stock of *D. melanogaster*, preferably Oregon-R. A wild-type stock of *D. simulans*, preferably S12a (Umea stock centre).

For Experiment 1, the following mutant strains of *D. melanogaster* are required:

Df(1)sc^{B57} w s/FM7
Df(1)sc^2 w s/FM7c

For Experiment 2 the following mutant strains of *D. melanogaster* are required:

y; pr cn Dp(1;2)sc^{19}/CyO
HSSC-2

For Experiment 3, the following mutant strain of *D. melanogaster* is required:

Compound-1DX y f/Y

These stocks can be requested from the authors or from international stock centres (e.g., Umea). For further description of the mutations and rearrangements, see Lindsley and Zimm (1992).

PROCEDURES FOR ESTABLISHING CROSSES AND ANALYSIS OF THEIR RESULTS

Maintenance of the laboratory and the fly stocks is as described in Chapter 14. A description of how to distinguish male from female flies is given in Chapter 11 (exactly the same procedures can be applied to *D. simulans*). All crosses must be carried out according to the following protocol because the effects of temperature and overcrowding in the vials can bias results. Temperature, in particular, strongly influences results; thus, flies should be raised at the temperature specified in the experiment and all crosses should be performed at this temperature. Care must be taken not to confuse the *D. melanogaster* and *D. simulans* stocks because they are virtually identical. All crosses are between *D. melanogaster* females and *D. simulans* males because the reverse cross is very difficult to perform. The hybrid progeny of these crosses are all female, because a copy of the *D. simulans* X chromosome is required for viability, hence male hybrids from the cross in this direction are not viable (Lemeunier et al., 1984; Hutter, 1997). Hybrid females carry one X chromosome (with the Achaete-Scute Complex) from each parent species. Virgin female *D. melanogaster* must be collected over a period of no longer than seven days and maintained at the temperature specified in the experiment. Virgin flies can be collected by emptying culture vials and collecting flies which emerge over the following 6 hours. Male *D. simulans* flies should be between 7 and 14 days old. Four simultaneous replicas of each cross should be set up, each containing 25 females and 25 males, and allowed to mate and lay for a period of seven days before being transferred to a fresh vial. Wherever possible, for each replica, only the first 25 progeny

Table 26.1.

	PS	ANP	PNP	ASA	PSA	APA	PPA	ADC	PDC	ASC	PSC
Dfsc^{B57}											
Bristles											
absent	37	28	117	52	199	49	45	94	45	29	52
Total											
bristles	200	200	200	200	200	200	200	200	200	200	200
% ABSENT	18.50	14.00	58.50	26.00	99.50	24.50	22.50	47.00	22.50	14.50	26.00
M.S.E.	2.75	1.83	2.22	0.82	0.50	4.99	3.95	3.00	2.06	1.89	5.48

TOTAL % AVERAGE NOTAL BRISTLE LOSS: 33.95

Balancer *FM7*											
Bristles											
absent	4	3	13	17	105	27	0	17	8	1	2
Total											
bristles	200	200	200	200	200	200	200	200	200	200	200
% ABSENT	2.00	1.50	6.50	8.50	52.50	13.50	0.00	8.50	4.00	0.50	1.00
M.S.E.	1.15	0.96	1.71	2.87	2.63	1.50	0.00	3.40	0.82	0.50	0.58

TOTAL % AVERAGE NOTAL BRISTLE LOSS: 8.95

test		0.0082	0.0036	0.0007	0.0069	0.0003	0.1367	0.0107	0.0071	0.0067	0.0075	0.0160

Cross: *Dfsc*^{B57}/*FM7* D. melanogaster female × D. simulans males.
Conditions: Four individual crosses of equal numbers of *simulans* males and *melanogaster* virgin females set up (1A, 1B, 1C, and 1D) in medium sized tubes at 25 °C. All crosses had 25 males and 25 virgin females. Females had been raised at 18 °C. Crosses were left for one week to mate and lay, then, flies were removed.
Analysis: The first twenty five hybrid progeny carrying the *Dfsc*^{B57} X-chromosome were selected and notal bristle loss was recorded for each hemithorax, and the first 25 hybrid progeny carrying the *FM7* chromosome were selected and notal bristle loss was recorded for each hemithorax.

to eclose should be scored for bristle loss on both heminota (therefore, for each cross $4 \times 25 \times 2 = 200$ heminota are scored). Progeny can be collected and stored in 50% ethanol until it is convenient to score them. Flies should be manipulated under the microscope with the aid of fine forceps. They may be kept in 50% ethanol at room temperature for an indefinite period. For statistical purposes, data sets from the individual replicates should be scored separately. Bristle loss must be recorded for each of the eleven bristles on each heminotum individually. These data are best recorded by constructing a grid with the abbreviated names of the eleven macrochaetes (see Table 26.1) along the top, and numbers 1 through 50 down the side, representing the 50 heminota to be scored from that replicate of the cross. If possible, data should then be logged in a spreadsheet package such as Microsoft Excel, because this greatly facilitates data handling and subsequent statistical analysis. Bristle loss should be expressed as a percentage of the 50 heminota scored in each of the four replications of the cross, and the final percentage bristle loss is the mean of these four values. Scoring the four replicates separately allows the mean standard error for the cross to be calculated, as well as

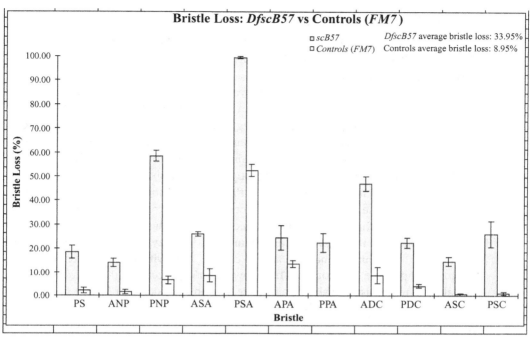

Figure 26.2. The data recording sheet obtained by the authors for a single cross of *Df(1)sc^B57* *D. melanogaster* females with *D. simulans* males (Experiment 1) (see also Table 26.1). The cross was replicated four times (1A, 1B, 1C, 1D), and 25 hybrid progeny from each replicate carrying the deficiency were analysed for loss of the eleven notal macrochaetes on both heminota. Hybrid progeny carrying the balancer chromosome were used as controls and analysed for bristle loss separately. The total loss of each bristle and the Mean Standard Error (M.S.E.) were calculated and expressed as a percentage of the 200 heminota scored, and the overall bristle loss was also determined. A Student's *t*-test was calculated for each bristle to establish whether there was a significant effect on bristle loss caused by the deficiency. The results were graphed plus or minus the M.S.E. with bristle loss on the Y axis and the bristle on the X axis. After Skaer and Simpson (2000).

statistical tests of significance. The mean standard error is an expression of our confidence in the mean that has been calculated; the lower its value, the greater the confidence that the mean is of significance. The formula for the mean standard error is the standard deviation of the four replicates, divided by the square root of the number of replicates, which in all cases will be $\sqrt{4} = 2$. Data should be represented graphically with each bristle on one axis and percent bristle loss on the other axis showing plus and minus error bars of the calculated mean standard error. For tests of significance between data sets, a two-tailed *t* test should be employed. A significant difference is accepted when the *t* test returns a p-value of 0.05 or less. An example data sheet and graph from Experiment 1 (authors' results) is illustrated in Table 26.1 and Figure 26.2 (notal bristles). Expected results are very unlikely to exactly replicate those quoted in this chapter, due to differences in fly stocks and environmental conditions. However, if procedures are followed with a degree of rigour, statistically significant results should be able to be replicated.

OUTLINE OF THE EXPERIMENTS

EXPERIMENT 1. THE EFFECT ON HYBRID BRISTLE LOSS OF INTRODUCING DELETION CONSTRUCTS OF THE Achaete-Scute LOCUS FROM THE *D.* melanogaster PARENT

There are two deletions of the AS-C employed in this experiment: $Df(1)sc^{B57}$ is a deletion of the entire AS-C, while $Df(1)sc^2$ is an internal deletion within the AS-C which removes an enhancer responsible for driving expression of the bristle formation genes *achaete* and *scute* in the region of the presumptive notum that gives rise to the scutellar bristles (ASC and PSC). Neither deletion has a phenotype in the heterozygous condition but it has been established through clonal analysis that flies homozygous $Df(1)sc^{B57}$ lose all their notal bristles, while flies homozygous for $Df(1)sc^2$ lose the ASC and PSC bristles. The deletions are "balanced" by the chromosomes *FM7* and *FM7c*. These homologous chromosomes prevent recombination thereby allowing the deletion to be maintained in the stock, but are themselves homozygous sterile. The balancer genes carry the dominant marker *Bar eye* causing flies with the balancer to exhibit a 'kidney-shaped eye' phenotype. This allows hybrid progeny of the crosses carrying the balancer to be distinguished from those carrying the deletion (which do not possess the *Bar eye* phenotype). They also carry the recessive marker *yellow* so that flies that are homozygous for the balancer, and therefore sterile, may be distinguished in the parental stock when setting up crosses. The recessive marker genes *white(w)* and *sable(s)* are also found on the deletion chromosomes, but need not be considered in this experiment.

The two crosses should be set up according to the conditions described above using virgin *D. melanogaster* females heterozygous for the deletion and the balancer, and *D. simulans* males. All flies should be reared, maintained and crossed at 25°C.

P1	*D. melanogaster* female		*D. simulans* male
	Deletion/Balancer	×	X⁺ Y
F1 Hybrid Females	Deletion/+ or Balancer/+		

Both types of progeny should be collected and scored as described earlier.

Internal control. The progeny carrying the balancer chromosome act as an internal control for the hybrid progeny carrying the deletion. It has been established that these balancers do not affect bristle loss in the hybrids (Skaer and Simpson, 2000).

EXPERIMENT 2. THE EFFECT OF SUPPLYING HYBRIDS WITH AN EXTRA COPY OF THE AS-C VERSUS SUPPLYING AN EXOGENOUS SOURCE OF SCUTE AT DIFFERENT PERIODS DURING HYBRID DEVELOPMENT

In this experiment, all flies should be reared, maintained, and crossed at 18°C. The mutant $y;\ pr\ cn\ Dp(1;2)sc^{19}/CyO$ bears an extra copy of the entire *D. melanogaster* AS-C on the second chromosome. This duplication is balanced by *Curly of Oster (CyO)*, a dominant marker that causes flies carrying it to exhibit wings which are curled back over themselves. *D. melanogaster* flies carrying one copy of this duplication exhibit occasional extra bristles.

Internal control. A cross between – *y; pr cn Dp(1;2)sc¹⁹/CyO D. melanogaster* females and *D. simulans* males should be set up and scored as described above, and the balancer carrying *CyO* progeny should be scored as an internal control as in Experiment 1.

The *HSSC-2* mutant strain of *D. melanogaster* carries a construct consisting of an extra copy of the *scute* gene under the control of the *hsp70* promoter on the second chromosome. When the temperature is raised to 37°C the promoter is activated and the developing animal is subjected to ubiquitously raised levels of the *Scute* protein. This stock is not balanced and is maintained in a homozygous condition. The cross should be set up as described above using *HSSC-2 D. melanogaster* females and *D. simulans* males, but the number of vials should be quadrupled, i.e. 16 replicates. When hybrid pupae start to appear in the vials in significant numbers, eight of the vials should be cleared of pupae, and placed in a water bath heated to 37°C for a period of 3 hours. The vials are then returned to 18°C for a further 24 hours after which pupae are collected. If less than 25 individuals per vial pupate during this period, collect pupae from the extra 4 vials until 100 pupae have been collected. The same vial must not be heat shocked more than once, and if the extra vials are required, data should be scored separately. Transfer the collected pupae to fresh vials, return them to 18°C to complete development, and score the progeny when they emerge.

Clear the remaining 8 vials of pupae as above and leave them for a further 24 hours. Remove the pupae that appear during this period, transfer them to a fresh vial, and maintain them at 18°C for an additional 24 hours. Then, heat shock again for a three-hour period as above. After this return them to 18°C to complete pupal development, and score on emergence.

Control. As the *HSSC-2* strain is maintained in a homozygous condition, internal controls are not available in this case. As a substitute control, conduct the above protocol, but instead of using *HSSC-2* females replace them with females of a wild-type *D. melanogaster* strain such as Oregon-R. A further control may be performed in which *HSSC-2* vials are not subjected to a heat shock. This will establish the background level of bristle loss in these hybrids.

EXPERIMENT 3. THE EFFECT OF HYBRID SEX ON BRISTLE LOSS

As described above, all crosses so far give rise to female hybrids owing to the difficulty of crossing *D. melanogaster* males with *D. simulans* females to produce male hybrid progeny. This practical problem can be negotiated by using a *D. melanogaster* compound X stock. Females of this stock have a pair of fused X chromosomes, but in addition, also carry a Y chromosome. When crossed to *D. simulans* males, the *melanogaster* Y will segregate with the *simulans* X to generate male hybrid progeny. It has been established that male hybrids generated in this manner exhibit the same bristle loss as those resulting from performing the difficult cross of *D. simulans* females to *D. melanogaster* males (Skaer and Simpson, 2000).

P1   *D. melanogaster* female   *D. simulans* male
                                        XX Y   ×   X Y
     F1 Hybrid Males   X Y

The cross should be set up as described above.

Control. Compare the results to a control cross of *D. melanogaster* Oregon-R females crossed to *D. simulans* males (which will yield female hybrids). Each cross should be conducted at 18°C.

EXPECTED RESULTS

EXPERIMENT 1. THE EFFECT ON HYBRID BRISTLE LOSS OF INTRODUCING DELETION CONSTRUCTS OF THE Achaete-Scute LOCUS FROM THE *D.* Melanogaster PARENT

The expected frequencies of bristle loss in hybrids carrying the $Df(1)sc^{B57}$ chromosome are numerically and graphically represented in Figure 26.2. The expected frequencies of bristle loss for hybrids carrying the $Df(1)sc^2$ chromosome are graphically represented in Figure 26.3. In the $Df(1)sc^{B57}$ hybrids, all eleven bristles are lost at a greater frequency than in the *FM7* control hybrid flies, and the values of the *t*-tests demonstrate that this greater loss is significant in all cases except in the APA bristle (see Figure 26.2, p = 0.13). Overall, bristles are lost at a frequency of 33.95% in the $Df(1)sc^{B57}$ hybrids versus 8.95% in the *FM7* control hybrids. From the graph in Figure 26.3, it can be seen that only the ASC and PSC bristles in hybrids carrying the $Df(1)sc^2$ chromosome are lost at a significantly higher frequency than in the controls.

The results of Experiment 1 allow us to conclude that when the *melanogaster* copy of the AS-C is removed in female hybrid flies, leaving only the *simulans* copy of the AS-C, there is a significant increase in bristle loss. Similarly, if the *D. melanogaster* enhancer responsible for driving expression of the AS-C genes in the region that gives rise to the ASC and PSC bristles is removed, as occurs in $Df(1)sc^2$ hybrids, the loss of those bristles is significantly exacerbated. Thus, a reduction in the product levels of the AS-C genes results in increased loss of bristles in the hybrid.

EXPERIMENT 2. THE EFFECT OF SUPPLYING HYBRIDS WITH AN EXTRA COPY OF THE AS-C VERSUS SUPPLYING AN EXOGENOUS SOURCE OF SCUTE AT DIFFERENT PERIODS DURING HYBRID DEVELOPMENT

Following the conclusion in Experiment 1 that reducing copies of AS-C proteins causes an increase in bristle loss, it might be expected that supplying hybrids with an extra copy of the AS-C via the $Dp(1;2)sc^{19}$ *D. melanogaster* stock would alleviate bristle loss. As can be seen from Figure 26.4a, this proves not to be the case: bristles are lost with

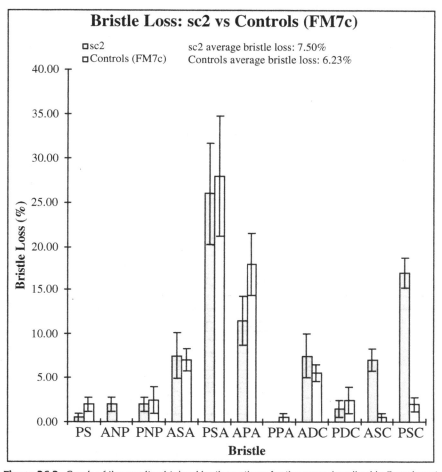

Figure 26.3. Graph of the results obtained by the authors for the cross described in Experiment 1.

approximately the same frequency. Therefore simply providing an extra template for increasing AS-C gene product levels is insufficient to alleviate bristle loss. However, hybrids supplied via the heat shock construct with an exogenous supply of the AS-C protein, Scute, do exhibit a significant rescue in bristle loss by comparison with controls (see Figure 26.4b). Notably this reduction in bristle loss is observed only when the heat shock is applied early, during the period in which endogenous Scute is active and the SOPs are being formed. Therefore, an extra copy of the AS-C does not reduce loss of bristles in the hybrid, while an augmented supply of the AS-C protein, Scute, does alleviate bristle loss as long as it is provided at the correct time in development. This allows us to conclude that a deficit of AS-C genes is likely to be the cause of bristle loss in the hybrid. Further, it suggests that, unlike in *D. melanogaster*, the extra copy of the AS-C provided by the duplication does not increase transcription of the AS-C genes in the hybrid.

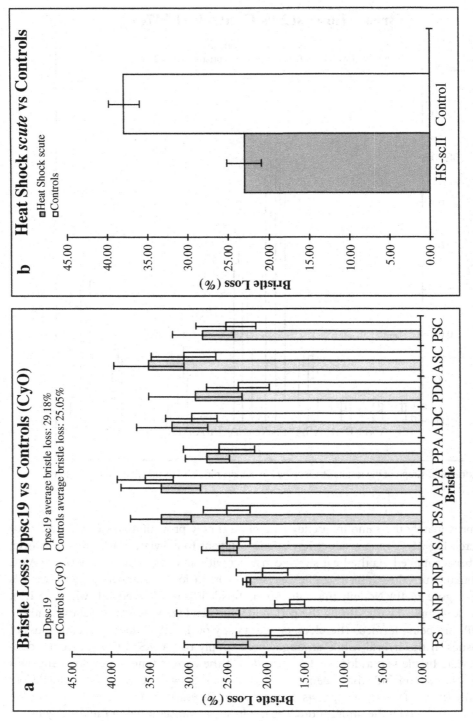

Figure 26.4. Graphs of the results obtained by the authors for the crosses described in Experiment 2.

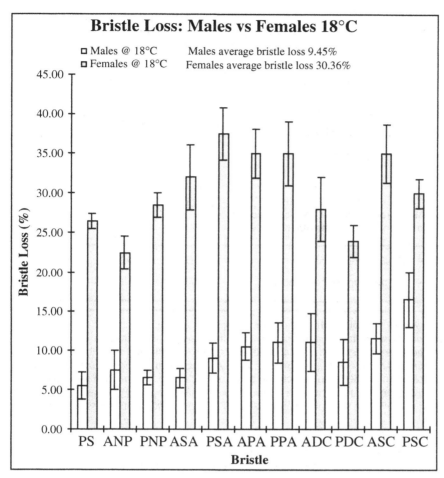

Figure 26.5. Graph of the results obtained by the authors for the cross described in Experiment 3. After Skaer and Simpson (2000).

EXPERIMENT 3. THE EFFECT OF HYBRID SEX ON BRISTLE LOSS

The effect of sex on hybrid bristle loss can be seen in Figure 26.5. In females, bristles are lost at a greater frequency; this is true for every bristle and is statistically significant for all bristles except the APA. Overall, males lose bristles at 9.45% while females lose bristles at 30.36%. Notably, males possess one X chromosome, hence one copy of the AS-C, while females have one X chromosome from each parent, and consequently both a *simulans* and *melanogaster* copy of the AS-C.

DISCUSSION

From the three experiments described in this chapter, we have concluded that altered (lowered) transcription of the AS-C genes is implicated in bristle loss in hybrids, and that restoring levels of the AS-C protein, Scute, alleviates bristle loss. However,

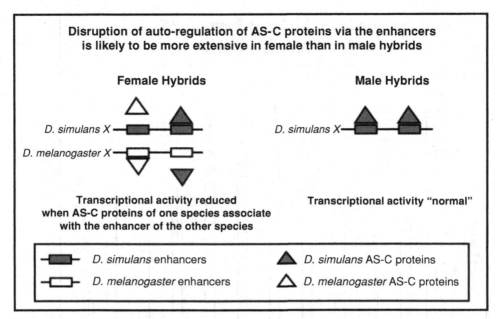

Figure 26.6. Model indicating the mechanism proposed to explain the loss of bristles in *melanogaster/simulans* hybrids. The hypothesis is that transcriptional activators encoded by the *melanogaster* genome bind less well to, or activate transcription less efficiently from, the regulatory elements of the *simulans achaete-scute* genes and vice versa. Autoregulation of *achaete-scute* via the sensory organ precursor enhancers (SOP) and Achaete (Ac) and Scute (Sc) will be less problematic in males with a single X-linked *achaete-scute* locus, because both activators and target regulatory sequences are from the same species. Reprinted from Skaer and Simpson (2000). Copyright (2000), with permission from Elsevier.

supplying hybrids with an extra copy of the AS-C does not appear to raise levels of AS-C proteins in the hybrid as bristle loss is not rescued. It is not immediately obvious why there are profound differences in loss of bristles between male and female hybrids. To more fully explain these results, further knowledge of AS-C transcription is required. In the Introduction it was mentioned that genes of the AS-C are initially transcribed in response to autosomal transcription factors that bind various regulatory elements in the AS-C. Transcription is then reinforced in SOPs via a feedback loop within the AS-C. The probable reason why levels of AS-C transcription are diminished in hybrids is because transcription factors from one species are binding regulatory elements from the other species, which either blocks, or leads to inefficient transcription of the AS-C genes. This is why supplying an extra copy of the AS-C via the duplication (in Experiment 2) does not lead to increased production of the AS-C proteins. The reason why males and females hybrids show differences in bristle loss is more complex. As stated, males have one copy of the AS-C, while females have two: one from each parent species. The feedback loop of AS-C transcription which reinforces levels of the AS-C proteins in the SOPs is therefore different in male and female hybrids; in the males all elements of the loop are derived from the single AS-C whereas in females the AS-C proteins of one species can bind the regulatory DNA in the AS-C of the other species which leads to inefficient transcription. This situation is illustrated in Figure 26.6. The AS-C proteins of

D. simulans may not work as efficiently on the regulatory DNA of *D. melanogaster*, and vice versa, because of small divergences, most likely in the regulatory DNA, between the two species. The function of the AS-C genes and the phenotype (i.e. the bristle pattern) remains the same between the two species, despite these divergences, because of a phenomenon called developmental homeostasis. This occurs when the expression of a gene has a significant functional importance in the development of the organism, and is thus under strong stabilising selection. Hence when nucleotide changes occur in the DNA that regulate such genes, compensatory changes at other sites in the regulatory DNA, which balance the initial changes, are selected for. This creates a "turnover" effect in the regulatory sequence, and ensures that while changes are occurring in the sequence that regulates a gene's expression, the expression of the gene itself remains unaltered.

TIME REQUIRED FOR THE EXPERIMENTS

Setting up crosses is rapid, although stocks must be amplified to such a degree that sufficient virgin females are being produced. This will require ordering stocks at least one month prior to the planned experiments conducted at 25°C and 2 months for the experiments conducted at 18°C (and earlier still if the group is large). Collection of eclosing progeny and the time delay involved in the temperature shift experiment (2) may cause scheduling problems, and should be taken into account when planning the practical timetable. Staff may be required to aid at these steps. Analysis of progeny is time consuming and students should work in pairs, one at the microscope, the other recording data. Logging data for one cross (i.e. 200 heminota) will take one pair of students approximately 2 hours.

POTENTIAL SOURCES OF FAILURE

In all experiments care must be taken when constructing crosses and handling stocks. Females must be virgin, and stocks must not be contaminated. Incorrect identification of the dominant markers on the balancer chromosomes will cause controls and experimental hybrids to be confused. Specified temperatures must be observed because these strongly influence results. Clear recognition of the eleven macrochaetes is important – other macrochaetes are present on the thorax (a pair of anterior bristles, the humerals, and laterally, the sternopleurals) and it should be demonstrated to students which bristles must be scored and which ignored. In Experiment 2 the 37°C heat shock must be no longer than three hours or larvae will die.

TEACHING CONCEPTS

Hybridisation and speciation. The experiments in this chapter illustrate the issue of the species concept, and how hybridisation between even very recently diverged species leads to incorrect gene regulation in the hybrid progeny. This is a driving force for the evolution of reproductive isolating mechanisms as hybridisation leads to infertile progeny.

Developmental homeostasis (phenotypic evolution versus genomic evolution). The conclusions convey the idea of the uncoupling of genomic and phenotypic evolution: changes at the level of the nucleotide sequence are not mirrored by concomitant changes in the phenotype of the animal. This is because the phenotype is under stabilising selection and the further nucleotide changes that compensate for the initial changes are selected for. Hence, there is genetic turnover but phenotypic stasis.

Gene regulation and control. The experiments provide a framework in which students may visualise how genes are regulated by transcription factors binding enhancer elements, and how a breakdown in gene regulation can lead to insufficient protein production with consequences for cell fate determination.

ALTERNATIVE EXERCISES

The effect of temperature on bristle loss. Experiments 1 and 3 are conducted at 18°C and 25°C respectively. Repeat the experiments at the reversed temperatures, and compare to results at the initial temperature. As previously mentioned, the effect of temperature on bristle loss in the hybrids is profound and yet there are remarkable differences between the sexes in bristle loss at the two different temperatures. Why do you think this might be?

QUESTIONS FOR FURTHER ANALYSIS
➤ In Chapter 23 the transgenic line *neuralized-lac Z* is used which permits the staining of SOP in imaginal discs. How could you prove that bristle loss is due to SOP loss?
➤ Bristle loss is strongly temperature dependent. How could you investigate this aspect of bristle loss further? Could this temperature sensitivity be linked to events that occur during different periods of development? For additional information see Skaer and Simpson (2000).

REFERENCES
Cubas, P., de Celis, J.-F., Campuzano, S., and Modolell, J. (1991). Proneural clusters of *achaete-scute* expression and the generation of sensory organs in the *Drosophila* imaginal wing disc. *Genes Dev.*, 5, 996–1008.

García-Bellido, A., and Santamaría P. (1977). Developmental analysis of the *achaete-scute* genes of *Drosophila melanogaster*. *Genetics*, 88, 469–86.

Ghysen, A., and Dambly-Chaudière, C. (1988). From DNA to form: The *achaete-scute* complex. *Genes Dev.*, 2, 495–501.

Huang, F., Dambly-Chaudière, C., and Ghysen, A. (1991). The emergence of sense organs in the wing disc of *Drosophila*. *Development*, 111, 1087–95.

Hutter, P. (1997). Genetics of hybrid inviability in *Drosophila*. *Adv. in Genet.*, 36, 157–85.

Jan, Y. N., and Jan, L. Y. (1995). Maggot's hair and bug's eye: Role of cell interactions and intrinsic factors in cell fate specification. *Neuron*, 14, 1–5.

Lemeunier, F., David, J. R., Tsacas, L., and Ashburner, M. (1984). The *melanogaster* species subgroup. In *The Genetics and Biology of Drosophila*, eds. M. Ashburner, H. L. Carson, and J. N. Thompson, vol. 3e, pp. 275–94. London: Academic Press.

Lindsley, D. L., and Zimm, G. (1992). *The Genome of Drosophila melanogaster*. San Diego: Academic Press Inc.

Skaer, N., and Simpson, P. (2000). Genetic analysis of bristle loss in hybrids between *Drosophila melanogaster* and *D. simulans* provides evidence for divergence of *cis*-regulatory sequences in the *achaete-scute* complex. *Dev. Biol.*, 221, 148–67.

Villares, R., and Cabrera, C. V. (1987). The a*chaete-scute* gene complex of *D. melanogaster*. Conserved domains in a subset of genes required for neurogenesis and their homology to *myc*. *Cell*, 50, 415–24.

SECTION XII. COMPUTATIONAL MODELLING

27 Theories as a tool for understanding the complex network of molecular interactions

H. Meinhardt

INTRODUCTION

Development is controlled by a highly complex network of interactions in which non-linear feedback loops play an important role. Our intuition is not reliable to predict the properties of such complex networks. In many branches of natural science theoretical approaches are an integral part of understanding the behaviour of complex systems. In this chapter, using the formation of an organizing region as an example, a theoretical approach will be used to determine what type of interactions allow pattern formation, and how one can formulate a hypothetical interaction into a set of differential equations convenient for computer simulations. Finally, the postulated interaction is compared with known experimental observations.

HOW TO GENERATE AN ORGANIZING REGION

In Chapters 1 and 5, the freshwater polyp hydra was discussed as a model organism. The hydra hypostome had long been known to be an organizing region (Browne, 1909, Lenhoff, 1991). Even after dissociation into individual cells, upon re-aggregation complete animals are formed (Gierer et al., 1972), showing that pattern formation is possible even if the initial situation is more or less uniform. We proposed that pattern formation is accomplished by local self-enhancement and long-ranging antagonistic effects (Gierer and Meinhardt, 1972, Meinhardt, 1982). A simple and molecularly feasible realization would consist of two interacting substances. One substance, called the activator, has an autocatalytic feedback on its own synthesis. It controls the production of the second substance, the inhibitor. The inhibitor spreads more rapidly and suppresses the self-enhancement of the activator. If the region in which the reaction can take place is sufficiently large a homogeneous distribution is unstable. Locally high activator concentrations will emerge (Figure 27.1). In fields that have an extension comparable with the range of the activator, polar patterns are generated with a high concentration at a marginal position. In larger fields, either isolated maxima or stripe-like distributions

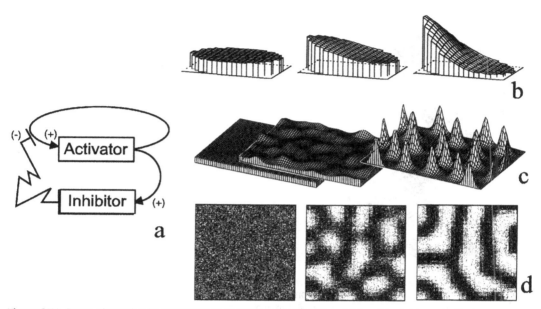

Figure 27.1. Pattern formation by autocatalysis and long-range inhibition. (a) A molecule termed activator is assumed that catalyses its own production and that of its highly diffusing antagonist, the inhibitor (Equations 1a, b). (b–d) Simulations: shown are the initial, an intermediate and a final concentration. (b) In fields that are small in comparison to the range of the activator, preferentially a single maximum is formed. The maximum can act as an organizing region because the resulting graded distributions are convenient to supply positional information for the surrounding cells. (c) In a region large compared with the range of the inhibitor, several maxima can be formed. (d) If the self-enhancement saturates (replacing a^2 by $a^2/(1 + ka^2)$ in Equation 1), stripe-like distributions emerge. Because of the saturation, the maxima reach an upper bound. Their limited height leads to an enlargement of the activated region. However, due to the lateral inhibition, activated cells prefer to have non-activated neighbors. This condition is satisfied with stripes: all activated cells have activated neighbors and, nevertheless, non-activated cells are close by.

can be generated. The mechanism proposed accounts for the extensive self-regulation that has been observed experimentally. After removal of the region of high activator concentration (the organizing region), the remnant inhibitor will decay in the remaining cells until the activator production sets in again (see Figure 27.2). Thus, this model can account for one of the most fundamental problems in developmental biology: the generation and maintenance of a pattern from an initially uniform situation. Although the genetic information is the same in all cells, a predictable number of cells will behave differently because they are exposed to a different signal. It is worthwhile to note that pattern formation from uniform initial situations is by no means a privilege of living systems. The formation of sand dunes in a desert is an example. Dune formation is based on the same mechanism. The homogeneous distribution of sand is unstable: a small local perturbation becomes amplified because more sand collects behind a wind shelter, and so on. Of course, an embryo is far more complex than a sand dune. But this example illustrates that de novo pattern formation by self-enhancement coupled with antagonistic effects of long range is an everyday experience.

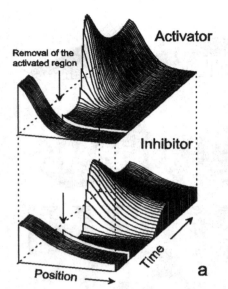

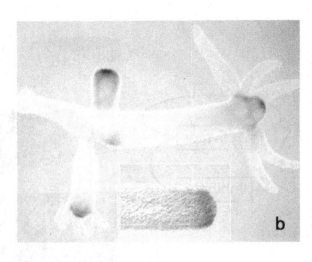

Figure 27.2. Regeneration. (a) Simulation: after removal of the activated region (arrow) a new activator maximum is formed after the decay of the remnant inhibitor (after Meinhardt, 1982). (b) In hydra, after head removal, a new gradient of *β-catenin* becomes established (Hobmayer et al., 2000). The insert shows the re-appearance of *β-catenin* transcription 1h after head removal (figure kindly supplied by Bernd Hobmayer). The molecular basis of the long-ranging inhibition is still unclear. It has been shown that the hydra patterning corresponds to an ancestral axis formation. The foot corresponds to the forebrain/heart patterning in vertebrates; the hypostomal opening is the most posterior structure. Nearly the whole body of the ancestral organism became converted into the brain. The trunk became later inserted during evolution in a narrow region between the tentacles and the posterior opening (Meinhardt, 2002). A budding and fully extended hydra can be upto 5 mm long. (b) Reprinted with permission from *Nature* (Hobmayer et al., 2000). Copyright (2000) McMillan Magazines Limited.

FORMULATION OF A PATTERN-FORMING INTERACTION

In developmental biology, models are usually formulated in a nonmathematical way. There is no check on whether the postulated and the real properties agree. For example, in many textbooks the selection of a neural precursor cell in *Drosophila* is assumed to be governed by next-neighbor lateral inhibition. In the drawings, the central cell of a cluster of equivalent cells always "wins" (i.e., it becomes neural) because it is known from experiments that this is the case. However, a mathematical formulation of such a model would reveal a different behavior: cells at the margin of the cluster would become neural because these cells are exposed to the lateral inhibition of fewer cells. Thus, a mathematical formulation would reveal that this model is wrong or incomplete. Further genetic evidence suggest that genes, such as *extramacrochaetae*, might be involved in defining the exact position of the final bristle (Cubas and Modolell, 1992).

A model has to describe the time course of the concentration profiles of the substances involved. This is possible using equations that describe the concentration *changes* at a particular position in a short time interval as a function of the other substances. Adding these concentration changes to the given initial concentrations yields to the concentration at a somewhat later time. This is the starting concentration to calculate the next change, and so on. Thus, repetitions of such calculations lead to the concentration profiles as functions of time. Three factors are expected to play a major role in the concentration change: the rate of production, the rate of removal (or decay), and the loss or gain due to an exchange with neighboring cells – for instance, by diffusion.

A possible equation that describes the concentration change of the activator a and inhibitor concentration b in a pattern-forming interaction is as follows (Gierer and Meinhardt, 1972, Meinhardt, 1982; Meinhardt and Gierer, 2000):

$$\frac{\partial a}{\partial t} = \frac{sa^2}{b} - r_a a + D_a \frac{\partial^2 a}{\partial x^2} + b_a \tag{1a}$$

$$\frac{\partial b}{\partial t} = s'a^2 - r_b b + D_b \frac{\partial^2 b}{\partial x^2} + b_b. \tag{1b}$$

where t is time, x is the spatial coordinate, D_a and D_b are the diffusion coefficients, and r_a and r_b the decay rates of a and b. The first term in (1a) describes the self-enhancement of the activator. It must be nonlinear (a^2) to overcome the activator removal by the normal decay ($-r_a a$). The term s, the source density, describes the ability of the cells to perform the autocatalysis (i.e., the competence). The net exchange by diffusion is proportional to the change of the concentration change (second derivative; $\partial^2 a/\partial x^2$) for the following reason: the net exchange is, of course, zero if all cells have the same concentration. However, the net exchange is also zero if a linear concentration gradient exists because any cell would lose exactly the same amount to its lower neighbor as it gains from its higher neighbor. A small activator-independent activator production b_a can initiate the system at low activator concentrations – for instance to accomplish regeneration. After removal of an established maximum (organizing region), this baseline activator production leads to a low level of activator concentration from which the autocatalysis can start again (see Figure 27.2). The expression b_b denotes the basal (= activator-independent) production rate of the inhibitor b.

In Equations 1a and 1b, the concentrations are assumed to vary continuously as functions of position. For computer simulation, the available space and the time have to be subdivided into discrete entities. Equation 1a can be approximated by

$$da_{i,t} = s\,a_{i,t}^2/b_{i,t} - r_a a_{i,t} + D_a(a_{i-1,t} + a_{i+1,t} - 2a_{i,t}) + b_a, \tag{1c}$$

where $i = 1, 2. \ldots$ is the number of the cell in a linear array of cells and $da_{i,t}$ the change of the activator concentration in the cell i in a short time interval dt. This allows the calculation of the activator and, analogously, the inhibitor change in each cell in a given time unit. Adding these changes to the actual concentrations leads to the new profiles at the later moment; $a_{i,t+1} = a_{i,t} + da_{i,t}$. The total time course is calculated from a repetition of many such iterations. In all simulations shown, the boundaries are assumed to be impermeable. This is achieved by assuming virtual cells beyond the boundaries that have the same concentrations as the cells at the borders. The net exchange between two cells that have the same concentration is, of course, zero. A simple program is provided on the Web page of the book. The program runs on a PC and can be modified. More complex computer programs for other developmental situations are given elsewhere (Meinhardt, 2003).

A condition for the formation of stable patterns is that the diffusion of the activator (or any other mode of spread) is much slower than that of the inhibitor, i.e., $D_a \ll D_b$ must be satisfied. Further, the activator has to have a longer time constant than the inhibitor, $r_a > r_b$, otherwise oscillations would occur.

Several molecular systems are known that are close to the proposed scheme. For instance, *Nodal* is required for mesoderm formation and for the left–right patterning in vertebrates. It has an autoregulatory feedback on its own production which is regulated by *lefty-2*, a longer range factor (Chen and Schier, 2002). In organizer formation in higher organisms, the *Wnt/β-catenin* pathway plays an essential role (see Figure 27.2). This pathway is well conserved. *β-catenin* RNA isolated from the Hydra injected into an early *Xenopus* embryo can trigger a second embryonic axis (Hobmayer et al., 2000). Evidence for autoregulation exists for this system too, but the long-ranging component is yet unknown. For hydra, elegant transplantation experiments have revealed the ranges of the activating and inhibitory influences (Technau et al., 2000). The observed ratio agrees precisely with the theoretical expectation. An autocatalytic and an inhibitory component have also been observed in the signaling of the blue–green algae *Anabaena* (Buikema and Haselkorn, 2001; Yoon and Golden, 1998), which forms long chains of cells in which every 7th cell becomes a nitrogen-fixing cell. Likewise, in *Drosophila*, self-enhancement and long-range inhibition are involved in the patterning of dorsal appendages of the egg (Wasserman and Freeman, 1998) and in the selection of sensory mother cells (see Culí and Modolell, 1998, and Chapter 22).

FORMATION OF PERIODIC STRUCTURES: BRISTLE FORMATION AS AN EXAMPLE

As mentioned in Figure 27.1, the mechanism leads to a periodic pattern in space if the range of the inhibitor is smaller than the total field. Several observations can be described by this mechanism. Wigglesworth (1940), for instance, found that in the bug *Rhodnius prolixus* new bristles become inserted after each mold and that the newly formed bristles emerge at largest possible distance from the older one. If we assume that the existing bristles are generated by high activator concentration, the distances between the existing maxima increase due to growth. In regions between existing maxima the inhibitor concentration becomes lower and lower until a new activator maximum is triggered from a low level of baseline activator production (term b_a in Equation 1a). With the local onset of activator production, the production of inhibitor also rises steeply, restricting the height and extension of the emerging maximum. Therefore, although the inhibitor has a very shallow distribution before a new signal is triggered, the new maximum becomes indistinguishable from the older maxima because of the self-regulating properties of the proposed mechanism (Figure 27.3).

In the wing disk of *Drosophila*, rows of small sensory bristles (microchaetae) are formed. The row-like arrangement is genetically fixed but not the position of the

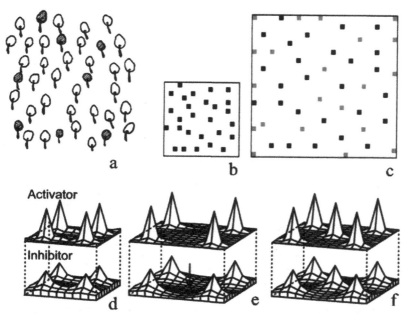

Figure 27.3. Insertion of new structures in a growing field. (a) Observation of Wigglesworth (1940): After the transition from the 4th to the 5th instar, new bristles (shaded) become inserted. (b, c) Simulation: with growth, the distance between maxima can become so large that new maxima emerge (grey in Fig. 27.3c). (d–e) Simulation showing the insertion of a new peak in more detail. Due to growth, the inhibitor concentration at a location between the four established peaks becomes so low (arrow in Fig. 27.3e) that the onset of autocatalysis can no longer be suppressed. The emerging maximum obtains the same shape as the previously established signalling centres.

individual bristles. The simulation in Figure 27.4 shows that very few positional cues are required to generate such a pattern. In the model if some cells have a better capability to produce the activator and the inhibitor (competence or source density, factor s in Equations 1a, b), the cells with the higher capability have a better chance to win. This is especially pronounced at a boundary region where cells with higher and lower competence are juxtaposed (Figure 27.4). A transient stripe-like activation occurs first at the high side of this border. Subsequently, the stripe disintegrates into individual spots due to the lateral inhibition. The inhibition that spreads from this initial stripe suppresses activation in the direct vicinity but enables the formation of other stripes at some distance. These stripes will also disintegrate into isolated maxima. Thus, a slightly different competence is sufficient to generate the arrangement of bristles into rows without the need to position every individual bristle by a complex genetic machinery.

Several components known to be involved in bristle formation have properties compatible with the proposed mechanism. Crucial are the genes *achaete* and *scute* (see Chapters 23 and 26). The enhancer region of *scute* has a binding site for its gene product, satisfying therefore the condition of local autocatalysis (Culí and Modolell, 1998). Lateral inhibition is accomplished by the *Delta-Notch* system (see Chapter 22). However, in the model it is assumed that the inhibitor has a range of several cell diameters.

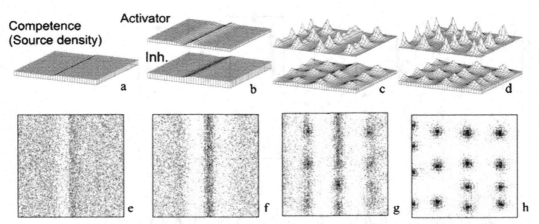

Figure 27.4. Generation of activated spots in rows: a simplified model for bristle initiation. Sensory bristles (microchaetae) on the thorax of *Drosophila* are arranged in rows. The details of the pattern are not genetically fixed (see Chapter 26). Model: an activator–inhibitor mechanism is assumed (Equation 1a, b). (a) The competence, i.e., the ability of the cells to perform the autocatalysis in the pattern-forming reaction (s in Equation 1a, b) is somewhat higher in some regions (right half). (b–d) Activation takes place first in a stripe-like fashion on the high side close to the border between the region of higher and lower competence. Due to the lateral inhibition, this stripe decays into several maxima (c). Their precise location depends on random fluctuations. Due to the long-range inhibition, more stripes of activation appear at some distances to the primary stripe. They also disintegrate into separate maxima (d). Thus, according to the model, very little genetic information is required to obtain local signals arranged in rows. (e–h) The same simulation in a plot as expected to show up in a staining experiment illustrates the highly dynamic character of the pattern-forming reaction. The density of pixels indicates the local concentration.

The Delta protein, functioning as the inhibitor, remains attached to the cell membrane. The behavior of the system, however, suggests that not only the direct neighbors are inhibited. How this inhibition spreads over several cell diameters is not yet known. In the tergites, *engrailed* represses bristle formation in p compartment and a6 region but not in a5 (Chapter 16). This difference may be sufficient to organize the different arrows in a posterior-to-anterior wave of sensory organ precursors (SOP) singling out (Marí-Beffa, unpublished results). Moreover, during tergite development, *extramacrochaetae* is probably also involved in preventing intercalary SOP singling out (Marí-Beffa et al., 1991) as observed in *Rhodnius prolixus* (see above).

CONCLUSION

Mathematically formulated models together with computer simulations provide a powerful tool for learning what the minimum requirements are to achieve a particular step in development. These models disregard several aspects, for instance that the signaling between cells is a complex event requiring ligands, receptors, and cascades to the nucleus. But to understand the underlying logic, the assumption of diffusion is a good first approximation. Models have been developed for many other situations that are beyond the scope of the present chapter, for the initiation of legs and wing in insects and vertebrates (Meinhardt, 1983*a*, *b*), for the generation of net-like structures

(Meinhardt, 1976, 1982), for the orientation of chemotactic cells by minute external signals (Meinhardt, 1999), and for the orientation of main body axes in a developing organism (Meinhardt, 2001, 2002). Animated simulations for these models are available at http://www.eb.tuebingen.mpg.de/depth/meinhardt/home.html. Computer programs to simulate such reactions on a PC are given elsewhere (Meinhardt, 2003). Nevertheless, these models are far from being complete. For instance, cell movements and the control of cell proliferation are two important aspects that are not yet integrated.

ACKNOWLEDGEMENTS

I wish to express my sincere thanks to Professor Alfred Gierer. Much of the basic work described in this chapter emerged from a fruitful collaboration over many years.

REFERENCES

Browne, E. N. (1909). The production of new hydrants in *Hydra* by insertion of small grafts. *J. Exp. Zool.*, 7, 1–23.

Buikema, W. J., and Haselkorn, R. (2001). Expression of the *Anabaena hetR* gene from a copper-regulated promoter leads to heterocyst differentiation under repressing conditions. *Proc. Natl. Acad. Sci. USA*, 98, 2729–34.

Chen, Y., and Schier, A. F. (2002). Lefty proteins are long-range inhibitors of squint-mediated nodal signaling. *Curr. Biol.*, 12, 2124–8.

Cubas, P., and Modolell, J. (1992). The *extramacrochaetae* gene provides information for sensory organ patterning. *EMBO J.*, 9, 3385–93.

Culí, J., and Modolell, J. (1998). Proneural gene self-stimulation in neural precursors – An essential mechanism for sense organ development that is regulated by *Notch* signaling. *Genes Dev.*, 12, 2036–47.

Gierer, A., Berking, S., Bode, H., David, C. N., Flick, K., Hansmann, G., Schaller, H., and Trenkner, E. (1972). Regeneration of hydra from reaggregated cells. *Nature New Biology*, 239, 98–101.

Gierer, A., and Meinhardt, H. (1972). A theory of biological pattern formation. *Kybernetik*, 12, 30–9 (available at http://www.eb.tuebingen.mpg.de/depth/meinhardt/home.html).

Hobmayer, B., Rentzsch, F., Kuhn, K., Happel, C. M., Cramer von Laue, C., Snyder, P., Rothbacher, U., and Holstein, T. W. (2000). WNT signalling molecules act in axis formation in the diploblastic metazoan *Hydra. Nature*, 407, 186–9.

Lenhoff, H. M. (1991). Ethel Browne, Hans Spemann and the discovery of the organizer phenomenon. *Biol. Bull.*, 181, 72–80.

Marí-Beffa, M., de Celis, J. F., and García-Bellido, A. (1991). Genetic and developmental analyses of chaetae pattern formation in *Drosophila* tergites. *Roux's Arch. Dev. Biol.*, 200, 132–42.

Meinhardt, H. (1976). Morphogenesis of lines and nets. *Differentiation*, 6, 117–23.

Meinhardt, H. (1982). *Models of Biological Pattern Formation.* London: Academic Press (available at http://www.eb.tuebingen.mpg.de/depth/meinhardt/home.html).

Meinhardt, H. (1983*a*). A boundary model for pattern formation in vertebrate limbs. *J. Embryol. Exp. Morphol.*, 76, 115–37.

Meinhardt, H. (1983*b*). Cell determination boundaries as organizing regions for secondary embryonic fields. *Dev. Biol.*, 96, 375–85.

Meinhardt, H. (1999). Orientation of chemotactic cells and growth cones: Models and mechanisms. *J. Cell Sci.*, 112, 2867–74.

Meinhardt, H. (2001). Organizer and axes formation as a self-organizing process. *Int. J. Dev. Biol.*, 45, 177–88.

Meinhardt, H. (2002). The radial-symmetric hydra and the evolution of the bilateral body plan: An old body became a young brain. *BioEssays*, 24, 185–91.

Meinhardt, H. (2003). *The Algorithmic Beauty of Sea Shells*, 3rd ed. Heidelberg, New York: Springer-Verlag.

Meinhardt, H., and Gierer, A. (2000). Pattern formation by local self-activation and lateral inhibition. *BioEssays*, 22, 753–60.

Technau, U., von Lane, C. C., Rentzsch, F., Luft, S., Hobmayer, B., Bode, H. R., and Holstein, T. W. (2000). Parameters of self-organization in *Hydra* aggregates. *Proc. Natl. Acad. Sci. USA*, 97, 12127–31.

Wasserman, J. D., and Freeman, M. (1998). An autoregulatory cascade of EGF receptor signaling patterns the *Drosophila* egg. *Cell*, 95, 355–64.

Wigglesworth, V. B. (1940). Local and general factors in the development of "pattern" in *Rhodnius prolixus*. *J. Exp. Biol.* 17, 180–200.

Yoon, H. S., and Golden, J. W. (1998). Heterocyst pattern-formation controlled by a diffusible peptide. *Science*, 282, 935–8.

APPENDIX 1. ABBREVIATIONS

11-UTP	11-uridine triphosphate
5′-UTR	5′-untranslated region
a1-6	cuticular types 1 to 6
A1-A8	first to eighth abdominal segment
AA	arachidonic acid
ABC-kit	avidin-biotin complex Kit
ABD-120	actin binding domain construct-120
abd-A	*abdominal-A* (gene)
Abd-B	*Abdominal-B* (gene)
ABRC	Arabidopsis Biological Resource Center
ac	*achaete* (gene)
AC	anchor cell
Ach	*Achaetous* (mutation)
Act5C	promoter of actin gene 5C
ADC	anterior dorsocentral (bristle)
ADHN	anterior dorsal histoblast nest
ADHs	alcohol dehydrogenases
A–E	wing interveins A to E
AEL	after egg laying
AER	apical ectodermal ridge
AG	*AGAMOUS* (gene)
ALP (AP)	alkaline phosphatase
AMC	antenno-maxillary complex
Amp	ampicillin
an 1-2	animal 1-2 (tier)
ANP	anterior notopleural (bristle)
ANR	anterior neural ridge
anti-HRP	anti-horseradish peroxidase
ANZ	anterior necrotic zone
AO	area opaca
ap	*apterous* (gene)
AP1	*APETALA 1* (gene)

AP2	*APETALA 2* (gene)
AP2/ERF	APETALA 2/Ethylene response factor
AP3	*APETALA 3* (gene)
APA	anterior postalar (bristle)
Arp 2/3	actin-related protein 2/3
ASA	anterior supraalar (bristle)
AS-C	*achaete-scute complex*
ASC	anterior scutellar (bristle)
Ath	anterior thalamus
ATP	adenosine triphosphate
Ava II (*Bme* 18I)	restriction enzyme from *Bacillus megaterium* 18
aVD	abdominal ventral denticle
AWM	anterior wing margin
Bam HI	restriction enzyme from *Bacillus amyloliquefaciens* H
bcd	*bicoid* (gene)
BCIP	5-bromo-4-chloro-3-indolyl-phosphate
bHLH	basic helix-loop-helix
BL	body length
BMPs	Bone Morphogenetic Proteins
BSA	bovine serum albumin
bwk	*bullwinkle* (gene)
bx	*bithorax* (mutations)
bxd	*bithoraxoid* (mutations)
cAMP	cyclic adenosine monophosphate
Cb	cerebellum
CCD	charge-coupled device (camera)
cDNA	complementary deoxyribonucleic acid
CF	cephalic fold
CF-SW	calcium-free sea water
cGy	centigrey
CHAPS	(3-(3-cholamidopropyl)dimethylammonio)-1-propane-sulfonate)
CNS	central nervous system
CPS	cephalopharyngeal skeleton
CRAC	cytosolic regulator of adenylyl cyclase
Cre-lox P	Cre recombinase/lox P sites (recombination system)
Cy	*Curly* (mutation)
CyO	*Curly of Oster* (balancer)
D	dorsal (compartment)
D0,1,2	complete (0) to slight (2) dorsalized phenotype
DA	dorsal (respiratory) appendage
DAB	3,3'-diaminobenzidine
D_{a-b}	diffusion coefficients
DAG	diacylglycerol
DC	dorsocentral (bristles)
DEAB	diethylamino benzaldehyde
DEPC	diethylpyrocarbonate
derf	*friend of echinoid* antisense
DF	dorsal fold

Df(2L)TW119	deficiency covering dl
DH	dorsal hairs
DIC	differential interference contrast
DIG	digoxygenin
Dl	*Delta* (gene)
dl	*dorsal* (gene)
Dll	*Distal-less* (gene)
DMEM	dulbecco's modification of Eagle's medium
DMSO	dimethyl sulfoxide
DNEGFR	dominant-negative form of the epidermal growth factor receptor
dNTP	deoxynucleotide triphosphate
dpc	days post coitum
DpM2	duplication *Dp(3;Y;X) M2*
Dra I (*Aha* III)	restriction enzyme from *Deinococcus radiophilus* I
dsh	*disheveled* (gene)
dsRNA	double-stranded RNA
E(spl)-C	*Enhancer-of-split*-complex
EB	embryoid body
EC	embryonic carcinoma (cell)
Eco RI	restriction enzyme from *Escherichia coli* I
EDTA	Ethylenediaminetetraacetate acid
EG	embryonic germ (cell)
EGF	Epidermal growth factor (protein)
EM	embryo medium
emc	*extramacrochaetae* (gene)
en	*engrailed* (gene)
En1/2	*engrailed 1 or 2* (genes)
ES	embryonic stem (cell)
EthBr	ethidium bromide
EtOH	ethanol
Ets	erythroblast transformation specific
f	*forked* (mutation)
FB	forebrain
FCS	fetal calf serum
FGF8	Fibroblast growth factor 8 (protein)
FITC	fluorescein isothiocyanate
Fk	Filzkörper
FL	feeder layer (culture)
FLP/FRT	Flipase/Flp recombination target
FM6/7	*First multiple 6/7* (balancer chromosome I)
fred	*friend of echinoid* (gene)
ftz-lacZ	*fushi tarazu-lacZ*
GATA-1	Zinc-finger transcription factor that binds to the (T/A)GATA(A/G) DNA sequence
GCb	grafted cerebellum
Gf (GOF)	gain-of-function
GFP	green fluorescent protein
GMR	Glass Multimer Reporter

Grk	*Gurken* (gene)
GS	goat serum
GSK-3β	glycogen synthase kinase 3 beta (protein)
Gtops	*Girardia tigrina opsin* (gene)
Gtsix-1	*Girardia tigrina six-1* (gene)
H	*Hairless* (gene)
H	haltere
H7	1-5 (Isoquinolinesulfonyl)-2-methylpiperazine
HA	head activation
Ha	heart anlage
HB	hindbrain
hb-lacZ	*hunchback-lacZ*
hh	*hedgehog* (gene)
HH10	Hamburger and Hamilton stage 10
HI	head inhibition
him-6	*high incidence of males-6* (gene)
Hind III	restriction enzyme from *Haemophilus influenzae*
HM	hydra medium
HO	head organizer
Hox	homeobox containing (gene)
hpf	hours post-fertilization
Hu	*Humeral* (mutation)
HyAlx	*Hydra Alx* (gene)
Hybra1	*Hydra Brachury 1* (gene)
iab-2-8	*infraabdominal-2 to -8* (mutations)
ICM	inner cell mass
IM	inner mesoderm
IMDM	iscove's modification of DMEM
iMES	induced mesencephalon
Inh	*inhibitor*
inv	*invected* (gene)
IPTG	isopropyl-β-D-thiogalactopyranoside
Is	isthmus
IsO	isthmic organizer
ksr-1	*kinase suppressor of activated RAS-1* (gene)
kuz	*kuzbanian* (gene)
L1-L5	wing veins L1 to L5
L3	sensilla of the vein L3
lacZ (β-gal)	*β-galactosidase* (gene)
LB	Luria Broth
LBA	LB+agar
LEC	larval epidermal cell
Ler	Landsberg *erecta*
lf (LOF)	loss-of-function
Lha	left heart anlage
LIF	leukemia inhibitory factor
Llpm	left lateral plate mesoderm
LM	lateral mesoderm
LPM	lateral plate mesoderm

M	*Minute* (mutation)
MAB	maleic acid buffer
Mac	macromere
MADS	<u>M</u>CM1, <u>A</u>GAMOUS, <u>D</u>EFICIENS and <u>S</u>RF (serum response factor) box
MAPK (MPK-1)	MAP kinase
MASH	*murine achaete scute homolog*
MB	midbrain
MBL	Marine Biological Laboratory
MBS	Modified Barth Saline
MC	mitomycin C
MEK (MEK-2, MAPKK)	MAP kinase kinase
MES	mesencephalon
Mes	mesomere
MH	mouth hooks
Mic	micromere
MMR	Marc's Modified Ringers
mRNA	messenger ribonucleic acid
MS-222	3-aminobenzoic acid ethyl ester
MSP-130	mesenchyme specific protein 130
MTG	monothioglycerol
Muv	multivulva
mwh	*multiple wing hair* (mutation)
N	*Notch* (gene)
NASC	Nottingham Arabidopsis Resource Center
NB	notal bristles
NBT	nitroblue tetrazolium
NEAA	nonessential amino acids
nej	*nejire* (gene)
neu	ErB-2 oncogene
neur-lacZ	*neuralized-lacZ*
NGF	nerve growth factor
Ni-NTA	Ni^{2+}-nitriloacetic acid
NP	notopleural (bristles)
otd	*orthodenticle* (gene)
Otx-2	*Orthodenticle-like homeobox gene-2*
P3-8.p	vulva percursor cells
PABA	Para-aminobenzoic acid
Pax-2	*Paired-box containing gene 2*
PB-1	phosphate-buffered medium
pBs	pBlueskript KS+ plasmid
PBS	phosphate-buffered saline
PBS-T (PBT, PBSw)	phosphate-buffered saline-Tween 20
pbx	*postbithorax* (mutation)
pc	pigment cells
pc	polar cells
PCR	polymerase chain reaction
PDC	posterior dorsocentral (bristle)
PDHN	posterior dorsal histoblast nest

PFA	paraformaldehyde
PGC	primordial germ cells
ph	pharynx
PH	pleckstrin homology domain
phc	photoreceptor cells
PI	*PISTILLATA* (gene)
PI(3,4,5)P3	phosphatidylinositol-3,4,5-triphosphate
PI(4,5)P2 or PIP2	phosphatidylinositol-4,5-biphosphate
PI3K	phosphatidylinositol 3-kinase
PKC	Protein kinase C (protein)
PMC	primary mesenchyme cells
PMG	posterior midgut
PNP	posterior notopleural (bristle)
pnr	*pannier* (gene)
pNR	procephalic neurogenic region
PNS	peripheral nervous system
PNZ	posterior necrotic zone
PPA	posterior postalar (bristle)
pRB	retinoblastoma tumour suppressor
PS	presutular (bristle)
PSA	posterior supraalar (bristle)
PSC	posterior scutelar (bristle)
psn	*presenilin* (gene)
ptc	*patched* (gene)
PTGS	Post-transcriptional gene silencing
Pth	caudal/posterior thalamus
pwn	*pawn* (mutation)
PZ	progress zone
Q¢PN	quail but not chick perinuclear antigen
r	rhabdomeres
R	Roentgen
r3-5	rhombomeres 3-5
RA	retinoic acid
r_{a-b}	decay rates
RALDH2	RA aldehyde-dehydrogenase-2
RARs	retinoic acid receptors
rf	reduction-of-function
Rh	rhombencephalon
RNAi	RNA interference
RNAse	ribonuclease
RTK	receptor tyrosine kinase
RT-PCR	reverse transcriptase-polymerase chain reaction
RXRs	retinoid X receptors
s	source density
Sb	*Stubble* (mutation)
sc	*scute* (gene)
SC	scutellar (bristles)
Scr	*Sex comb reduced* (gene)
SEM-5	Sex muscle abnormal-5 (protein)

SEP1-3	*SEPALLATA 1-3* (genes)
Ser	*Serrate* (gene)
sha	*shavenoid* (mutation)
Shh	*Sonic hedgehog* (gene)
Six-1	*Sine oculis-1* (gene)
SM6a	*Second multiple 6a* (balancer)
SMC	sensillum mother cell
smo	*smoothened* (gene)
sna	*snail* (gene)
SNP	single nucleotide polymorphism
so	sensory organ
sog-lacZ	*short gastrulation-lacZ*
SOP	sensory organ precursor (cell)
SOS-1	Son of sevenless orthologue-1 (protien)
Sp	spiracles
spz	*spätzle* (gene)
SSC	saline-sodium citrate (buffer)
SSEA-1	stage specific embryonic antigen-1
ssRNA	single-stranded RNA
Su(H)	*Suppressor of Hairless* (gene)
T	tuft
T1-T3	first to third thoracic segment
TAIR	The Arabidopsis Information Resource
Tb	*Tubby* (mutation)
TBE	Tris-(hydroxy methyl)aminomethane, boric acid & EDTA
Tcf	T cell factor (protien)
TdT	terminal deoxynucleotidyl transferase
Tet	tetracycline
TGF-β	Transforming growth factor beta
Th	threshold
Tl	*Toll* (gene)
TM	Tris-malate-(buffer)
TM2/3/6B/8	*Third multiple 2/3/6B/8* (balancer chromosomes III)
TP	transduction pathway
TPA	12-*O*-tetradecanoylphorbol-13-acetate
tRNA	transfer ribonucleic acid
TUNEL	TdT-mediated dUTP nick end-labelling (method)
tVD	thoracic ventral denticles
twi-lacZ	*twisted-lacZ*
UAS	upstream activating sequence
ubGFP	GFP under an ubiquitous promoter
Ubx	*Ultrabithorax* (gene)
unc-5	*uncoordinated-5* (gene)
V	Ventral (compartment)
V1,2	strongly ventralized phenotype
VD	ventral denticles
veg1/2	vegetative 1/2 (tier)
VF	ventral fold
vn	*vein* (gene)

vNR	ventral neurogenic region
VPC	vulval precursor cell
VT	vesicular trafficking
Vul	vulvaless
w (W)	wing
w	*white* (mutations)
WISH	whole mount in situ hybridization
WM	wing margin
WT	wild-type
Xba I	restriction enzyme from *Xanthomonas badrii*
Xho I	restriction enzyme from *Xanthomonas holcicola*
y	*yellow* (mutations)
YFP	yellow fluorescent protein
zen	*zerknüllt* (gene)
ZLI	zona limitans intrathalamica
ZPA	zone of polarizing activity

APPENDIX 2. SUPPLIERS

3M Company
General Office
3M Center
St. Paul, Minnesota 55144-1000
USA
Tel: 1-651-733-1110
http://www.3m.com

Ambion (Europe) Ltd
Spitfire Close
Ermine Business Park
Huntingdon
Cambridgeshire PE29 6XY
UK
Tel: 44-1480-373-020
Fax: 44-1480-373-010
http://www.ambion.com

ASCO Laboratories
52 Levenshulme Road
Manchester
Lancashire M18 7NN
UK
Tel: 44-161-224-5184
Fax: 44-161-224-5184
http://www.kellysearch.com

BDH
Hunter Boulevard
Magna Park
Lutterworth
Leicester LE17 4XN
UK
Tel: 44-800-223-344 Ext 3753
Fax: 44-120-266-4769
http://www.bdh.com

Becton Dickinson, S. A.
1 Becton Drive
Franklin Lakes, New Jersey 07417
USA
Tel: 1-201-847-6800
http://www.bd.com

Bio-Rad Laboratories Headquarters
1000 Alfred Nobel Drive
Hercules, California 94547
USA
Tel: 1-510-724-7000
Fax: 1-510-741-5817
http://www.bio-rad.com

Biosynth
1665 West Quincy Avenue
Suite 155
Naperville, Illinois 60540
USA
Tel: 1-800-270-2436
Fax: 1-800-276-2436
http://www.biosynth.com

Biozym Diagnostik GmbH
Postfach 180
D-31833 Hessisch Oldendorf
Germany
Tel: 49-5152/52430
Fax: 49-5152/524320
http://www.biozym.com

Brand GmbH
Postfach 1155
D-97861 Wertheim
Germany
Tel: 49-9342-808-0
Fax: 49-9342-808-236
http://www.brand.de

Carl Roth GmbH & Co.
Schoemperfenstr. 1-5
D-76185 Karlsruhe
Germany
Tel: 49-721-5606-0
Fax: 49-721-56-06-149
http://www.Carl-Roth.de

Carl Zeiss, Inc.
One Zeiss Drive
Thornwood, New York 10594
USA
Tel: 1-914-747-1800
Fax: 1-914-682-8296
http://www.zeiss.de

Carolina Biological Supply Company
2700 York Road
Burlington, North Carolina 27215-3387
USA
Tel: 1-800-334-5551(US);
    336-584-0381(international order)
Fax:1-800-222-7112
http://www.carolina.com

Chemicon International, Inc.
28820 Single Oak Drive
Temecula, California 92590
USA
Tel: 1-800-437-7500
Fax: 1-800-437-7502
http://www.chemicon.com

Chroma Technology Corp.
P.O. Box 489
10 Imtec Lane
Rockingham, Vermont 05101
USA
Tel: 1-800-824-7662
Fax: 1-802-428-2525
http://www.chroma.com

Developmental Studies Hybridoma Bank
Department of Biological Sciences
The University of Iowa
28 Biology Building East
Iowa City, Iowa 52242
USA
Tel.: 1-319-335-3826
Fax: 1-319-335-2077
http://www.uiowa.edu/dshhwww

Difco Laboratories
P.O. Box 331058
Detroit, Michigan 48232-7058
USA
Tel: 1-313-462-8500
Fax: 1-313-462-8517
http://www.rapidmicrobiology.com

Difco Laboratories Ltd.
P.O. Box 14B
Central Avenue
West Molesey
Surrey KT8 2SE
UK
Tel: 44-181-9799951
Fax: 44-181-9792506

DocFrugal Scientific Corp.
9865 Mesa Rim Road
Suite 206
San Diego, California 92121
USA
Tel: 1-800-789-5550
Fax: 1-888-789-0444
http://www.flystuff.com

Drummond Scientific Company
500 Parkway
Broomall, Pennsylvania 19008
USA
Tel: 1-610-353-0200
Fax: 1-610-353-6204
http://www.drummondsci.com

Dumont
CH-2924 Montignez
Switzerland
Tel: 41-324752121
Fax: 41-324752123
http://www.outlis-dumont.com

Electron Microscopy Sciences
P.O. Box 251
321 Morris Road
Fort Washington, Pennsylvania
    19034
USA
Tel: 1-215-646-1566
Fax: 1-215-646-8931
http://www.emsdiasum.com

EMD Biosciences, Inc.
CALBIOCHEM
10394 Pacific Center Court
San Diego, California 92121
USA
Tel: 1-800-854-3417
Fax: 1-800-776-0999
http://www.calbiochem.com

EMD Biosciences, Inc.
Novagen Brand
441 Charmany Drive
Madison, Wisconsin 53719
USA
Tel: 1-608-238-110
Fax: 1-608-238-1388
http://www.novagen.com

Epicentre
726 Post Road
Madison, Wisconsin 53713
USA
Tel: 1-608-258-3080
Fax: 1-608-258-3088
http://www.epicentre.com

Eppendorf AG
Netheler-Hinz-GmbH
Barkhausenweg 1
D-22331 Hamburg
Germany
Tel: 49-40-53-8010
Fax: 49-40-53-801-556
http://www.eppendorf.com

Fine Science Tools
373-G Vintage Park Drive
Foster City, California 94404-1139
USA
Tel: 1-800-521-2109
Fax:1-800-532-2109
http://www.finescience.com

Fisher (Scientific)
2761 Walnut Ave.
Tustin, California 92780
USA
Tel: 1-800-766-7000
Other Tel: 1-412-490-8300
Fax: 1-800-926-1166
http://www.fishersci.com

Fisher Scientific UK Ltd.
Bishop Meadow Rd.
Loughborough
Leicestershire LE11 5RG
UK
Tel: 44-150-923-1166 Tel: 44-1-202-669700
Other Tel: Fisher Scientific Europe:
    49-89-242180
Fax: 44-150-923-18
http://www.fisher.co.uk

Fluka
Industriestrasse 25
CH-9471 Buchs SG
Switzerland
Tel: 41-81-755-2511
Fax: 41-81-755-2815
http://www.sigmaaldrich.com/Brands

Gene Tools, LLC
One Summerton Way
Philomath, Oregon 97370
USA
Tel: 1-541-929-7840
Fax: 1-541-929-7841
http://www.gene-tools.com

Gibco Invitrogen Corporation
1600 Faraday Avenue
P.O. Box 6482
Carlsbad, California 92008
USA
Tel: 1-760-603-7200
Fax: 1-760-602-6500
http://www.invitrogen.com

Greiner und Gassner GmbH
Dachaver Str. 233
D-80637 Munich
Germany
Tel: 49-89-1577640
Fax: 49-89-154701
http://ww.eco-select.de/firm/0589.htm

Harvard Apparatus, Inc.
84 October Hill Road
Hollinston, Massachusetts 01746
USA
Tel: 1-508-893-8999
Fax: 1-508-429-5732
http://www.harvardapparatus.com

ICN (MP) Biomedicals, Inc.
15 Morgan
Irvine, California 92618-2005
USA
Tel: 1-800-854-0530
Fax: 1-800-334-6999
http://www.icnbiomed.com

ICN Pharmaceuticals Ltd.
Cedarwood. Chineham Business Park
Crockford Lane
Basingstoke
Hampshire RG24 8WD
UK
Tel: 44-1256-374-620
Fax: 44-1256-374-621
http://www.icnbiomed.com

Invitrogen Corporation
3 Fountain Dr.
Inchinnan Business Park
Paisley PA4 9RF
Scotland
UK
Tel: 44-0141-814-6100
Fax: 44-0141-814-6260
http://www.invitrogen.com

Jackson Immuno Research Laboratories Inc.
872 West Baltimore Pike
P.O. Box 9
West Grove, Pennsylvania 19390
USA
Tel: 1-800-367-5296
Fax: 1-610-869-0171
http://www.jacksonimmuno.com

JAX Research Systems
The Jackson Laboratory
610 Main Street
Bar Habor, Maine 04609
USA
Tel: 1-800-422-6423
Fax: 1-207-288-6150
http://Jaxmice.jax.org

Jencons-PLS
Unit 6 Forest Row Business Park
Station Road
Forest Row
East Sussex RH18 5DW
UK
Tel: 44-1342-826836
Fax: 44-1342-826771
http://www.jencons.co.uk

Labor
Bruno Kummer-Freiburg
Katharinenstraße 16
D-79104 Freiburg/Breisgan
Germany
Tel: 49-761-387700
Fax: 49-761-273085
http://www.kummer-laborgeraete.de

Labscientific, Inc
114 West Mt. Pleasant Avenue
Livingston, New Jersey 07039
USA
Tel: 1-800-886-4507.
Fax: 1-973-992-0827
http://www.labscientific.com

Leica Microsystems Inc.
2345 Waukegan Road
Bannockburn, Illinois 60015
USA
Tel: 1-800-248-0123
Fax: 1-847-405-0164
http://www.leica.com

MatTek Cooperation AC
200 Homer Avenue
Ashland, Massachusetts 01721
USA
Tel: 1-508-881-6771
Fax: 1-508-879-1532
http://www.mattek.com

Merck
Merck KgaA.
Frankfurter Strasse 250
D-64293 Darmstadt
Germany
Tel: 49-6151-72-0
Fax: 49-6151-72-2000
http://www.chemdat.de

Merck and Co. Inc
One Merck Drive
P.O. Box 100
Whitehouse Station, New Jersey
   08889-0100
USA
Tel:1-908-423-1000
http://www.merck.com

Merck Chemicals Ltd.
Merck House
Poole/Dorset BH15 1TD
UK
Tel: 44-1202-669-700 (661616).
Fax: 44-1202-666530
http://www.merckscltd.co.uk

Molecular Probes
29851 Willow Creek Rd.
Eugene, Oregon 97402
USA
Tel: 1-541-465-8300
Fax: 1-541-344-6504
http://www.probes.com

Narishige Scientific Instrument Lab.
27.9 Minamikaraayama 4.chom
Setagaya.ku
Tokyo 57.0062
Japan
Tel: 81-3-3308-8233
Fax: 81-3-3308-2005
http://www.narishige.co.jp

Nasco
901 Janesville Avenue
P.O. Box 901
Fort Atkinson, Wisconsin 53538-0901
USA
Tel: 1-800-558-9595
Fax : 1-920-563-8296
http://www.rapidmicrobiology.com

Nikon
Deutschland GmbH
Tiefenbroicher Weg 25
D-40472 Düsseldorf
Germany
http://www.nikon.de

Nunc GmbH & Co. KG
Hagenauer Strasse 21a
D-65203 Weisbaden
Germany
Tel: 49-611-18674-0
Fax: 49-611-18674-74
http://www.nalgenunc.com

Oxoid Limited
Wade Road
Basingstoke
Hampshire RG24 8PW
UK
Tel: 44-1256-841144
Fax: 44-1256-463388
http://www.oxoid.com

Oxoid Inc.
800 Proctor Avenue
Ogdensburg, New York 13669
USA
Tel: 1-613-226-1318
Fax: 1-613-226-3728
http://www.oxoid.ca

Pelikan Holding AG
Zugerstrasse 76b
0CH-6340 Baar
Switzerland
Tel: 41-41-768-5090
Fax: 41-41-768-5095
http://www.pelikan.de

Promega Corporation
2800 Woods Hollow Road
Madison, Wisconsin 53711
USA
Tel: 1-608-274-4330
Fax: 1-608-277-2516
http://www.promega.com

Qbiogene, Inc.
2251 Rutherford Road
Carlsbad, California 92008
USA
Tel: 1-800-424-6101
Fax: 1-760-918-9313
http://www.qbiogene.com

Qiagen GmbH
QIAGEN Strasse 1
D-40724 Hilden
Germany
Tel: 49-02103-29-12000
Fax: 49-02103-29-22000
http://www.quiagen.com

Qiagen Inc.
28159 Avenue Stanford
Valencia, California 91355
USA
Tel: 1-661-702-3000
Fax: 1-800-718-2056
http://www.qiagen.com

Roche Diagnostics Corporation
Roche Applied Science
P.O. Box 50414
9115 Hague Road
Indianapolis, Indiana 46250-0414
USA
Tel: 1-800-428-5074
Fax: 1-317-576-4240
http://www.roche.com

Roche Diagnostics GmbH
Roche Applied Science
Roche Molecular Biochemicals
Sandhofer Str. 116
D-68305 Mannheim
Germany
Tel: 49-621-759-8568
Fax: 49-621-759-4136
http://www.roche-applied-science.com

Roche (Boheringer Manheim)
F. Hoffmann-La Roche Ltd
Group Headquarters
Grenzacherstrasse 124
CH-4070 Basel
Switzerland
Tel: 41-61-688 1111
Fax: 41-61-691 9391
http://www.roche.com

Sanders Brine Shrimp Company
3850 South 540 West
Ogden, Utah 84405
USA
Tel: 1-801-393-5027
Fax: 1-801-621-3825
http://www.sandersbshrimp.com

Sci-Mart, Inc.
3221 Woodson Rd.
St. Louis, Missouri 63114
USA
Tel: 1-800-434-8850
Fax: 1-314-427-0121
http://www.scimart.com

SERVA Electrophoresis GmbH
Carl-Benz-Str. 7
P.O. Box 10 52 60
D-69115 Heidelberg
Germany
Tel: 49-6221-13840-0
Fax: 49-6221-13840-10
http://www.serva.de

Sigma-Aldrich Corp. (Fluka)
P.O. Box 14508
St. Louis, Missouri 63178
USA
Tel: 1-800-325-3010
Fax: 1-800-325-5052
http://www.sigmaaldrich.com

Sigma-Aldrich Company Ltd.
Fancy Road
Poole Dorset BH12 4GH
UK
Tel: 44-0800-447788
Fax: 44-1202-715-460
http://www.sigmaaldrich.com/order

Sigma-Aldrich Chemie GmbH
Eschenstrasse 5
82024 Munich
Germany
Tel: 49-89-65130
Fax: 49-8965131169
http://www.sigmaaldrich.com

Singer Instrument Co. Ltd.
Roadwater, Watchet
Somerset TA23 ORE
UK
Tel: 44-1984-640226
Fax: 44-1984-641166
http://www.singerinst.co.uk

Small Parts, Inc.
13980 N.W. 58th Court
P.O. Box 4650
Miami Lakes, Florida 33014-0650
USA
Tel: 1-800-220-4242
Fax: 1-800-423-9009
http://www.smallparts.com

Stoelting Co.
Physiology Division
620 Wheat Lane
Wood Dale, Illinois 60191
USA
Tel: 1-630-860-9700
Fax: 1-630-860-9775
http://www.stoelingco.com

Strathmann Biotec(h)
Feodor-Lynen-Str. 5
D-30625 Hannover
Germany
Tel: 49-04-0-55 90 50
Fax: 49-04 0-55 90 51 00
http://www.strathmann-biotec-ag.de

Stromberg's Chicks
P.O. Box 400
Pine River, Minnesota 56474
USA
Tel: 1-800-720-1134
Fax: 1-218-587-4230
http://www.strombergschickens.com

Sutter Instrument Company
51 Digital Drive
Novato, California 94949
USA
Tel: 1-415-883-0128
Fax: 1-415-883-0572
http://www.sutter.com

Tetko Inc.
111 Calumet St.
Depew, New York 14043
USA
Tel: 1-716-683-4050
Fax: 1-716-683-4053
http://www.tetko.com

Vector Laboratories
30 Ingold Road
Burlingame, California 94010
USA
Tel: 1-800-227-6666
Fax: 1-650-697-0339
http://www.vectorlabs.com

VWR
10105 Carroll Canyon Road
San Diego, California 92131
USA
Tel: 1-800-932-5000
Tel: 1-858-695-7600
http://www.vwr.com

VWR International GmbH
Frankfurter Str. 133
D-64293 Darmstadt
Germany
Tel: 49-61-51-72-3000
Fax: 49-61-51-72-3333
http://www.vwr.com

VWR Scientific
VWR International Ltd.
Merck House
Poole
Dorset BH15 1TD
UK
Tel: 44-1202-664421
Fax: 44-1202-665599
http://www.vwrsp.com

World Precision Instruments, Inc.
175 Sarasota Center Boulevard
Sarasota, Florida 34240
USA
Tel: 1-941-371-1003
Fax: 1-941-377-5428
http://www.wpiinc.com

INDEX